T0263873

NIELS BOHR
COLLECTED WORKS
VOLUME 7

NIELS BOHR (WITH ALBERT SCHWEITZER), COPENHAGEN 1959.

NIELS BOHR
COLLECTED WORKS

GENERAL EDITORS

ERIK RÜDINGER
UNTIL 1989

FINN AASERUD
FROM 1989

THE NIELS BOHR ARCHIVE, COPENHAGEN

VOLUME 7
FOUNDATIONS OF
QUANTUM PHYSICS II
(1933–1958)

EDITED BY

JØRGEN KALCKAR

THE NIELS BOHR INSTITUTE, COPENHAGEN

1996

ELSEVIER
AMSTERDAM · LAUSANNE · NEW YORK · OXFORD · SHANNON · TOKYO

ELSEVIER SCIENCE B.V.
Sara Burgerhartstraat 25
P.O. Box 211, 1000 AE Amsterdam
The Netherlands

Library of Congress Catalog Card Number: 70-126498
ISBN Collected Works: 0 7204 1800 3
ISBN Volume 7: 0 444 89892 1

This book is printed on acid-free paper
Transferred to digital print on demand, 2005
Printed and bound by Antony Rowe Ltd, Eastbourne

GENERAL EDITOR'S PREFACE

This is the first volume of the Niels Bohr Collected Works that has been prepared from beginning to end under the auspices of the Niels Bohr Archive, which received official status as an independent institution under the Danish Ministry of Education in 1985, the centennial of Bohr's birth. It is also the first volume for which I share responsibility, having taken over Erik Rüdinger's duties as Director of the Archive and General Editor of the Collected Works in 1989. Work on the two-volume "Foundations of Quantum Physics" (Volumes 6 and 7) was begun as a joint venture between General Editor Rüdinger and Editor Jørgen Kalckar, who share responsibility for Volume 6. Kalckar's prior experience in working with "Foundations" has not only been of great help in completing the present volume but has also allowed me to acquaint myself with the tasks peculiar to the General Editor. Even more importantly, work on this and future volumes is greatly facilitated by Rüdinger's conscientious over-all planning, before my appointment, of the whole series. Rüdinger's prior efforts are thus of great assistance in preparing the contents of the two remaining volumes and have furthermore rendered superfluous the task of defining and formulating editorial specifics, which are clearly laid out in Rüdinger's General Editor's Preface to Volume 5, published in 1984.

By contrast, the collaboration with the publisher has changed substantially. Thus, at the time of my arrival, new personnel took over responsibility for the Collected Works at Elsevier. What is more, the publication process itself changed when the Press decided to produce Volume 7 from electronic files prepared at the Archive. This decision necessitated the development of new working procedures at the Archive in close collaboration with Elsevier, notably represented by Publisher Joost Kircz and Desk Editors Henk Pruntel and Fer Mesman. I look forward to our continued fruitful collaboration in preparing the last two volumes in the series.

I join in Kalckar's thanks in his Foreword to the personnel at the Archive, who continue to work enthusiastically on the Collected Works. Finally, I would like to thank the members of the Archive's board for their approval and support.

Finn Aaserud
The Niels Bohr Archive
May 1995

FOREWORD

The present volume of the Niels Bohr Collected Works is a direct continuation of Volume 6[1]. Its content falls quite naturally into two parts: Part I presents the extension of the analysis of complementary relationships to the foundations of *relativistic* quantum physics; Part II covers thirty years of continued efforts by Bohr to achieve, through further analysis of basic issues, an ever clearer understanding of the rôle of fundamental concepts in the quantal description, and a deepening of insights into its content[2]. Why, one might ask, the necessity of this protracted labour, this tireless work running, as it were, like a pedal bass accompanying Bohr's fundamental contributions to nuclear physics (Collected Works, Vol. 9), to the problems of penetration of charged particles through matter (Vol. 8) and – in the years during and after the Second World War – his intense occupation with the political consequences of the nuclear arms race he so early foreboded and tried to forestall (Vol. 11).

A tentative answer might be found in Bohr's reluctance to consider any sphere of truly fundamental significance as a "closed" subject and in his deep awareness of the impossibility of ever exhausting the infinite manifold of shadings embodied in that strange concept to which we refer by the word "reality". We meet, I think, in Bohr's intellectual temperament a fundamental duality: on the one hand, the strong urge towards the grand synthesis, an overall view of "reality", already so characteristic a trait in his early youth, as alluded to in the General Introduction that opens Volume 6; on the other hand, equally strong, his feeling of responsibility to pay most scrupulous attention to every minute

---

[1] The general principles for the editing of Volumes 6 and 7 are laid down in *Foreword to Volumes 6 and 7* (Vol. 6, pp. V--VII).

[2] The essays of Bohr that deal mainly with the rôle of complementarity in domains outside the realm of physics are included in Volume 10.

detail of that "reality". This duality demands from the reader careful consideration of the significance of every phrase, every cunning innuendo, in Bohr's elaborately constructed periods and paragraphs – as well as a keen awareness of what is *not* said. Indeed, while Bohr could ponder for a long time how to express most clearly what he wanted to say, he was never in doubt as to what he definitely did *not* wish to say.

A peculiar feature that may strike the reader of these essays is that, apart from his briskly completed, terse and direct reply to Einstein, Podolsky and Rosen in 1935, Bohr apparently never found it worth while to publish accounts of his solutions to any of the other paradoxes suggested by Einstein[3], until the invitation to contribute to the volume celebrating Einstein's 70th birthday in 1949 provided an obvious challenge to do so. In the face of evidence, therefore, we must conclude that in this Birthday Volume Bohr, in his polite manner, slightly exaggerated the impact of Einstein's objections on the clarification of his own thoughts.

It is true that, in recent times, salient features of Bohr's analysis have been elucidated by exploitation of more transparent mathematical schemes. Still, the original arguments couched in Bohr's own idiom will hardly be superseded. Indeed, to my mind, the true marvel of the series of papers reproduced in Part II is the strength of inspiration and skill with which Bohr turns any *pièce d'occasion* into an ever-renewed, lapidary synthesis of a whole life's ponderings on the epistemological insights gained through the development of physics in our century. Throughout this volume we perceive Bohr struggling to live up to what a true connoisseur once during conversation referred to as "his responsibility as tacitly elected spokesman for the quantal community".

* * *

Ten years ago, the general structure and representation of the present volume were conceived by the former director of the Niels Bohr Archive, Erik Rüdinger, and me, and the entire planning worked out as a joint undertaking. It took place through many long, ardent and friendly debates that I have never ceased to miss. In those distant days, I also thankfully enjoyed the collaboration with the former secretary at the Archive, Helle Bonaparte, whose proficiency in English and German was an invaluable assistance. In addition, her optimism,

---

[3] In Volume 6, it was pointed out that the second version of the Como Lecture (1927), published in 1928 *after* the debates with Einstein at the Solvay Conference, does not contain as much as one single revision that could be interpreted as a result of these discussions (cf. Vol. 6, p. [32]).

however sarcastically expressed, as regards the final outcome of the whole enterprise – in spite of its often somewhat dubious appearance – provided a most welcome encouragement.

* * *

The preparation of the manuscript itself was carried out with remarkable efficiency by the members of the Niels Bohr Archive, headed by Finn Aaserud. Recalling lessons of sage saws, that it is best not to swap horses when crossing a stream, I was myself, obviously, rather concerned when a new director of the Niels Bohr Archive inherited Erik Rüdinger's position as General Editor. I could have spared myself such worries! I should add that through all these years of preparatory work, Finn Aaserud has relieved me of any concern whatsoever with respect to practical matters. I am deeply indebted to him for this generosity and for his and Felicity Pors's competent help in matters of language and style.

At this place it is a special pleasure also to pay tribute to Hilde Levi, who provided all the photographic material. Abraham Pais offered many a piece of shrewd advice and kept up moral discipline through his constant admonitions to go ahead at full speed. Martin B. Sørensen lent, together with Judith Hjartbro, most valuable assistance in tracing references and material for the facsimili, and shared with Finn Aaserud the labour of preparing the index. Helle Beckmann Johansen assisted in checking many formulae and the secretary at the Niels Bohr Archive, Anne Lis Rasmussen (during a period assisted by Virma Rasmussen), patiently typed and retyped the introductions and the many letters and helpfully took part in the proof-reading. To each and all: my thanks.

* * *

The major part of my own work was carried out under the skies of Italy. I am greatly indebted to the Catania Istituto di Fisica Teorica, Sicily, for its magnanimity, personified as it were in the shape of Marcello Baldo. I cherish the many memories of inspiring discussions with him, from which I have learned so much. I owe a special gratitude for the hospitality of the Danish Academy in Rome, in particular to the warm and friendly helpfulness of Karen Ascani, its librarian and administrative secretary. Without the many peaceful months that I have spent there, since my first visit in 1985, I should never have gathered the strength to fulfil a task that I felt to be far beyond my natural powers. In this connection, the generous support of the Foundation of the Niels Bohr Institute and the Niels Bohr Fund of the Royal Danish Academy of Sciences and

Letters is gratefully acknowledged. I also feel in great debt to colleagues at the Institute and members of its board for their lenient tolerance and forbearing consideration through the decade that I have devoted to this work. I can only hope that they will feel that the result is to some extent worthy of their patience.

* * *

The momentum needed to complete my intermittent and hesitating work on the manuscript for the Introductions, as well as with the translation of the substantial correspondence included, was provided by Jens Lindhard: first, he declined to discuss with me any topic in physics, however alluring, before the manuscript was finished; second, after I had achieved this feat – as I naïvely imagined – in Rome during the summer of 1992, the following months witnessed many days of heated discussions, here and in Aarhus, resulting in an almost complete recasting of the form and content of the Introductions.

At the very last stage of the work, several important sections of the manuscript were thoroughly reformulated during intense debates with Aage Bohr and Ole Ulfbeck. The resulting number of improvements is vast. Still, the critical reader might nevertheless find errors and shortcomings for which I obviously claim sole responsibility.

* * *

I could not end this Foreword, however, without turning back to Aage Bohr and Jens Lindhard, two links in a precious chain, of which by now there is but little left. To them I wish not only to convey my deep gratitude, but also to express how strangely moving I experienced these painstaking discussions, line by line, conducted in a mood of earnest guised in laughter, so fully in harmony with Bohr's own style of working – in happy days, long past.

Jørgen Kalckar
The Niels Bohr Institute
May 1995

CONTENTS

PART I: RELATIVISTIC QUANTUM THEORY

PART II: COMPLEMENTARITY: BEDROCK OF THE QUANTAL DESCRIPTION

PART III: SELECTED CORRESPONDENCE

Correspondents

INVENTORY OF RELEVANT MANUSCRIPTS
IN THE NIELS BOHR ARCHIVE

INDEX

ABBREVIATED TITLES OF PERIODICALS

Acad. Copenhague, Math. Phys. Com.	Matematisk–fysiske Meddelelser udgivet af Det Kongelige Danske Videnskabernes Selskab (København)
Ann. d. Phys.	Annalen der Physik (Leipzig)
Ber. Sächs. Akad. math.–phys. Kl.	Berichte über die Verhandlungen der Sächsischen Akademie der Wissenschaften zu Leipzig, mathematisch–physikalische Klasse
Berl. Ber.	Sitzungsberichte der Königlichen Preussischen Akademi der Wissenschaften zu Berlin
Copenhagen Acad. Math.–phys. Comm. Dan. Mat.–fys. Medd.	Matematisk–fysiske Meddelelser udgivet af Det Kongelige Danske Videnskabernes Selskab (København)
Handbuch der Phys. Hb. d. Physik	Handbuch der Physik (Berlin 1933)
Helv. Phys. Acta.	Helvetica Physica Acta (Basel)
J. de Physique	Le Journal de physique et le radium (Paris)
Journ. Chem. Soc.	Journal of the Chemical Society (London)
Journ. Frankl. Inst.	Journal of the Franklin Institute (Philadelphia)
Kgl. Danske Vid. Sels. Math.–fys. Medd.	Matematisk–fysiske Meddelelser udgivet af Det Kongelige Danske Videnskabernes Selskab (København)
Kon. Ned. Akad. v. Wet.	Proceedings, Koninklijke Nederlandse Akademie van Wetenschappen (Amsterdam)

Leipziger Ber.	Leipziger Berichte
Mat.–Fys. Medd. *Mat.–Fys. Medd. Dan.* *Vidensk. Selsk.* *Mat.–Fys. Med. of the Kgl.* *Danske Videnskabernes* *Selskab*	Matematisk–fysiske Meddelelser udgivet af Det Kongelige Danske Videnskabernes Selskab (København)
Naturwiss.	Die Naturwissenschaften (Berlin)
Naturwiss. Rundschau	Naturwissenschaftliche Rundschau (Stuttgart)
Overs. Dan. Vid. Selsk. *Overs. Dan. Vidensk.* *Selsk. Virks.*	Oversigt over Det Kongelige Danske Videnskabernes Selskabs Virksomhed (København)
Phil. Mag.	Philosophical Magazine (London)
Phil. Sci.	Philosophy of Science (East Lansing)
Phys. Rev.	The Physical Review (New York)
Phys. Zs.	Physikalische Zeitschrift (Leipzig)
Proc. Cambr. Phil. Soc.	Proceedings of the Cambridge Philosophical Society
Proc. Roy. Soc.	Proceedings of the Royal Society (London)
Prog. Theor. Phys.	Progress of Theoretical Physics (Kyoto)
Rev. Mod. Phys.	Reviews of Modern Physics (New York)
Sächs. Akad. *Sächs. Akad. der Wiss.* *Verh. d. Sächs. Ak.*	Berichte über die Verhandlungen der Sächsischen Akademie der Wissenschaften zu Leipzig, mathematisch–physikalische Klasse
Verh. der Kgl. Dän. *Gesellsch. d. Wiss.*	Oversigt over Det Kongelige Danske Videnskabernes Selskabs Forhandlinger (København)
Z. Phys. *Z. Physik* *Zs. f. Phys.*	Zeitschrift für Physik (Braunschweig)

ABBREVIATIONS

AHQP	Archive for History of Quantum Physics
AIP	American Institute of Physics, College Park, Maryland
Bohr MSS	Bohr Manuscripts, AHQP
BR	Bohr–Rosenfeld, *Zur Frage der Messbarkeit der elektromagnetischen Feldgrössen* (see p. [7], ref. 6)
BSC	Bohr Scientific Correspondence, AHQP
Mf	Microfilm
MS	Manuscript
NBA	Niels Bohr Archive, Copenhagen
PWB II	Pauli Wissenschaftlicher Briefwechsel, Bd. II (see p. [10], ref. 9)

ACKNOWLEDGEMENTS

For some of the published works in this volume, the editors and the publisher were unfortunately unable to trace the copyright holders and thus to enter a formal request for permission to reproduce the material. These works were nevertheless considered sufficiently important to be reprinted without further delay. The effort to identify the copyright holders will be continued.

N. Bohr and L. Rosenfeld, "Zur Frage der Messbarkeit der elektromagnetischen Feldgrössen", Mat.–Fys. Medd. Dan. Vidensk. Selsk. **12**, no. 8 (1933), is reprinted by permission of the Royal Danish Academy of Sciences and Letters.

N. Bohr and L. Rosenfeld, "On the Question of the Measurability of Electromagnetic Field Quantities" in "Selected Papers of Léon Rosenfeld" (eds. R.S. Cohen and J.J. Stachel), D. Reidel Publishing Company, Dordrecht 1979, pp. 357–400, is reprinted by permission of the publisher, Kluwer Academic Publishers.

N. Bohr, "Sur la méthode de correspondence dans la théorie de l'électron", Report of the 7th Solvay Meeting in 1933, "Structure et propriétés de noyeaux atomiques", Gauthier-Villars, Paris 1934, pp. 216–228, and the contribution by N. Bohr to "Discussion du rapport de M. Dirac", *ibid*, pp. 214–215, are reprinted by permission of the publisher, Dunod Editeur.

N. Bohr, "Problems of Elementary-Particle Physics", Report of an International Conference on Fundamental Particles and Low Temperatures held at the Cavendish Laboratory, Cambridge, on 22–27 July 1946, Volume 1, Fundamental Particles, The Physical Society, London 1947, pp. 1–4, is reprinted by permission of the publisher, Institute of Physics Publishing.

N. Bohr, "Discussion with Einstein on Epistemological Problems in Atomic Physics" in "Albert Einstein: Philosopher–Scientist" (ed. P.A. Schilpp), The Library of Living Philosophers, Vol. VII, Evanston, Illinois 1949, pp. 201–241, is reprinted by permission of P.A. Schilpp.

N. Bohr, "Quantum Physics and Philosophy – Causality and Complementarity" in "Philosophy in the Mid-Century, A Survey" (ed. R. Klibansky), La Nuova Italia Editrice, Firenze 1958, pp. 308–314, is reprinted by permission of the publisher, La Nuova Italia Editrice s.p.a.

N. Bohr, "On Atoms and Human Knowledge", Dædalus **87** (1958) 164–175, is reprinted by permission of Dædalus, Journal of the American Academy of Arts and Sciences.

The following abstracts and article by N. Bohr from Overs. Dan. Vidensk. Selsk. Virks.: "Om den begrænsede Maalelighed af elektromagnetiske Kraftfelter" (with L. Rosenfeld), Juni 1932 – Maj 1933, p. 35; "Analyse og Syntese indenfor Atomfysikken", Juni 1941 – Maj 1942, p. 30; "Naturvidenskabens Erkendelsesproblem", Juni 1950 – Maj 1951, p. 39; "Atomerne og den menneskelige erkendelse", Juni 1955 – Maj 1956, pp. 112–124, are reprinted by permission of the Royal Danish Academy of Sciences and Letters.

The following articles by N. Bohr are reprinted by permission from Nature, © Macmillan Magazines Limited: "The Limited Measurability of Electromagnetic Fields of Force" (with L. Rosenfeld), **132** (1933) 75 (abstract); "Quantum Mechanics and Physical Reality", **136** (1935) 65.

The following articles from the Physical Review are reprinted by permission of The American Physical Society: N. Bohr and L. Rosenfeld, "Field and Charge Measurements in Quantum Electrodynamics", **78** (1950) 794–798; N. Bohr, "Can Quantum-Mechanical Description of Physical Reality be Considered Complete?", **48** (1935) 696–702; A. Einstein, B. Podolsky and N. Rosen, "Can Quantum-Mechanical Description of Physical Reality Be Considered Complete?", **47** (1935) 777–780.

W. Heisenberg, "Über die mit der Entstehung von Materie aus Strahlung verknüpften Ladungsschwankungen", Ber. Sächs. Akad. math.–phys. Kl. **86** (1934) 317–322, is reprinted by permission of the publisher, Akademie Verlag GmbH.

ACKNOWLEDGEMENTS

L. Landau and R. Peierls, "Extension of the Uncertainty Principle to Relativistic Quantum Theory" in "Collected Papers of L.D. Landau" (ed. D. ter Haar), Pergamon Press, Oxford 1965, pp. 40–51, is reprinted in extract by permission of the publisher, Elsevier Science Ltd.

J. Lindhard, "Indeterminacy in Measurements by Charged Particles" (manuscript) is included in extract by permission of J. Lindhard.

L. Rosenfeld, "On Quantum Electrodynamics" in "Niels Bohr and the Development of Physics", Pergamon Press, London 1955, pp. 70–95, is reproduced by permission of the publisher, Elsevier Science Ltd.

L. Rosenfeld, "Niels Bohr in the Thirties" in "Niels Bohr, his life and work as seen by friends and colleagues" (ed. S. Rozental), North-Holland Publ. Co., Amsterdam 1967, pp. 114–136, is reproduced by permission of the publisher, Elsevier Science B.V.

J.A. Wheeler, Abstract of N. Bohr and L. Rosenfeld, "Field and Charge Measurements in Quantum Electrodynamics" in A. Pais, "Developments in the Theory of the Electron", Institute for Advanced Study and Princeton University, Princeton, New Jersey 1948, pp. 42–45, is reproduced by permission of J.A. Wheeler.

PART I

RELATIVISTIC
QUANTUM THEORY

INTRODUCTION

by

JØRGEN KALCKAR

1. ON THE QUESTION OF THE MEASURABILITY OF ELECTROMAGNETIC FIELD
QUANTITIES (1933)

In 1927 Bohr and Heisenberg analysed, on the basis of indeterminacy relations, complementary features in processes of observation within non-relativistic
quantum mechanics. In the paper to be discussed here, on the problem of measurements of local fields, Bohr and Rosenfeld extended the analysis to systems
of infinitely many degrees of freedom and to situations where retardation effects and the relativity of simultaneity must be taken explicitly into account.
Through detailed investigation, Bohr and Rosenfeld exhibit the basis for the
physical interpretation of quantum electrodynamics[1].

In preliminary form, the quantization of the free radiation field, i.e. the concept of photons, was introduced by Einstein in 1905 and constituted the germ
of a general quantum theory of fields. The recognition of the radiation field as
an infinite system of quantized, so-called virtual, harmonic oscillators provided
a rational basis for the photon concept that proved decisive for the later development of quantum electrodynamics. In 1913 Bohr initiated what we now call
the "old" quantum theory, which was able successfully to account for a great
variety of experimental evidence. Nevertheless, Bohr emphasized from the very
beginning the preliminary character of this description. Indeed, detailed studies
revealing its limited range of validity led the way to the creation by Heisenberg,
Schrödinger and Dirac of the consistent formalism of quantum mechanics for
material particles in 1925–1926. The "zero-point vibrations" for a single har-

---

[1] A most lucid summary of the salient points in this paper is found in the first section of an
unpublished manuscript by Bohr on *Field and Charge Measurements*, dating from the thirties. The
first section is quoted in this Introduction, p. [26], and it is reproduced in its entirety on p. [195].

[3]

monic oscillator, implied by the uncertainty relations, soon directed attention to the resulting "vacuum fluctuations" of the fields represented as an infinite manifold of such oscillators. Furthermore, the commutation relations between the generalized position and momentum variables for the virtual harmonic oscillators implied commutation relations between complementary electromagnetic field quantities, with associated indeterminacy relations.

Indeed, the first step towards a comprehensive theory of quantum electrodynamics was taken by Born, Heisenberg and Jordan already in 1925, when matrix mechanics was born. Dirac's paper "Foundations of Quantum Electrodynamics", presenting so many of the essential tools for field quantization, belongs to the same year as the uncertainty relations. Finally, the publication in 1928 of Bohr's Como Lecture coincided, within months, with Dirac's publication of the relativistic wave equation for the electron, with the quantization of fermion fields by Jordan and Wigner, and – most important in the present context – with the derivation of the relativistically invariant commutation relations for free electromagnetic fields by Jordan and Pauli.

Still, considerable difficulties remained within the formalism of quantum electrodynamics because of the various kinds of divergencies and the presence of "negative energy states", which could give rise to doubts as regards the consistency of the description and its very foundations. It must therefore have been reassuring that the derivation by Klein and Nishina in 1929, on the basis of the Dirac equation, of the cross-section for Compton scattering of light by free electrons promised to establish a certain degree of contact with the realm of experience, through its agreement with the classical formula of Thomson in the correspondence limit. Indeed, in 1931 Lise Meitner and collaborators succeeded in performing the first direct experimental test of the Klein–Nishina formula.

At the time when Bohr's Faraday Lecture was published, not only had Heisenberg and Pauli rounded off, as it were, the formalism of relativistic field quantization, but the existence of the positron had been predicted and tentative experimental evidence published by C.D. Anderson. That same year, 1932, C. Møller derived, on the basis of the Dirac equation, the relativistic cross-section for elastic electron–electron scattering.

* * *

Let us now turn to our main theme. From the material available in the Niels Bohr Archive, it seems that Heisenberg was the first to raise the issue of the connection between commutators and the interpretation of the associated indeterminacy relations within quantum electrodynamics. In the early summer of

Werner Heisenberg, Copenhagen 1936.

1929 Heisenberg lectured in Chicago[2], and during his stay there he wrote a letter to Bohr where he sketched such a generalization of their earlier analysis within non-relativistic quantum theory. The relevant part of the letter runs as follows:

"In connection with my lectures here I have also contemplated how one could elucidate the indeterminacy relations for the wave amplitudes. I got the idea of a possible scheme, but I am not quite satisfied with it; I believe that it is only half correct; it would, however, on occasion interest me to hear your opinion on this question in general. My own thoughts run something like this:

As a matter of course, any measurement would yield not E and H at an exact *point* but average values over perhaps very small spatial regions. Let Δv be the volume of the spatial region, then the commutation relations

Heisenberg to Bohr,
16 June 29
German text on p. [437]

---

[2] These lectures are published in W. Heisenberg, *Die physikalischen Prinzipien der Quantentheorie*, Hirzel Verlag, Leipzig 1930. Reprinted in Hochschultaschenbücher, Mannheim 1958, and in W. Heisenberg, *Gesammelte Werke/Collected Works*, Springer-Verlag, Berlin 1985. English edition: *The Physical Principles of Quantum Theory*, University of Chicago Press, Chicago 1930. Reprinted by Dover, New York 1949 (1967).

[5]

between E and the potentials Φ look like this[3]

$$E_i\Phi_k - \Phi_k E_i = \delta_{ik}\, 2hci\frac{1}{\Delta v}, \tag{1}$$

where E_i and Φ_k are now to be interpreted as average values over the spatial volume Δv. Consequently one would expect indeterminacy relations of the form

$$\Delta E_i\Delta\Phi_k \gtrsim \delta_{ik}\frac{hc}{\Delta v} \quad\text{or}\quad \Delta E_x\Delta H_y \gtrsim \frac{hc}{\Delta v\Delta\ell} \tag{2}$$

(and cyclic permutations), if $\Delta v = (\Delta\ell)^3$ is assumed to be a cube.

One method of deduction might be as follows: The energy in Δv must be

$$\text{Energy} = \frac{\Delta v}{8\pi}(E^2 + H^2). \tag{3}$$

If E and H were known exactly one would, for small Δv, face a contradiction to the corpuscle picture, since the energy remains composed of discrete light quanta. The largest light quanta to be detected at all are those with wavelengths not much smaller than $\Delta\ell$. In order to resolve the contradiction, the indeterminacy of the left hand side of (3) must be of the same order of magnitude as this largest detectable hv, i.e.

$$\Delta E_x\Delta H_y \gtrsim \frac{hc}{\Delta v\Delta\ell}.$$

Another, even less solid derivation that I have considered is the following[4]:

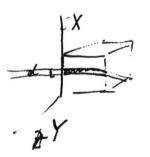

In order to measure the fields E_x and H_y, let us send two electron beams of width d through the cube $(\Delta\ell)^3$, one in the positive, the other in the negative Z-direction. The angle of deflection is given by

$$\frac{1}{p_z}\,e(E_x \pm \frac{p_z}{\mu c}\,H_y)\cdot\frac{\mu\Delta\ell}{p_z},$$

so that the accuracy with which H_y and E_x can be determined amounts to

$$\Delta E_x = \frac{\lambda}{d}\cdot\frac{p_z}{e}\cdot\frac{p_z}{\mu\Delta\ell} = \frac{h}{de}\cdot\frac{p_z}{\mu\Delta\ell}\,;\qquad \Delta H_y = \frac{h}{de}\cdot\frac{p_z}{\mu\Delta\ell}\cdot\frac{\mu c}{p_z}. \tag{4}$$

---

[3] These are of course the equal time commutators at a given instant of time; the averaging pertains solely to the spatial domain.

[4] In his book (both languages) Heisenberg has chosen to discuss the relation $\Delta E_x \cdot \Delta H_z$; consequently, the electrons enter along the y-axis, and the figure is changed accordingly.

[6]

So far, however, we have neglected the fact that the electrons travelling in the beams themselves modify the field noticeably, and even with a partially unknown amount, because it is not known *where* in the beam the electron travels. The ensuing indeterminacy in the average values of E_x and H_y over the cube $(\Delta\ell)^3$ amounts to[5]

$$\Delta E_x \sim \frac{ed}{(\Delta\ell)^3}; \qquad \Delta H_y \sim \frac{ed}{(\Delta\ell)^3} \cdot \left[\frac{p_z}{\mu c}\right].$$ (5)

From (4) and (5) it follows then that

$$\Delta E_x \Delta H_y \gtrsim \frac{hc}{\Delta v \Delta \ell}.$$

I believe that the essential features here are correct, but the utilization of the average values of E and H in the cube is still not quite solid. But would you in principle consider such a discussion reasonable?"

Actually the relation (2) is wrong as shown later by Bohr and Rosenfeld in their more detailed study (BR[6], pp. 12–14, this volume pp. [68]–[70]). In fact, the antisymmetry of the expression

$$[E_j(\vec{x}_1), H_k(\vec{x}_2)] = -2i\hbar c\, \varepsilon_{jk\ell} \frac{\partial}{\partial x_1^\ell} \delta(\vec{x}_1 - \vec{x}_2)$$

implies the vanishing of the commutator between the average values of E_x and H_y for coinciding spatial regions. Even though in his letter to Bohr, Heisenberg admits that his derivations are "not quite solid", they are repeated almost verbatim in his book (chapter III, §§1 and 2), perhaps because it is a reprint of his lecture notes.

The immediate incentive to the work by Bohr and Rosenfeld arose, however, from a manuscript circulated by Landau and Peierls in 1930 and eventually published under the title "Erweiterung des Unbestimmtheitsprinzips für die relativistische Quantentheorie"[7].

---

[5] The factor $\frac{p_z}{\mu c}$ is lacking in Heisenberg's letter but is correctly inserted in his book.

[6] This abbreviation is used here and in the following for reference to pages and formulae in the paper by Bohr and Rosenfeld, *Zur Frage der Messbarkeit der elektromagnetischen Feldgrössen*, this volume p. [55]. The English version is found on p. [123].

[7] L. Landau and R. Peierls, *Erweiterung des Unbestimmtheitsprinzips für die relativistische Quantentheorie*, Z. Phys. **69** (1931) 56–69. Reprinted in English translation in *Quantum Theory and Measurement* (eds. J.A. Wheeler and W.H. Zurek), Princeton Series in Physics, Princeton 1983; and in *Collected Papers of L.D. Landau* (ed. D. ter Haar), Pergamon Press, Oxford 1965. The crucial sections are reprinted in this volume, Appendix to Part I, p. [229].

Lev Landau.

The general background was as follows:

On the basis of quantization of the electromagnetic radiation field, Jordan and Pauli (1928) and Pauli and Heisenberg (1929) derived, as already mentioned, the commutation relations between components of a free electromagnetic field, quoted in the Bohr–Rosenfeld paper as equations (1) and (2). They imply indeterminacy relations between such quantities, which are of the usual reciprocal type and in accordance with the causality inherent in special relativity (cf. BR, eq. (8)). In flat contradiction to this result, Landau and Peierls claimed to have demonstrated that it was impossible to determine even a *single* field component by means of a charged point particle.

The crux of their argument refers to an allegedly uncontrollable energy–momentum loss through radiation emission during the very act of measurement, while the charge is accelerated by the external field it is supposed to determine.

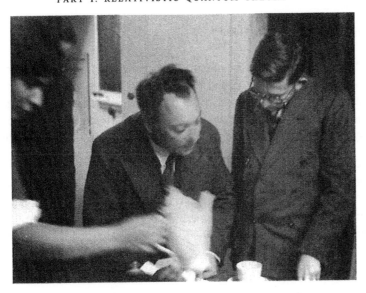

Hilde Levi, Wolfgang Pauli and Rudolf Peierls, Copenhagen 1936.

On this basis, they arrived at an absolute lower limit for the accuracy of definition of electric or magnetic fields, ΔE or ΔH, given by

$$\Delta E \sim \Delta H \gtrsim \frac{\sqrt{\hbar c}}{(c\Delta t)^2}$$

where Δt represents the latitude in temporal fixation of the measuring process. According to these expressions, only static fields ($\Delta t \rightarrow \infty$) are measurable within the quantal domain[8].

Later, in the introduction to their paper, Bohr and Rosenfeld emphasize that if it were indeed true "that the measurability of field quantities is subjected to further restrictions which go essentially beyond the presuppositions of quantum field theory", it would "deprive this theory of any physical basis". In order to resolve this dilemma, Bohr and Rosenfeld utilize an extended test body, sufficiently heavy to permit its acceleration to be neglected, for any given value of its charge and of the momentum transfer from the field to be measured.

Actually, the derivation by Landau and Peierls was criticized early by Pauli, who during the summer 1931 must have felt it was time he took Peierls to task.

---

[8] Cf. in this connection the letter from Heisenberg to Pauli, 12 March 1931, reproduced in Part III, p. [440].

[9]

He barks[9]:

Pauli to Peierls,
3 July 31
PWB II, letter [281]
German

"Today I had a look at your physics again and this time at the paper by you and Landau. When I had to give a colloquium on this paper I noticed that your arguments as regards the inequalities for the field strengths are erroneous. (Perhaps the inequalities themselves are correct, however, and you may improve the derivation). Obviously, it is wrong that the radiation energy $\frac{e^2}{c^3} \frac{(v'-v)^2}{\Delta t}$ represents an *uncertain* energy change. (Note also that the whole consideration pertains only to *macroscopic* test bodies with charge e for which $e^2/hc > 1$). It may be that the radiation energy also contains some uncertainty in the time development, but in first approximation the radiation energy certainly represents a *definite* change. Hence the equation

$$(v'-v)\Delta P > \frac{e^2}{c^3} \frac{(v'-v)^2}{\Delta t}$$

is certainly also wrong as an uncertainty relation. This is already clear from the fact that it does not contain h and if correct would postulate a fundamental momentum uncertainty for charged particles in the classical theory![10]

Please write me your opinion about this question. In this case, by the way, I am in favour of publication. Either you (together with Landau) must publish an improved derivation of the inequalities for the field strengths, or you must confess in public that these inequalities (for macroscopic test bodies of high charge) are so far without foundation."

It adds to the confusion that Landau and Peierls do not at all distinguish between the *energy* and the *momentum* of the radiation hypothetically emitted during the act of measurement. In connection with Pauli's rejection of the "uncertainty" suggested by Landau and Peierls because it does not contain h, it may be observed that neither does the corresponding expression (BR, eq. (42)) derived by Bohr and Rosenfeld: We have there to do with the classical average field produced by the macroscopic test body, but the displacement D_x of this body is a quantal variable complementary to the momentum of the test body.

---

[9] The full letter in German is reproduced in Wolfgang Pauli, *Wissenschaftlicher Briefwechsel mit Bohr, Einstein, Heisenberg u.a., Band II: 1930–1939* (eds. K. v. Meyenn, A. Hermann and V.F. Weisskopf), Springer-Verlag, Berlin 1985. In the following, any letter from this volume is referred to as PWB II followed by the number of the letter.

[10] Cf. in this connection the comments in the appendix on p. [229].

Dated as late as 18 January 1933, we find the following amazing letter from Pauli to Heisenberg. At this time the Bohr–Rosenfeld paper was almost finished and Bohr must obviously have known that the inequalities (1) in Pauli's letter below, were wrong, contrary to Pauli's remark in parentheses in the letter. However, Pauli might not have followed the progress of the work in detail during this period. (This is confirmed by Bohr's letter to Pauli, 25 January 1933, quoted in Part III, p. [463]).

"Now I have yet another particular question which I have discussed with Peierls for a long time without reaching a solution. That is the question of the *uncertainty relations for the field quantities*. In the later part of the article[11], the proofs of which are still to arrive, there is a passage which is not correct and has to be altered. According to Landau and Peierls the following uncertainty relations should be valid for the *individual* field components[12],

Pauli to Heisenberg, 18 Jan 33
PWB II, letter [304]
German

$$|\Delta \vec{E}| > \frac{\sqrt{hc}}{(c\Delta t)^2} \; ; \qquad |\Delta \vec{H}| > \frac{\sqrt{hc}}{(c\Delta t)^2}. \tag{1}$$

(Bohr believes these inequalities to be *correct*, as he has definitely emphasized to me after repeated inquiries). [*sic*]

The argumentation for these inequalities follows according to Landau and Peierls via the inequalities

$$e|\Delta \vec{E}|\Delta t > |\Delta \vec{p}| \tag{2}$$

$$|\Delta \vec{p}| \, \Delta t > \frac{h}{v' - v} \tag{3}$$

$$|\Delta \vec{p}| \, \Delta t > \frac{e^2}{c^3}(v' - v). \tag{4}$$

I have no objections to (2), (3), but the inequality (4) is of a different character. It should be valid for momentum determinations of *charged macroscopic* bodies. Only for $e^2 > hc$, i.e. for highly charged bodies, does it give anything new in comparison with (3). (Then there would always be *many* light quanta emitted by the momentum measurement). In this case the inequality (4) should be established as a consequence of the *radiation* from the body.

---

[11] W. Pauli, *Die allgemeinen Prinzipien der Wellenmechanik*, Handbuch der Physik **24**, no. 1 (1933) 83–269, pp. 256 ff.

[12] The following equations are numbered as in the original document.

I raise an *objection* to this idea. One may after all enclose the charged test body in a very large box that is impenetrable to light, and *measure* with arbitrary accuracy the recoil of the box as well as the energy it has absorbed[13]. Thus the change in momentum of the test body through radiation does not seem to be *uncertain*, as assumed by Landau and Peierls, but on the contrary accurately *measurable*.

Thus it remains open whether the inequalities (1) for the field strengths could not after all be circumvented by the utilization of highly charged macroscopic test bodies. Neither Peierls nor I have succeeded in invalidating my objection (although intuitively I hold (1) to be sensible).

What do you think of this? Furthermore: I remember that the considerations in your book implied not only an inequality for $|\Delta \vec{E}_x \cdot \Delta \vec{H}_y|$, but also for $\Delta \vec{E}$ and $\Delta \vec{H}$ separately. However, I do not know any longer what these inequalities look like and how they came about. *Could you write to me about this?*"

* * *

It is obvious that the treatise by Bohr and Rosenfeld should neither be regarded as a rejoinder to Heisenberg's analysis nor to Landau and Peierls's paper. An essential distinction is the recognition that a consistent analysis involves fields averaged over finite space–time domains and that the measurements thus require test bodies with corresponding extensions. (In this connection it is important to note their critical remark (BR, p. 26) on the arbitrariness in identifying the indeterminacy Δx in the position of a point charge with the total spatial size L of the domain).

The Bohr–Rosenfeld paper presents a systematic analysis of those new features of complementary relationships exhibited by observations in measuring situations concerning systems of infinitely many degrees of freedom, when special relativity must be taken explicitly into account. The simplest case to address is a measurement of a single electric or magnetic component of a free radiation field – averaged over a given space–time region – by means of an extended and heavy, electrified or magnetized test body. In this situation, only the constants h and c are involved which do not suffice to define any scale with dimension of a length. Consequently, the atomic constitution of the test body can be neglected and its charge treated as a continuous classical distribution. Furthermore, it is possible to idealize the body so that it possesses only a

---

[13] This possibility is excluded by the arguments in BR, p. 26, this volume p. [82].

[12]

Léon Rosenfeld, Copenhagen 1931.

single degree of freedom, represented by its displacement as a whole, and non-relativistic quantum mechanics suffices to account for this translatory motion.

Even in this comparatively simple situation, Bohr and Rosenfeld demonstrate that particularly contrived and highly idealized compensation devices must be brought into action in order that the method of idealized measurement provide an optimal determination of the field components, in accordance with the indeterminacy relations (BR, eq. (8))[14]. But the most subtle part of the discussion concerns the rôle of the non-compensable field fluctuations in connection with the question of reproducibility of average field measurements over one and the same spatial region, repeated during partly overlapping time intervals[15].

[14] Cf. in this connection the illuminating analysis of a particularly simple situation by W. Heitler in his monograph, *Quantum Theory of Radiation*, Clarendon Press, Oxford 1954, 3rd edition, chapter II, §9.
[15] Cf. the letter from Bohr to Heisenberg, 13 March 1933, quoted in Part III, p. [443].

[13]

In the Selected Correspondence at the end of this volume, further letters, exchanged between Bohr, Heisenberg and Pauli during this period, are quoted. From these the reader may gain a vivid impression of the initial confusion, as well as the quite formidable effort required to resolve the paradoxes and thereby to analyse and elucidate the conceptual basis of quantum electrodynamics.

∗ ∗ ∗

A problem of crucial importance for the field measurements is to achieve a rigid displacement of the extended test body at a definite instance, even if the spatial extension L (in units of c) is very large compared to the time interval T over which the field component is averaged; at the same time the total momentum of the test body must remain well defined. Bohr and Rosenfeld solve this problem with a remarkably cunning device (cf. BR, pp. 30–34, this volume pp. [86]–[90]), which establishes exactly the same correlation between relative distance and total momentum of quantal objects as the one we encounter in the Einstein–Podolsky–Rosen experiment (cf. p. [250]). This circumstance is brought out with special clarity in the following exchange of letters between Weisskopf, Rosenfeld and Bohr:

Weisskopf to Rosenfeld, 2 Dec [33]
German text on p. [512]

[Zurich,] December 2, [1933]

Dear Rosenfeld,

I am just now conducting a seminar on worms[16], and there is a passage that I do not understand. It concerns the measurement of the total momentum of the test-body system, on p. 30 in the 4th proof.

There you describe the method of momentum measurement by means of the Doppler effect, and you claim that each test-body element is subjected to the same displacement, the absolute magnitude of which, however, remains unknown. Now, I do not understand how, in the arrangement described, all test-body elements do in fact suffer the same displacement. Because it is possible through a simple modification of the experimental arrangement to measure the momentum of every individual test-body element: You just have to place a system of mirrors along the path of the reflected rays, by means of which the reflected light from each test-body element is

---

[16] From A. Pais I have learned the background for this peculiar phrase. The story runs something like this: "Once upon a time there was a German professor in zoology who was not only the leading expert with respect to worms ('Würmer'), but absolutely obsessed by this topic. It happened that he was invited to give a lecture on: 'Ornithology'. He began the lecture with the sentence: Many birds live from eating worms! – whereupon he continued to discuss nematology, his pet subject."

[14]

sent in different directions, so that their spectrum *after separation* could be determined with *arbitrary accuracy*. Obviously, it is impossible that the introduction of such a system of mirrors, which merely deflects the reflected radiation, could influence the incoming light hitting the test-body elements. Therefore, since one may determine the momentum of each individual test-body element, their *relative* displacements must remain uncertain. If I accidentally measure all the reflected rays together, as Bohr does, it does of course not make any change in the displacement, but, as far as I can see, it also does not represent the optimal method for measuring the total momentum.

I am absolutely sure that you and Bohr are right, but I should be grateful for further explanation. If you present the question to Bohr, please tell him in advance that it is not meant as an objection, but is merely pedagogical.

With most cordial greetings to all
Copenhageners and Belgian ladies
in your surroundings,

Cordially yours,
Weisskopf

[Copenhagen,] December 5, [19]33

Bohr to Weisskopf,
5 Dec 33
German text on p. [513]

Dear Weisskopf,

Rosenfeld just showed me your friendly letter which gave me quite a fright, not because we have a bad conscience, but because I feel that our endeavours, our struggle to couch the issue in words, have met with even less success than I feared. Indeed, the exposition in the section you refer to seemed to us particularly clear, and almost too elaborate.

The point is that, in general, loss of knowledge of the position of a body through determination of its momentum does not at all arise from restrictions with regard to the tracing of details in the collision processes, but solely from the impossibility of fixing the development of these processes relative to a definite space–time frame of reference. In our case, this latter circumstance is disclosed by the fact that the absolute instant of time for the opening of the diaphragm that serves to cut off the incoming beam of radiation cannot be determined together with the amount of energy which is exchanged between the shutter and the radiation passing through the diaphragm. In contrast, given a sufficiently heavy shutter, the process of opening may be followed in arbitrary detail. The accuracy that we achieve by

Victor Weisskopf, Copenhagen c. 1938.

measuring the total momentum depends entirely on the fact that we only need to employ a single diaphragm. If a determination of the momentum of each individual test-body element were required, we would have to use as many independent diaphragms as there are test-body elements. Thereby these bodies would not only suffer different displacements, but the accuracy of the determination of the total momentum would be reduced by $1/\sqrt{N}$, where N is the number of test-body elements. Incidentally, the reason why – as we have shown – it can be achieved that all displacements of the test-body elements are the same, and, at the same time, the determination of the total momentum corresponding to the displacements remains as exact as possible, is that the total momentum is conjugate to the position of the centre-of-mass, but commutes with the relative displacements of the test-body elements. Besides, these relative displacements may in fact be determined and even regulated afterwards, by means of a suitable telegraphic procedure, without changing the total momentum. As we have emphasized in section 5, we can freely choose between an arrangement of mirrors and a telegraph arrangement.

I hope that this reply will satisfy you and I should be grateful if you would let me know as soon as possible. By the way, I need hardly say that I am only grateful for any such fright – whether it be of a physical or a psychological nature. Rosenfeld and I suffered many similar experiences while going through the manuscript for the last time, and we have tried to save the reader from these through a number of small alterations in the last paragraph.

With the kindest greetings from both of us, also to Pauli – I was in every respect very happy with his most recent letter.

Yours

[Niels Bohr]

As Bohr explains, the total momentum of the composite test body commutes with the relative positions of the elements

$$[P, x_{nm}] = 0,$$

$$P = \sum_m p_m, \quad x_{nm} = x_n - x_m.$$

It is therefore possible to determine the total momentum in a measurement in which the displacements of the test-body elements, though indeterminate, are the same for all the constituents of the body. This possibility is closely analogous to the arrangement discussed in connection with the so-called "Einstein–Podolsky–Rosen Paradox" (cf. Part II, p. [250]).

* * *

The remarks in the preceding section of this introduction did not address the commutation relations between field quantities since it merely concerned the conditions for measurability of a single field component. It might therefore be worthwhile to emphasize that the extension of the analysis by Bohr and Rosenfeld to elucidate the complementary relationships between the determination of space–time averages of *two* field components was not at all motivated by apparent paradoxes like those suggested by Einstein in connection with the indeterminacy relations between position and momentum in non-relativistic quantum mechanics. Moreover, field measurements appear to receive but little attention in modern quantum electrodynamics, and in current textbooks the work by Bohr and Rosenfeld is merely referred to in passing, at most, in the first chapter. Still, I believe that this profound analysis – the *only* one, so far – of "relativistic complementarity" not only discloses new aspects of the interplay between the relativistic and the quantal descriptions, but also illuminates and emphasizes in a novel manner fundamental features of the measurability problem in *non*-relativistic quantum mechanics.

***

No early drafts of the Bohr–Rosenfeld paper have been found in the Niels Bohr Archive. A glimpse of the prevailing mood is provided, however, through the following pages extracted from Rosenfeld's reminiscences about the progress of the work, which he contributed to the collection of essays dedicated to Niels Bohr on the occasion of his seventieth birthday[17].

ON QUANTUM ELECTRODYNAMICS
L. Rosenfeld

... et discors concordia fetibus apta est.
OVID, *Metam*. I, 433

When I arrived at the Institute on the last day of February, 1931, for my annual stay, the first person I saw was Gamow. As I asked him about the news, he replied in his own picturesque way by showing me a neat pen drawing he had just made*. It represented Landau, tightly bound to a chair

---

* I am afraid this work of art has been allowed to disintegrate before its historical value could be realized. [The substitute is from G. Gamow, *Thirty Years That Shook Physics*, Doubleday, New York 1966.]

[17] *Niels Bohr and the Development of Physics* (ed. W. Pauli with the assistance of L. Rosenfeld and V. Weisskopf), Pergamon Press, London; McGraw–Hill Book Co., New York 1955.

Gamow's later reconstruction of his Landau–Bohr drawing.

and gagged, while Bohr, standing before him with upraised forefinger, was saying: "Bitte, bitte, Landau, muss ich* nur ein Wort sagen!" I learned that Landau and Peierls had just come a few days before with some new paper of theirs which they wanted to show Bohr, "but" (Gamow added airily) "he does not seem to agree – and this is the kind of discussion which has been going on all the time." Peierls had left the day before, "in a state of complete exhaustion," Gamow said. Landau stayed for a few weeks longer, and I had the opportunity of ascertaining that Gamow's representation of the situation was only exaggerated to the extent usually conceded to artistic fantasy.

There was indeed reason for excitement, for the point raised by Landau and Peierls [1] was a very fundamental one. They questioned the logical consistency of quantum electrodynamics by contending that the very concept of electromagnetic field is not susceptible, in quantum theory, to any physical determination by means of measurements. The measurement of a field component requires determinations of the momentum of a charged

---

* This is a familiar danicism of Bohr's for "darf ich".

test-body; and the reaction from the field radiated by the test-body in the course of these operations would (except in trivial cases) lead to a limitation of the accuracy of the field measurement, entirely at variance with the premises of the theory. In fact, the quantization of the field only entails reciprocal limitations of the measurements of pairs of components, arising from their non-commutability, but no limitation whatsoever to the definition of any single field component. On the other hand, one had to face another inescapable consequence of the field quantization, the occurrence of irregular fluctuations in the value of any field component; the existence of this fluctuating "zero-field" (as it was called because it persists even in a vacuum) was known to be responsible for one of the divergent contributions to the self-energy of charged particles, but its meaning was very obscure. Landau and Peierls, somewhat illogically, tried to bring it in relation with their alleged limitation of measurability of the field, and this only further confused an already tangled issue.

MEASURABILITY OF ELECTROMAGNETIC FIELDS[*]

Bohr's state of mind when he attacked the problem reminded me of an anecdote about Pasteur [3]. When the latter set about investigating the silkworm sickness, he went to Avignon to consult Fabre. "I should like to see cocoons," he said, "I have never seen any, I know them only by name." Fabre gave him a handful: "he took one, turned it between his fingers, examined it curiously as we would some singular object brought from the other end of the world. He shook it near his ear. 'It rattles,' he said, much surprised, 'there is something inside.' " My first task was to lecture Bohr on the fundamentals of field quantization; the mathematical structure of the commutation relations and the underlying physical assumptions of the theory were subjected to unrelenting scrutiny. After a very short time, needless to say, the roles were inverted and he was pointing out to me essential features to which nobody had as yet paid sufficient attention.

His first remark, which threw decisive light on the problem, was that field components taken at definite space–time points are used in the formalism as idealizations without immediate physical meaning; the only meaningful statements of the theory concern averages of such field components over finite space–time regions. This meant that in studying the measurability of field components we must use as test-bodies finite distributions of charge

---

* This section contains an analysis of the paper on the subject by Bohr and Rosenfeld [2], to which the reader is referred for further details.

and current, and not point charges as had been loosely done so far. The consideration of finite test-bodies immediately disposed of Landau and Peierls' argument concerning the perturbation of the momentum measurements by the radiation reaction: it is easily seen that this reaction is so much reduced, for finite test-bodies, as to be always negligible.

On the other hand, the construction and manipulation of extended test-bodies proved a most perplexing affair.

. . .

At first, all went well. We had a lively surprise when we found out that the only case of coupled measurements for which a reciprocal indeterminacy relation had been explicitly written down and discussed in the literature, namely that of an electric field and a perpendicular magnetic field in the same volume element, was one in which unlimited accuracy had to be expected from the correctly integrated commutation law. It was immediately clear that in such a case the mutual perturbations of our extended test-bodies would cancel out and that we could fulfil the theoretical prediction. More exciting was the realization that the reaction of the test-body upon itself in the course of the measurement of any single field component could be automatically compensated, at least so far as it can be classically computed. In fact, this reaction, being proportional to the displacement of the test-body, can be matched exactly by an elastic spring mechanism of known strength.

I must not by-pass the problem of the actual momentum measurement of the extended test-bodies. To take account of relativistic effects, these bodies must be imagined to consist of a large number of independent elements, and it was far from obvious how the total momentum of all these elements could be obtained without multiplying the error far beyond the optimum limit which we required. Moreover, one had to make sure that the relativity requirements could be met without further restriction to the measurability of the momentum. This necessitated a much more detailed analysis of the measuring process than one was wont to in ordinary quantum mechanics. Bohr succeeded in showing that the measurement of the total momentum can even be performed in such a way that the displacements of the elements, though uncontrollable within a finite latitude Δx, are all equal, and that the determination of the total momentum is only limited by the uncertainty of the common displacement Δx to the extent $\hbar/\Delta x$ indicated by the indeterminacy relation. The interest of this result transcends its immediate application to the problem of field measurement: it affords a specially clear example of a measuring process which can be entirely described in a purely classical way, and in which the origin of the reciprocal indeterminacies is thus directly traced to the impossibility of specifying the dynamical characteristics of the system without loosening its connection with the space–time frame of reference*. The solution of this problem is one of the most striking products of Bohr's uncanny virtuosity in this subtle kind of analysis. It cost him some hard thinking, but was for him a source of intense enjoyment.

...

The realization of such a complete harmony between the formalism and its physical interpretation was felt as a fitting reward for our tribulations, and with great alacrity Bohr presented our conclusions and handed in the manuscript of our paper at a meeting of the Danish Academy on the 2nd December, 1932. The reading of the fourteen or so successive proofs only took about one more year. One point especially still gave us much trouble to the very last: what part has one to assign to the field fluctuations in the logical structure of the theory? This is a quite fundamental question, which I have kept to the last for some comment.

The self-reactions and mutual perturbations of the test-bodies can only be compensated by spring mechanisms to the extent of their classically evaluated average magnitudes. Their actual values, however, owing to the quantal character of the electromagnetic field, deviate from the classical averages to an extent which necessarily escapes our control: for such a control would involve a determination of the actual numbers of photons emitted in the various modes of oscillation and accordingly entail the loss of our knowledge of the phase relationships between these modes. Such statistical fluctuations accompanying any field defined by classical sources have, however, a universal character: in fact, they are just the zero-field fluctuations. This may be seen in the following way:

The radiation part of any field may be represented, in the usual way, in the form

$$F = \sum_i (a_i f_i + a_i^\dagger f_i^*), \qquad (10)$$

in terms of the annihilation and creation amplitudes $a_i, a_i^\dagger$ belonging to the different modes labelled by the index i. We may take formula (10) to represent the space–time average of some field component over a given space–time region; the f_i, f_i^* will then be the corresponding space–time integrals over the appropriate progressive phase functions. If the field is defined by

---

* It has apparently, like so many things, escaped the attention of those quixotic young physicists who, spell-bound by distorted echoes of a lore unfathomed, try to split the quantum with the rusty sword of mechanistic materialism.

a classical source distribution, the expectation values $\langle a_i \rangle, \langle a_i^\dagger \rangle$ are given in modulus and phase:

$$\langle a_i \rangle = \overline{N}_i^{1/2} e^{i\chi_i}, \qquad \langle a_i^\dagger \rangle = \overline{N}_i^{1/2} e^{-i\chi_i}; \tag{11}$$

in this notation, $\overline{N}_i$ represents the average number of photons of mode i.

Now, we can readily set up the expression for the state-vector describing this situation. In fact, since the emission processes which build up the radiation field are statistically independent events, the actual numbers N_i of photons in the different modes are distributed around the averages $\overline{N}_i$ according to the Poisson formula

$$p(N_i) = (N_i!)^{-1} \overline{N}^{N_i} e^{-\overline{N}_i} \tag{12}$$

and the state-vector is of the form

$$\Psi(N_1 N_2 \ldots) = \Pi \phi(N_i) \qquad \text{with} \quad |\phi(N_i)|^2 = p(N_i). \tag{13}$$

In order to obtain the expressions (11) for the average amplitudes, it suffices to fix the phases of the factors $\phi(N_i)$ by

$$\phi(N_i) = |\phi(N_i)| e^{iN_i\chi_i}. \tag{14}$$

Using the state-vector (13), (14), a straightforward calculation shows that the mean square fluctuation of F,

$$\langle F^2 \rangle - \langle F \rangle^2,$$

reduces *exactly* to the expression $\sum_i |f_i|^2$, quite independent of the $\overline{N}_i$'s and χ_i's.

The Poisson distribution of photons has lately been established by very elegant calculations utilizing the new methods of quantum electrodynamics [4]. In those days, however, people were not so learned. Judging from the surprise which the formula (12) excited in those who saw it, I think Bohr and, no doubt, Pauli must have been the only ones who knew that this property is an inherent part of the photon idea.

The property just discussed of the quantal field fluctuations gives the clue to the assessment of their significance for the consistency of the theory. In fact, it becomes clear that the fluctuations affecting the fields produced by the test-bodies during the measurements cannot be separated from those which equally affect the field which we want to measure. The impossibility of compensating or controlling them in any way, far from being an imperfection of the measuring device, is a property which it must necessarily possess to ensure that all the consequences of the theory are in principle

[23]

verifiable by measurement. The fact that the zero-field fluctuations are superposed on to the classical field distribution is indeed a well-defined theoretical prediction, and we see that we are able to suppress the perturbations arising from the manipulation of the test-bodies to an extent which
just leaves scope for the test of this prediction.

Now the existence of such fluctuations certainly means that the causal
connection of classical theory between the fields and their sources is lost
in quantum theory: this is just one aspect of the breakdown of determinism and its replacement by a wider form of statistical causality. This issue,
however, should not be confused with the question whether the field concept
can be consistently upheld in the quantal mode of description. Our analysis of field measurability shows that any measurement of a field average
will always give a definite answer, which will be reproducible (this means
that measurements of the same field component over two almost coinciding space–time regions will yield the same answer): but the result of any
such measurement may differ by any amount from its classical estimate.
The breakdown of the classical connection between field and source is a
feature of the quantum theory of fields just as fundamental as the reciprocal limitation of measurability of field components; both are in fact direct
and formally quite parallel consequences of the commutation laws which
express the quantization of the field.

The classical solution of the field equations, expressed in terms of the
retarded potentials corresponding to the given charge and current distribution, does not satisfy the commutation rules. Owing to the linear character
of the equations, however, it can always be supplemented by a solution F_0
of the type (10) of the homogeneous equations, which represents a pure radiation field due to far-away sources not included in the system. If there is
no such external source at infinity, the radiation field F_0 would in classical
theory reduce to zero, but in quantum theory it does not vanish identically.
The condition $a_i^\dagger a_i = 0$ expressing that the field does not contain any photon, implies that the expectation value of each component of F_0 vanishes
everywhere, but any space–time average of the field component will exhibit
fluctuations with a finite quadratic mean

$$\langle F_0^2 \rangle = \sum_i |f_i|^2.$$

This leads us again to our previous conclusion, but the present argument
shows how essentially it is linked to the characteristic noncommutability of
the quantal variables. Indeed, one may say that the existence of the zero-
field immediately follows from the additional requirement imposed on the

[24]

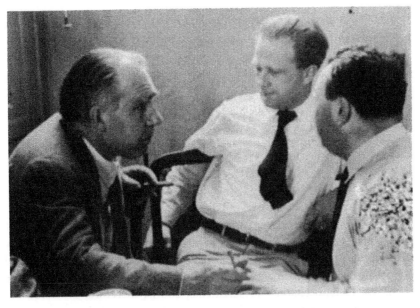

Niels Bohr, Werner Heisenberg and Wolfgang Pauli, Copenhagen 1936.

solutions of the field equations by the commutation laws.

At the end of our laborious inquiry, we had thus completely vindicated the consistency of quantum electrodynamics, at least in its simplest form. Our increased insight invited a re-assessment of the scope of the analysis we had just completed. We had set out on the suspicion of grave defects in the logical structure of the theory, and used the direct method of testing definitions of concepts by investigating the concrete measuring processes they embody. Knowing, however, that the formalism is free from logical flaws and that the physical interpretation of its symbols is in strict accordance with their relation to observable quantities, it becomes tautological to assert the possibility of constructing measuring devices capable of verifying all theoretical predictions. All that the actual design of such instruments can do is to illustrate the logical relationships and, perhaps, elucidate their meaning.

...

References

[1] L. Landau and R. Peierls; *Z. Physik* **69**, 56, 1931

[25]

[2] N. Bohr and L. Rosenfeld; *Dan. Mat.–fys. Medd.* **12**, No. 8, 1933
See also B. Ferretti, *Nuovo Cimento* **12**, 558, 1954

[3] J.H. Fabre; *Souvenirs entomologiques,* **IX**, Chapter XXIII

[4] W. Thirring and B. Touschek; *Phil. Mag.* **62**, 244, 1951
R. Glauber; *Phys. Rev.* **84**, 395, 1951
H. Umezawa, Y. Takahashi and S. Kamefuchi; *Phys. Rev.* **85**, 505, 1952
J. Schwinger; *Phys. Rev.* **91**, 728, 1953

...

* * *

I shall of course not attempt to summarize in further detail the main lines of argumentation in the paper by Bohr and Rosenfeld. Indeed, as already alluded to on p. [3], we are fortunate in having a succinct summary of its salient points in an unpublished manuscript by Bohr dating from a few years later. It deals primarily with charge measurements, and is, therefore, discussed in section 3. However, as a background, the introductory part of this manuscript outlines the issues associated with field measurements and, in particular, it elucidates the rôle of the quantal field fluctuations, produced by the test body, as being inseparable from the fluctuations in the field itself which is the object of the measurement. This part[18] is, therefore, reproduced below, as a guide for the reader.

*1. Field Theory.

The quantities defining the electromagnetic field at a given point of space–time are idealizations derived from the consideration of the forces acting on electrified or magnetized test-bodies filling up finite space regions around the given point during finite time intervals, by letting the space–time regions thus constituted shrink indefinitely towards the given point. While in classical theory this limiting process does not involve any difficulty of principle, it may give rise in the quantum theory of fields to certain paradoxes, connected with the appearance in the fundamental commutation rules for field components of the singular Delta-function. To avoid all such paradoxes it is necessary in quantum theory explicitly to consider, instead of the mentioned idealizations, mean values of field quantities over finite space–time regions; on account of the fact that the universal constants of field the-

---

[18] The manuscript is reproduced in full on p. [195].

ory, the quantum of action h and the velocity of light c, are not sufficient to define units of length or time, no limitation is imposed by the theory on the dimensions of the space–time regions considered.

An important consequence of this fact is that in the construction and use of test-bodies for the purpose of testing the predictions of the theory, any limitation arising from the atomic constitution of matter may entirely be disregarded. The idealized test-bodies will thus be assumed to approximate to any desired accuracy continuous distributions, over the given space regions, of electric or magnetic charge (the last case being obtained by considering each element of the test body as one end of a long flexible magnetized rod). If this distribution is uniform, the test body will be suited to measure the ordinary mean value of the form[19]

$$\overline{E}_x^{(G)} = \frac{1}{VT} \int\limits_{(T)} dt \int\limits_{(V)} dv \, E_x.$$
(1)

We shall first confine ourselves to this case, and later on discuss generalizations.

The mean value under consideration is obtained from the total momentum transferred to the test body by dividing this momentum by the product of time interval and total charge of the test body. The measurements required are thus determinations of the momentum of the test body in the direction considered at the beginning and end of the time interval. These momentum measurements involve uncontrollable displacements of the test body; the accuracy of the momentum determination Δp_x being connected with the displacement Δx by Heisenberg's relation

$$\Delta p_x \cdot \Delta x \geq h.$$
(2)

In order that the dimensions of the space region over which the mean value is taken be sufficiently well-defined, it is necessary to arrange the measuring procedure so as sufficiently to reduce the displacement Δx; this can, however, be effected without impairing the accuracy of the field measurement, since the latter depends on the product of Δx [and] the charge of the test body, which may be chosen arbitrarily large. Similarly the duration Δt of the momentum measurements may be assumed as small as desired in comparison with the time interval T.

Some complication, but no essential difficulty, arises from the consideration of the requirements of relativity. Thus, due to the retardation of all

---

[19] The numbering of the formulae is added by the editor for convenience of the reader.

forces, it is not permissible to treat the extended test body as a solid body, but we must imagine it subdivided in arbitrarily small, though finite parts, the momentum of each of which is measured independently. By making use of the fact, fundamental in the theory of measurements, that the relative course of the measuring process may always be accurately described in a classical way, it is nevertheless possible to manage to give all partial test bodies exactly equal displacements at exactly the same instant. The uncertainty Δx is due to the impossibility, when aiming at a momentum determination of accuracy Δp_x, of fixing the absolute position of the test bodies with respect to the fixed frame of reference.

It might be thought, however, and it has in fact been suggested (by Landau and Peierls), that a quite essential difficulty should be involved in the measurement of the momentum of a charged body, because the acceleration of this charge during the measurement gives rise to an electromagnetic field, the reaction of which on the test body affects the measured value of the momentum, by an amount depending on the uncontrollable displacement of the test body during the measurement. An estimation of the resulting uncertainty in the result of the momentum determination, carried out by treating the test body as a point charge, has led (Landau and Peierls) to the conclusion that it is so large as to deprive the definition of any field quantity of any unambiguous meaning. When due account is taken, however, of the finite extension of the test bodies, it is easily seen that the perturbation arising from the radiation of the test body during the measurement can always be arbitrarily reduced. By the use of the idealized measuring procedure outlined above, it is thus possible to define with unlimited accuracy the mean values of all field quantities over any finite space–time region. It must also be pointed out that the results of such measurements are strictly reproducible; the repetition of the measurement consists in carrying out a second determination by means of a test body (or system of test bodies) superposed on the first test body in the space region considered, and extending over a time interval infinitesimally displaced with respect to the time interval of the first measurement.

As regards the interpretation of the field quantities thus measured, it must in the first place be noticed that they consist of the sum of the field present before the measurement and of the field created within the region occupied by the test body by the body itself during the whole time interval. Now the latter field, as far as it can be calculated on classical theory, is proportional to the displacement of the test body and can thus be exactly compensated by means of suitable (elastic) contrivances. But this classical part of the field of the test body is only its average value; around this value there are statistical fluctuations due to the fact that the field in question partly consists of a certain number of light quanta, of which only the average value is unambiguously predictable. (*Or*: due to the fact that the field in question partly arises from individual emission processes, of which only the average effect is unambiguously predictable). The results of the measurements will thus unavoidably be affected by these fluctuations.

On the other hand, it is a characteristic consequence of the quantum theory of field[s] that all mean values of field quantities (except in the trivial case where the state of the system is defined just by the specification of these values) are always subject to statistical fluctuations, which do not vanish even in complete absence of charges and electromagnetic energy. These fluctuations become the larger, the smaller the dimensions of the space–time extensions are; in the limiting cases $cT \gg L$ and $cT \ll L$ respectively, they are of the order of magnitude

$$\frac{\sqrt{hc}}{L \cdot cT} \quad \text{and} \quad \frac{\sqrt{hc}}{L^2}. \tag{3}$$

Now it is readily estimated that, whereas the compensable part of the field of the test body gets larger and larger when the accuracy of the measurement is increased, the fluctuating part of this field tends to a finite limit which has exactly the same expression as the theoretical fluctuations in empty space just mentioned. This may be seen as follows:

The wavelength of the radiation mainly emitted during the displacement of the test body is of the order of magnitude of the linear dimensions L of the body. If F is the value of the field created during the displacement of the test body, the total energy emitted will be F^2V, corresponding to a number[20]

$$\frac{F^2V}{hc/L} = \frac{F^2L^4}{hc} \tag{4}$$

of light quanta. In the case $cT \ll L$, the mean value of the field being F, its mean fluctuation is of the order of magnitude $F/\sqrt{N}$; in the other limiting case, the mean value of the field is reduced in the ratio L/cT, and its fluctuation thus affected in the same manner. In both cases the above-mentioned theoretical expressions are obtained.

This coincidence finds its explanation in the theory in a most elementary way. In fact, it is a general and immediate consequence of the theory that

---

[20] Misprint L^2 in original.

in the case where the sources of the field are described classically (as it must be done for measuring instruments such as test bodies), the solution of the field equations and fundamental commutation rules consists of the superposition of the classical field as calculated from the source distribution, and a purely transversal quantum field, corresponding to a complete absence of light quanta. While this "zero field" of course contains no energy (if the expression for the energy density is defined so as to remove the zero-point energy), its electric and magnetic forces exhibit fluctuations around their average value zero, which are just the above-mentioned fluctuations in empty space. In this connexion it may also be recalled that the so-called spontaneous emission processes can formally be described as "induced" by the zero field just considered.

Somewhat more generally, if besides classically described sources there is a given distribution of light quanta, the field fluctuations will include, besides terms depending on this distribution, a term exactly equal to the fluctuation of the zero field. In the light of these remarks we are now able completely to solve the question of the interpretation of field measurements. We see, in fact, that the fluctuations of the field created by the measuring process itself cannot be separated from the zero field fluctuations already present before the measurement. The occurrence of such fluctuations does therefore not mean any limitation in the testing of predictions of the quantum theory of fields, but on the contrary is essential for the consistency of the theory, by preventing exactly to the extent wanted the establishing of a connexion on classical lines between the sources of the field and the results of measurements of field quantities.

Apart from the peculiar feature just discussed, where we have to do with an influence of the measuring process on the very quantity to be measured, we meet also in the quantum theory of field[s] with the ordinary features of complementary limitation when we consider measurements of two field quantities, i.e. mean values of two field components over given space–time regions. Also here a detailed investigation shows that the interaction of the test bodies during the measuring process, after having been compensated to the largest possible extent, gives rise to a reciprocal limitation of the accuracies of the results of the measurements, exactly equal to the limitation derived from the commutation relations. This completes the proof of the full harmony existing between the consequences of the quantum theory of fields and the possibilities of measurements, to the extent where the atomic properties of the measuring instruments may be neglected.

2. ON THE CORRESPONDENCE METHOD IN ELECTRON THEORY, SOLVAY 1933

This paper deals with the relationship between classical electron theory and the Dirac theory. Special attention is paid to the lower limit for space–time localization: the classical electron radius $\delta = e^2/mc^2$ $(\tau = \delta/c)$, and the Compton wavelength $\lambda = h/mc$ $(\theta = \lambda/c)$, respectively, and the complementary control of conservation of momentum and energy.

As pointed out by Bohr in a footnote to the title, this paper provides us with an elaborated version of remarks he contributed during the general discussion session. It was by no means meant as a finished treatise and should not be read as such. Indeed, in a letter to Pauli, Bohr writes:

"By the way, as you also realized, the remarks are only to be regarded as a sketch for a more thorough treatise in English on the nature and limitations of the correspondence viewpoint, which I hope to finish soon[21]. I only returned them to Langevin in their present form in order not further to delay the printing of the Solvay report."

> Bohr to Pauli,
> 15 March 34
> Danish text on p. [474]
> Translation on p. [477]

It seems conspicuous that several impulses received from Dirac's talk "Théorie du positron", delivered at the same Solvay meeting, have played a rôle in connection with Bohr's work on his paper. I believe that the sketch as we have it here gains in clarity if it is read in conjunction with the letters between Bohr, Heisenberg and Pauli[22] that are quoted at the end of this volume: Perplexity is shared by the three pioneers as they are confronted by the remaining paradoxes of quantum electrodynamics. Typical in this respect is the following letter from Bohr to Pauli:

Copenhagen, February 15, 1934

> Bohr to Pauli,
> 15 Feb 34
> Danish text on p. [469]

Dear Pauli,

Thanks for your nice letter which came as a most welcome occasion for a little outpouring of my heart. In fact I have during the last few days re-

---

[21] Never published.

[22] Unfortunately, during these years we have to guess Pauli's reactions at second hand from the responses by Bohr. In connection with his escape to Sweden in 1943 during the Nazi-German occupation of Denmark, a great number of German letters from the thirties were burned in great haste. These letters could have endangered those few trustworthy friends and colleagues of Niels and Harald Bohr, who still remained in Germany, because of hints at common efforts to aid the escape of Jewish physicists and mathematicians via Copenhagen to safety in the Anglo-Saxon sphere of the world. From Bohr's reply to Pauli, 24 February 1934 (Part III, p. [471]), we learn, however, that Pauli must have reacted promptly – with at least two "nice letters" – to the remarks quoted here.

INSTITUT INTERNATIONAL DE PHYSIQUE SOLVAY

SEPTIÈME CONSEIL DE PHYSIQUE — BRUXELLES. 22-29 OCTOBRE 1933

Photo Benjamin Couprie

28, avenue Louise, Bruxelles

E. HENRIOT   F. JOLIOT W. HEISENBERG   H. A. KRAMERS   N. F. MOTT   G. GAMOW P. BLACKETT   M. COSYNS   Aug. PICCARD   L. ROSENFELD

F. PERRIN   E. STAHEL   P. A. M. DIRAC   J. ERRERA   C. D. ELLIS   E. O. LAWRENCE

E. SCHRÖDINGER   N. BOHR   E.T.S. WALTON P. DEBYE   Ed. BAUER J.E VERSCHAFFELT J. D. COCKROFT

Mme I. JOLIOT   E. FERMI   B. CABRERA W. BOTHE   W. PAULI   E. HERZEN   R. PEIERLS

A. JOFFÉ   Mme CURIE   N. S. ROSENBLUM   M. de BROGLIE   Mlle L. MEITNER   J. CHADWICK

O. W. RICHARDSON   Lord RUTHERFORD   L. de BROGLIE

P. LANGEVIN   Th. DE DONDER

Absents : A. EINSTEIN et Ch.-Eug. GUYE

[32]

turned with great enthusiasm to my old primitive view on the problems of electron theory, having been quite uneasy for some time because of the difficulty concerning the infinite self-energy of the electron which you have so strongly emphasized. Indeed, it struck me that the formalism of quantum electrodynamics not only, as Rosenfeld and I have tried to show, is an absolutely consistent idealization independent of atomic theory, but that any application of it stands in a clearly exclusive relationship to the elementary particle problems. As we have shown, any field measurement within the domain of quantum theory requires test bodies with an electric charge $E \gg \sqrt{hc}$ which is distributed continuously over a domain whose linear dimension is $L \gg h/Mc$, where M is the mass of the test body. This means, however, that the domain of applicability of the theory lies entirely outside the atomic problems proper and is restricted to problems, in which – in contrast to the former – the coupling between the charged bodies and the fields is sufficiently large to ensure the complementary relationship between the concepts of field and photons that is presupposed in the formalism[23].

In the atomic problems proper, the characteristic feature of which is just the extremely small coupling between particles and radiation fields, the conditions for the applicability of the field concept are quite different. The very smallness of the value of ε^2/hc, and the resulting smallness of the electron radius compared to h/mc, entails not only that in the quantum mechanical problems proper, including the hole theory, the field concept may be applied in a purely classical way to a very high approximation[24], but also that in the radiation problems proper one may to the same approximation confine oneself to applying the old modest correspondence arguments in a consistent manner.

I am afraid that such an attitude may perhaps at first appear much too reactionary; but if I am not quite mistaken, it is really the only sober view on atomic problems that is possible at the moment. In fact, the point is that just as classical electron theory represents an idealization valid when all actions are large compared to h, so quantum electrodynamics is an idealization valid only when all charges are large[25] compared to ε. This does

---

[23] In the analysis of field measurements from 1933, it is important that the test body is not treated as an elementary particle.

[24] This seems difficult to understand since the processes discussed within the hole theory involve only one or two photons.

[25] This is true, it seems, only when a proper application of the concept of a local field is involved. In the scattering processes commonly discussed in quantum electrodynamics, only electrons or other corpuscles carrying the unit charge ε enter.

of course not at all mean that the latter theory cannot be applied in the discussion of such universal problems as the law of heat radiation, but merely that the situation within atomic theory is essentially different from that presupposed in each of the idealizations mentioned.

The idea that the field concept has to be used with great care lies also close at hand when we remember that all field effects in the last resort can only be observed through their effects on matter. Thus, as Rosenfeld and I showed, it is quite impossible to decide whether the field fluctuations are already present in empty space or merely created by the test bodies. Even though the fluctuations are an unavoidable part of quantum electrodynamics, it is therefore in the electron theory permissible completely to ignore these fluctuations and all other consequences of the field quantization in a consistent manner[26]. This is of course also evident from Einstein's simple derivation of the law of heat radiation on a correspondence basis.

This attitude has in particular meant a major relief for me as regards the relativistic measuring problem. As you know, I have for a long time thought it impossible through investigation of these problems to find any argument whatsoever against the application of space–time concepts in electron theory, even within domains small compared to h/mc. Nevertheless I was beginning to feel rather uneasy because of the suspicion that the necessity of field quantization could make the treatment of the measuring problems as well as the attempts at a further extension of the hole theory quite illusory[27]. As matters now stand, there is, however, in my opinion not the slightest reason to believe that in the measuring problem we may meet with any surprise whatsoever which is not implicitly hidden in the mathematical formulation of the hole theory, if only this fulfills the sufficient invariance requirements. Certainly, the detailed treatment of the possible measuring devices is in general rather complicated, but by now we have of course also all realized that the theoretical predictions, with the testing of which such measurements would be concerned, would generally be of a very compli-

[26] Taken at its face value, this remark seems wrong. It is contradicted not only by the conclusion of the paper by Bohr and Rosenfeld (1933) but also by Bohr's own lucid summary in the unpublished manuscript (p. [195]). Furthermore, T. Welton in *Some Observable Effects of the Quantum-Mechanical Fluctuations of the Electromagnetic Field*, Phys. Rev. **74** (1948) 1157–1167 and V.F. Weisskopf in *Recent Developments in the Theory of the Electron*, Rev. Mod. Phys. **21** (1949) 305–315 demonstrated how the field fluctuation could qualitatively account for the Lamb shift.

[27] This sentence appears incomprehensible, but here and in the previous footnote, we must remember that we do not know the manuscript to which these remarks refer.

cated nature. Only in the special case, when the problem may be regarded with sufficient accuracy as a one-body problem, is the situation simpler, and, as I indicated in Brussels, one may here easily prove that by performing sufficiently many experiments with sufficiently varied bundles of radiation, one may test all the fundamentally statistical predictions of the theory as far as the density functions are concerned.

In these very days I am attempting with Rosenfeld's help to work out an account of the consistent application of the correspondence argument in atomic theory, and thus I should be very grateful if in a couple of lines you would let me know whether from the very beginning you would entirely disagree with an attitude like the one I have indicated here.

With many kind regards from us all,

Your age-old
Niels Bohr

I have not succeeded in finding any draft in the Niels Bohr Archive that tallies with the description of the work given in the last paragraph of the letter.

It is of course true that Bohr hardly possessed any first-hand experience with the mathematical aspects of quantum field theory, but so much the more striking is his detailed understanding of its physical implications. As an illustration I quote the following extract of a letter to Pauli from March 1934:

"All of us were also interested in Weisskopf's beautiful paper[28], and I believe that there is complete agreement in this matter, although I am not quite clear as to Heisenberg's view and I should be happy to know your opinion about his remarks in a letter which I just received and of which I send a copy.

In spite of the general agreement in this matter I should still like to suggest a small, but perhaps not insignificant change in the didactic presentation of the situation. As I indicated in the short letter that I sent you a couple of days ago, after sending off the remarks, I have given much thought to the problem of line width and I believe that one must stress more clearly

Bohr to Pauli,
15 March 34
Danish text on p. [474]
Translation on p. [477]

---

[28] V.F. Weisskopf, *Über die Selbstenergie des Elektrons*, Z. Phys. **89** (1934) 27–39. This paper contains an error in the calculation, which was corrected by Weisskopf in *Berichtigung zu der Arbeit: Über die Selbstenergie des Elektrons*, Z. Phys. **90** (1934) 817–818, arriving at the correct result that according to the hole theory, the self-energy of the electron diverges only logarithmically. (Cf. the letters from Weisskopf to Heisenberg 21 September 1934, PWB II, letter [383a], and Heisenberg to Weisskopf 2 October 1934, PWB II, letter [383b].)

what is meant by a consistent treatment of it on the basis of the correspondence principle. Indeed, I do not see any deeper point in making a distinction between quantum electrodynamics and the kind of application of the field concept involved in Weisskopf's paper. All of these methods have with more or less justification been characterized as correspondence methods, while – in accordance with the deeper insight that we have gained through the recent development – I would like to employ this designation in a fundamentally more limited manner.

Of course I am quite aware of the dangers inherent in the more naive treatment of the problem of line width that was attempted in the old days, where one distinguished too crudely between atomic states and radiation; and I perfectly understand the demand for a more consistent treatment as it found expression in Dirac's theory. Since, as we now see, the latter path is barred for fundamental reasons, however, we are in my opinion forced to regard the problem of line width as defying closer analysis in a sense similar to the interference phenomena that occur for electrons passing through two holes in a diaphragm. Of course, the latter problem is of quite a different nature and the analogy mentioned is therefore purely dialectic [didactic?], but I still believe that the attitude is natural as well as consistent.

Thus, as far as a usual absorption line is concerned, where all features are fully accounted for experimentally, I do not see any difficulty for a completion of the simple correspondence treatment of the dispersion problem by using the classical dispersion formula with radiation damping. This formula constitutes, in fact, with due regard to the principles of energy conservation and superposition, a unity from which no part can be separated in a consistent manner.

In the case of an ordinary absorption line I suppose indeed that the very definition of the line width is associated with an experimental arrangement which depends on a measurement of the dispersion of strictly monochromatic light. In the case of spectral lines corresponding to transitions between two states with finite lifetimes, the classical analogy is of course rather more dubious, and in particular those experimental arrangements which would permit an unambiguous definition of the line width are much more complicated. Personally I believe, however, that the result – which I suppose in your parlance one would say had been obtained through swindle by means of Dirac's theory – is sufficiently simple to be considered correct.

I think that here, as in all other genuine correspondence considerations, including the fundamental Heisenberg equations and Dirac's wave equation, we must base the argumentation on consistency, and above all on simplicity. As far as the basis of physics is concerned, one may even be inclined to advance the postulate that nature is governed by laws exactly to the extent to which it is at all possible to define in a consistent manner. But I shall not continue this idle chatter until I have the opportunity of watching the expression on your face at close range.

...

P.S. After Klein and Rosenfeld have looked more closely into Weisskopf's paper, the question of a fundamental difference between his methods and results and the earlier conceptions has become very unclear. Perhaps you could explain this to us."

3. FIELD AND CHARGE MEASUREMENTS

Almost immediately after the publication of their paper in 1933, Bohr and Rosenfeld began work on an extension of the analysis from the measurability of components of the free radiation field to situations where creation and annihilation of real and virtual electron–positron pairs must be taken into account. In the former analysis only the constants h and c entered, from which no absolute scale of length or time interval could be constructed. In the latter case, however, where the charge e and mass m of the electron enter explicitly, the Compton wavelength $\lambda = h/mc$ provides, together with the dimensionless fine structure constant, scales for spatial and temporal extensions. One could then envisage that fields averaged over a given space–time volume $L^3 cT$ (measured in units of λ) would exhibit a novel kind of quantal fluctuations depending on e and m. According to Rosenfeld's recollections, quoted below, Bohr and Rosenfeld had at an early stage suspected the existence of similar fluctuations for the *charge* (in units of e) averaged over a given space–time domain, and they therefore welcomed Heisenberg's paper of 1934[29] in which he derived quantitative expressions for these fluctuations in two limiting cases.

In particular, Heisenberg demonstrated that, in contrast to the field fluctuations, the charge fluctuations remained infinite, even when averaged over a sharply defined space–time volume. This result was not entirely unexpected since it is closely related to the infinite production of pairs from the edge of a potential that within an interval of the order λ increases from zero to mc^2, early referred to as the "Klein paradox". Thus, in order to obtain finite values

[29] W. Heisenberg, *Über die mit der Entstehung von Materie aus Strahlung verknüpften Ladungsschwankungen*, Ber. Sächs. Akad. math.–phys. Kl. **86** (1934) 317–322. Reprinted in *Selected Papers on Quantum Electrodynamics* (ed. J. Schwinger), Dover, New York 1958, pp. 62–67. Reproduced in this volume, Appendix to Part I, p. [239]. Cf. also Heisenberg, *Bemerkungen zur Diracschen Theorie des Positrons*, Z. Phys. **90** (1934) 209–231 and **92** (1934) 692.

for the charge fluctuations, the space–time domain must be limited by a smooth cut-off, defined by a finite characteristic space–time scale that enters explicitly into the formal expression for the fluctuations.

In the Bohr Memorial Volume[30] Rosenfeld recalls:

"The progress (by no means smooth) of the investigation was followed with tense interest by both Heisenberg and Pauli, and their reactions illustrate quite strikingly the difference in their temperaments. Heisenberg, on a visit to Copenhagen, received our first conclusions, which, it must be admitted, were still far from clear, with strong scepticism. He flatly declared that he did not believe a word of my calculations, and at once started to repeat some of them on the blackboard; as he proceeded, and the results I had stated began to emerge, it was amusing to watch the growing expression of surprise on his face. As soon, however, as he had grasped the significance of the problem raised by these mysterious field fluctuations, he changed over from passionate incredulity to passionate zeal. Within a few weeks, he sent us two papers inspired by this discussion. One of them (which was never published) was a spirited attempt to avoid the troublesome inconsistencies which beset quantum electrodynamics by explicitly considering (as we had found necessary to do in our study) mean field intensities in space–time regions rather than their values at space–time points. The other paper[29] pointed for the first time to the occurrence of similar fluctuations of the mean electric charge or electric current contained in a space–time region; this was a most valuable hint to us that the problem we were struggling with was not confined to electromagnetic fields, but existed also for the wave-fields associated with charged particles.

Pauli's attitude was more cautious, but in its way equally inspiring; he was reluctant to express any opinion on the probable outcome of the investigation, but he constantly stressed the importance of the issue at stake. No one better than Pauli understood the earnestness of Bohr's endeavour, because he himself was as earnestly concerned with the philosophical and human problems which ultimately guided Bohr in his quest. When he heard of Heisenberg's calculation of the charge and current fluctuations, he encouraged us to tackle this problem as well. Quite rightly, he pointed out to Bohr that he was the only one to master the peculiar methods required for such analyses of ideal measurements; even to this day, no physicist can boast

---

[30] L. Rosenfeld, *Niels Bohr in the Thirties*, in *Niels Bohr, His life and work as seen by his friends and colleagues* (ed. S. Rozental), North-Holland Publ. Co., Amsterdam 1967, pp. 126–127.

of approaching Bohr's virtuosity in this respect. We did yield to Pauli's exhortations; and although this new investigation suffered many interruptions, and only reached the stage of publication in 1950, it did not take us very long to complete a first survey of the ground, which led us to conclusions rather similar to those we had obtained in electrodynamics. Pauli received these provisional results with expressions of great satisfaction, tempered by the fear (justified by the event) that the further course of the work would be slow. 'However', he added in his usual sarcastic fashion, 'since you have managed to publish the work on the electromagnetic field, it has become impossible to state with certainty that the other work will never be published.' This mild derision, of course, was only a cloak for really deep feelings of grateful admiration; Pauli did also tell Bohr, at that time, that the analysis of the measurability of electromagnetic fields had meant a great deal to him, by deepening his understanding of quantum theory and strengthening his faith in the soundness of its foundations."

* * *

I have found it convenient to divide the rest of this section into two parts. This division is preferred because of the extended temporal hiatus in Bohr and Rosenfeld's work, between the thirties when the work began, until the late forties, when it was resumed. In the meantime, quantum electrodynamics had developed into a systematic formal description which, in spite of seeming mathematical "inconsistencies", yielded results of remarkable accuracy that were in close agreement with experimental measurements of high precision.

OVERTURE IN THE THIRTIES

In the collections of the Niels Bohr Archive several folders are preserved containing an abundance of apparently scattered drafts of qualitative evaluations and pages of tentatively formulated text dealing with field and charge measurements in quantum electrodynamics. But fortunately, we also found among these remnants a rather coherent and most interesting manuscript[31] which is reproduced on p. [195].

---

[31] Judging from the material in the same folder I estimate that it dates from the mid-thirties. Some of the corrections are in Bohr's own handwriting, but several of these we have not been able to decipher.

Niels Bohr and Léon Rosenfeld, Tisvilde 1931.

In spite of the fact that the analysis is obviously incomplete, this manuscript appears to be much more in Bohr's own style – and therefore much clearer, as far as the basic physical argumentation is concerned – than is the sketchy and strangely formal style of the paper published in 1950. As already mentioned in connection with the paper on field measurements from 1933, the first section of this manuscript provides an excellent summary of the salient points of that analysis.

Before embarking on a study of the Bohr manuscript itself, it may be an advantage for the reader to consider the addendum by Wheeler in the well-known treatise on electron theory by Pais[32]. It summarizes with remarkable clarity the main points of a manuscript by Bohr, apparently circulated already well before 1939. Its title is: "Field and Charge Measurements in Quantum Theory".

---

[32] A. Pais, *Developments in the Theory of the Electron*, Institute for Advanced Study and Princeton University, Princeton, New Jersey 1948, pp. 42–45. Cf. footnote 3 in N. Bohr and L. Rosenfeld, *Field and Charge Measurements in Quantum Electrodynamics*, Phys. Rev. **78** (1950) 794–798, reproduced on p. [211].

FIELD AND CHARGE MEASUREMENTS IN QUANTUM THEORY
By N. Bohr and L. Rosenfeld

The manuscript of this paper, which will shortly appear in the Mat.-Fys. Med. of the *Kgl. Danske Videnskabernes Selskab*, was written already in 1939, but its publication has been delayed due to the events of later years. We are indebted to Professor Bohr and Professor Rosenfeld for the opportunity of reading the manuscript and making an abstract of it which is here included with their consent. – John A. Wheeler.

This paper analyzes the question how far measurements of electromagnetic field quantities and electric charges are limited by typical quantum mechanical fluctuation phenomena. The first half considers the consequences of the field fluctuations of quantum electrodynamics, and the second deals with the charge fluctuations of pair theory. The purpose of the analysis is to clarify the logical aspects of the definition and use of concepts compatible with the mathematical formalism, and thus to bring out some aspects of this formalism which might otherwise have escaped attention, and which are fundamental features of its logical consistency.

The paper begins with a review of the earlier communication of Bohr and Rosenfeld (see p. 13 [in Pais' booklet]) on the question of measurability of electromagnetic fields. It is emphasized again that suitably constructed test bodies – to the extent that one can look apart in the measurements from their atomic constitution – permit one in idealized experiments to determine field quantities with just the precision claimed by quantum electrodynamic theory. In particular, the average value of a field quantity over a finite space–time interval can be found with arbitrary accuracy. The same is true of determinations of values of two different field quantities over the same space–time interval. However, reciprocal uncertainty relations in general exist between the values of field quantities averaged over two different space–time intervals. About this complementary character of the two quantities in question the analysis of the idealized experiments yields the same conclusions as does the formalism of electrodynamics.

The earlier considerations of simple space–time averages of field quantities are extended. In particular the measurability of the electric charge is considered. For this purpose one considers the case where (1) the test bodies employed envelop a certain closed volume, (2) the thickness b of the test bodies tends to an infinitesimally small value, and (3) the partial test bodies are connected by way of suitable levers with a single external device for the

John A. Wheeler, 1949 (Courtesy AIP).
The portrait still hangs in Bohr's old office at the Institute.

measurement of force or momentum. The connections are made in such a way that one obtains the value of the normal component of the electric field integrated over the surface in question. It is shown that this integral can in principle be determined with an arbitrary accuracy so far as concerns any possible limitations imposed upon the measurement by the quantum theory of fields. It is therefore concluded that the determination via the theorem of Gauss of the electric charge enclosed within the surface offers no difficulty so long as one disregards the consequences of the phenomenon of pair creation for the measurement.

That the measurement of the charge contained within a region of space cannot be made with arbitrary accuracy is, however, the conclusion reached by Heisenberg (*Sächs. Akad.*, **86**, 317 (1934))[33] through an analysis in the light of pair theory of the fluctuations to be expected in such a region. From his paper it furthermore follows that the fluctuation of charge depends not only upon the magnitude of the volume itself but also upon the sharpness b with which the boundaries of this volume are defined. In the case where the product of the velocity of light and the interval T during which the mean value of the charge is taken is smaller than both the quantity b and the distance $\hbar/mc$, the mean square deviation $\overline{(\Delta e)^2}$ of the charge from its expectation value is given by the expression:

$$\overline{(\Delta e)^2} = \frac{(\text{electronic charge})^2 \times (\text{surface of enclosure})}{\begin{pmatrix} \text{thickness } b \text{ of test bodies} \\ \text{which define enclosure} \end{pmatrix} \times \begin{pmatrix} \text{velocity of light times} \\ \text{time of measurement} \end{pmatrix}} \tag{1}$$

The remainder of the paper of Bohr and Rosenfeld is devoted to a discussion of the physical significance of the charge fluctuations implied by this formula and a derivation of the formula directly from an analysis of the measuring process itself. It is thus shown that the possibilities of measurement contained within the framework of the theory of fields and pairs in principle permit a determination of the charge contained within an enclosure with just the accuracy specified by the equation in question, again subject to the condition that one may look apart from the atomic constitution of the measuring devices.

On the question of the origin of the charge fluctuations, attention is drawn to the earlier view expressed by Oppenheimer (*Phys. Rev.*, **47**, 144 (1935))[34], that these charge fluctuations are not only inseparable from but in fact due to the zero-point fluctuations in the electromagnetic field itself. The contrary conclusion is reached by Bohr and Rosenfeld. They show that the charge fluctuations arise entirely from the pair field. The absence of any contribution to the charge fluctuation from the zero-point fluctuations of the electromagnetic field is indeed already evident from their theorem about the exact measurability – within the framework of pure quantum electrodynamics – of the surface integral of the normal component of the electric field.

The uncertainty in the value of the Gaussian integral – and therefore in the value of the included charge – is found to be due to the effect of pairs unavoidably created by the measuring device itself. This phenomenon shows up when one considers in detail a device composed as mentioned above of a

---

[33] This volume, Appendix to Part I, p. [239].

[34] R. Oppenheimer, *Note on Charge and Field Fluctuations*, Phys. Rev. **47** (1935) 144–145.

large number of small charged test bodies arranged to surround completely the volume under study. Specifically, Bohr and Rosenfeld point out:

a. That the typical charged test body receives a normal displacement in the course of the measurement;

b. That the displacement of charge creates a supplementary electric field within the region occupied by the test body itself;

c. That this electric field will in general disturb the infinite distribution of negative energy electrons and bring into existence pairs of positive and negative electrons;

d. That the charge of these pairs will be so distributed as to produce an additional electric polarization;

e. That this electric polarization will produce a supplementary force on the test body which will disturb the result of the measurement;

f. That this disturbance is, however, proportional in a certain approximation to the magnitude of the displacement of the test object;

g. That the perturbation can therefore be compensated by suitable elastic devices;

h. That the perturbation consequently will not in itself impair the possibility of measuring with arbitrary accuracy the charge contained within the given volume element;

i. That the supplementary force created in this way, for example by N pairs of positive and negative electrons, will be subject however to unpredictable statistical fluctuations about its normal avarage value proportional to $N^{1/2}$;

j. That there will be on this account an unavoidable fluctuation in the value of the surface integral of the normal component of the electric force; and

k. That this fluctuation, when calculated, is found to correspond to an uncertainty in the charge contained within the given volume just equal to the figure of Heisenberg quoted above.

Although there is no direct connection between these charge fluctuations and the field fluctuations discussed in the earlier half of their paper, Bohr and Rosenfeld nevertheless point out the close analogy between the two kinds of fluctuations. In both cases the displacement of the measuring device produces a back reaction on this test object itself, in the one case by way of the radiation field which it creates, in the other case by way of the positive and negative electrons which it generates. In both cases, the average value of the reaction so developed is directly proportional to the magnitude of the displacement and can therefore either be compensated by a suitable elastic device or otherwise be taken into account in a predictable way. However, there are necessarily quantum fluctuations about this average value because the reaction is transmitted by a finite number of photons in the one case or a finite number of pairs in the other. It is these fluctuations which set a limit to the accuracy which can be obtained in certain field and charge measurements.

From their analysis of the measuring process Bohr and Rosenfeld conclude that existing theory – when used within its proper domain of application and thus in particular employed in a way which does not depend upon the atomic character of instruments of measurement – gives a logically self-consistent account of the possibilities of determining charges and electromagnetic field quantities.

* * *

Hardly *any* of these points – numbered (a) to (k) – are really discussed in the paper by Bohr and Rosenfeld finally published; the more remarkable is the close connection to the early manuscript under discussion here. Naturally I sent a copy of the Bohr manuscript to Wheeler in order to learn whether that could be the one to which he was referring. In answer to my enquiry, Wheeler wrote:

> "Now as to your own question: I honestly don't know where to look for any copy of the manuscript of the Bohr–Rosenfeld paper. You know the old statement that three moves are equal to one fire. I don't remember where I wrote that summary. I suspect that Bram Pais will have a far better memory on this point than I.
>
> I suspect that the points (a) to (k) that you refer to arose out of my own incessant ignorant questioning of B & R and associates as to what it was all about –"

Already in early conversations, however, Pais had told me that he had no recollection of ever having seen a draft of the published paper. Thus the existence of a more complete, and possibly lost, manuscript from the late thirties, remains so far an enigma.

Indeed, the following letter from Bohr to Heisenberg contains a remarkably definite statement as regards the imminent completion of the manuscript and its approaching publication, although its content is merely hinted at:

[45]

Bohr to Heisenberg,
22 May 35
Danish text on p. [450]
Translation on p. [453]

"... I shall say a few words about the measuring problem, of which Rosenfeld and I have just now arrived at a better understanding. I don't know whether you have seen a paper by Oppenheimer in the Physical Review[35] a short while ago, in which – on the basis of the in itself correct, not to say obvious, idea that all density measurements according to their very nature have to be traced back to field measurements – he has thought himself able to demonstrate a contradiction between your evaluation of the density fluctuations and Rosenfeld's and my evaluation of the field fluctuations. However, he has completely missed the main point of our paper, according to which the field fluctuations in themselves do not entail any fundamental limitation to field measurements, but only to the [possibility of] tracing the results back to the field sources. Thus, fluctuations do not at all imply a limitation of density measurements, since the fluctuating fields as a matter of principle are divergence free, at least as long as we remain within the domain of the usual quantum electrodynamics. The relationship between your density fluctuations and the problem of density measurements is, however, most intimately connected with the modifications of the field equations in the positron theory as treated by Euler and Kockel[36]. Indeed, a closer investigation reveals here a complete consistency of the formalism as regards the connection between the possibilities of definition and measurement. As Rosenfeld and I have painfully convinced ourselves, in order to demonstrate such a connection one has to be just as alert against pitfalls as in our old struggle with the problem of field measurements. For the moment we are working on a small paper for the Physical Review about these questions and I am going to send you the manuscript in a few days. [*sic*]

About all this I am sure we quite agree. But it is not at all clear to me whether I have understood the trend of your working plan, and in particular I do not see how you find arguments in the field and density fluctuations for primarily considering average values over finite, more or less sharply limited space–time regions. Indeed, the result of the paper by Rosenfeld and me was just that apart from the self-energy question, the present form of quantum electrodynamics was more satisfactory than one might be inclined to believe on the basis of earlier investigations of the question of measurement, and in particular that the consideration of average values was an

[35] R. Oppenheimer, *Note on Charge and Field Fluctuations* (ref. 34).

[36] H. Euler and B. Kockel, *Über die Streuung von Licht an Licht nach der Diracschen Theorie*, Naturwiss. **23** (1935) 246–247. Cf. also W. Heisenberg and H. Euler, *Folgerungen aus der Diracschen Theorie des Positrons*, Z. Phys. **98** (1935) 714–732.

integral part of the formalism and therefore by no means pointed to the necessity of changing the present basis for the correspondence description. After the latest papers by you and your collaborators and our own small new contribution to the measuring problems, it seems to me that we are facing a very similar situation in electron theory. Of course I quite agree with you that the entire basis has to be revised substantially before one can build a really satisfactory and coherent theory; but at the same time it appears to me that there is considerable evidence that an introduction of average values, such as you suggest, will hardly suffice. The characteristic feature of the well-known paradoxes concerning the radiation [emitted] in collisions between swift electrons is precisely the fact that in these problems we are already concerned with space–time averages and that therefore we fail to see how an explicit introduction of average values on the present basis could prove very useful.

I am therefore still more inclined to believe that the whole difficulty of electron theory is due to the fact that the problems of definition as well as of measurement are conceived in a much too classical manner, and that a solution to the riddles is only to be achieved by revising the basis entirely so that the existence of the quantum of charge becomes far more intimately connected with the existence of the quantum of action than in the methods used so far."

Heisenberg's reply[37] is reproduced in the Selected Correspondence at the end of this volume.

In 1937, writing from Paris on his way to his trip around the world, Bohr estimates the time schedule for the completion of the paper less optimistically, but he is rather explicit with respect to what he and Rosenfeld have "shown" and "demonstrated" in their "new work":

"... it might interest you that Rosenfeld ... and I have recently – in connection with a vain attempt to make our work on charge measurements ready for publication before my journey – taken up for renewed discussion the whole question of application of classical concepts in quantum electrodynamics and electron theory. It is a fact that both in our old and in our new work we were able to show that as far as measuring possibilities are concerned, these theories represent in a purely logical respect a consistent idealization. However, quite recently I have begun to suspect that the

Bohr to Heisenberg,
21 Jan 37
Danish text on p. [457]
Translation on p. [458]

---

[37] Heisenberg to Bohr, 30 May [1935]. Reproduced on p. [455].

[47]

preposterous consequences of such an idealization lead far beyond the borders of reality already with respect to the basic presupposition of identifying field quantities with their actions on test bodies of finite dimensions. As Rosenfeld and I demonstrated in our paper, the theoretical field and charge fluctuations are absolutely inseparable from that part of the reaction on the test body that cannot be compensated due to the individuality of the processes of radiation and pair creation. Obviously this circumstance allows the possibility of depriving all such fluctuations of any simple physical reality. If one accepts this point of view – which for the moment appears to me as the only reasonable one – it would mean that we are forced again, in this domain, to rely on simple correspondence and complementarity arguments. These days I have discussed this perspective in more detail with Rosenfeld and got the impression that, even for a man more erudite than I, it may perhaps be less deterring than it might appear at first glance."

Perhaps, some time in the future new material will appear, throwing light on the question of whether there ever existed a more complete manuscript from the thirties than the one reproduced in this volume (p. [195]).

COMPLETION (1949–1950)

The Conference on the Foundations of Quantum Mechanics held at Shelter Island on 1–3 June 1947 marked the beginning of a new epoch in the development of quantum electrodynamics: Lamb and Retherford reported that through very delicate experiments they had found that, in the fine structure spectrum of hydrogen, the energy of the $2s^{1/2}$ level was higher than the energy of the $2p^{1/2}$ level by an amount corresponding to approximately 1000 megacycles. This is in disagreement with the prediction of the Dirac theory according to which the two levels are degenerate. Their paper was received by the Physical Review on 18 June 1947[38].

Remarkably enough, already on 27 June this journal received from Bethe the theoretical explanation which for the first time demonstrated how, through the process of mass renormalization originally suggested by Kramers, it was possible to separate the formal infinities, encountered in such calculations from

---

[38] Cf. *Selected Papers on Quantum Electrodynamics* (ref. 29), which contains reprints of several of the important papers alluded to in this section, together with extensive references.

Niels Bohr, Princeton 1948 (Courtesy Princeton University).

well-defined theoretical predictions amenable to experimental tests. As a matter of fact, Bethe obtained the value 1040 megacycles for the Lamb shift in the non-relativistic approximation corresponding to a maximum frequency $\omega_{max} \sim mc^2/\hbar$.

However dubious the methods might appear from a formal point of view, the impressive agreement between the theoretical predictions and experimental results created quite a new confidence in the consistency of the basic structure of quantum electrodynamics. This optimistic atmosphere inspired a true avalanche of fundamental contributions – by Schwinger and Tomonaga as well as by Dyson, Feynman and many others – through which the entire formal edifice of relativistic quantum theory was reshaped. Thus, when Bohr and Rosenfeld resumed their work in 1949 they were building on much firmer ground than in the thirties. This very fact, however, entailed of course a certain lack of incitement to continue a detailed analysis of charge–current measurements: there were no profound paradoxes to confront – whatever difficulties there remained in quantum electrodynamics resided elsewhere. Characteristically, it was Pauli who, in spite of this circumstance, encouraged Bohr and Rosenfeld to go on by his emphasis on the fundamental importance of the whole enterprise.

[49]

Also Dyson took a vivid interest in the work.

There is a rich and interesting correspondence from this period between Bohr, Pauli, Rosenfeld and Dyson, of which a selection is reproduced in Part III. The point of departure is a letter to Rosenfeld of 5 July 1949 (p. [482]) in which Pauli comments on "the" old manuscript by Bohr and Rosenfeld from the thirties, as summarized in the points (a) to (k) by Wheeler in the addendum to Pais's booklet (quoted p. [41]). He criticizes a lack of distinction between "fluctuations" and "uncertainties". Furthermore, he asks Rosenfeld, sceptically, how they in their calculations had avoided the problem of an infinite "self-charge". Unfortunately Rosenfeld's reply of 7 July 1949 seems to be lost, but Pauli's letter of 11 July shows that he accepted the answer, at least partly (it "clarified a number of points").

In his letter to Pauli of 15 August 1949 (p. [486]) Bohr announces that the apparently well-established divergence of the charge fluctuations for sharply bounded space–time regions was wrong! Pauli rejects this result as impossible and after some further letters back and forth locates the error in Rosenfeld's calculation. Indeed, in the limiting case considered, a logarithmic divergence remains in accordance with the result first obtained by Jost (p. [500]). This debate ends with Bohr's letter to Rosenfeld of 2 September (p. [501]). The Dyson correspondence[39] is not included here. These letters find their proper place in a study of the history of quantum electrodynamics, rather than in a volume of Bohr's Collected Works.

4. ELEMENTARY PARTICLES, SUMMARY OF REFLECTIONS

Problems of Elementary-Particle Physics (1947)
Some General Comments on the Present Situation in Atomic Physics (1950)

A few lines must suffice to introduce these two brief communications by Bohr. Obviously, they were never intended to be proper contributions to the new developments. Still, they are testimonials of the eager interest with which Bohr followed novel fundamental ideas, and, as we learn from a letter of 21 August 1949, reproduced on p. [493], even Pauli was well satisfied with Bohr's summary at the Solvay meeting in 1948. Within the domain of quantum field theory, Pais served during this period as Bohr's chief informant and, through his

---

[39] Dyson to Pauli, March 1950; Pauli to Dyson, 27 March 1950; Dyson to Rosenfeld, 3 April 1950. Rosenfeld Papers, NBA.

Wolfgang Pauli, Niels Bohr and a "tippetop", Lund 1954.

renowned biographies of Einstein[40] and of Bohr[41], we receive a moving impression of the contrast between Einstein's aloofness and isolation within the community of physicists, and Bohr's insatiable curiosity to "learn" through vivid contacts with many of the most gifted colleagues among the younger generations.

Science, to repeat Bohr's own phrasing, is about "posing questions to nature in the form of experiments" and "to learn" through attentive listening to its answers. In that way we are taught the proper use of language, which enables us to formulate unambiguously novel fundamental insights and their inescapable epistemological consequences.

[40] A. Pais, 'Subtle is the Lord ...', The Science and the Life of Albert Einstein, Oxford University Press, Oxford 1982.
[41] A. Pais, Niels Bohr's Times, in Physics, Philosophy, and Polity, Clarendon Press, Oxford 1991.

[51]

I. THE LIMITED MEASURABILITY OF ELECTROMAGNETIC FIELDS OF FORCE

(WITH L. ROSENFELD)

OM DEN BEGRÆNSEDE MAALELIGHED AF
ELEKTROMAGNETISKE KRAFTFELTER
Overs. Dan. Vidensk. Selsk. Virks. Juni 1932 – Maj 1933, p. 35

THE LIMITED MEASURABILITY OF
ELECTROMAGNETIC FIELDS OF FORCE
Nature **132** (1933) 75

Communication to the Royal Danish Academy on 2 December 1932

ABSTRACT

NIELS BOHR gav en Meddelelse: *Om den begrænsede Maalelighed af elektromagnetiske Kraftfelter.*

Meddelelsen gør Rede for en Undersøgelse udført i Samarbejde med L. Rosenfeld, ved hvilken, i Overensstemmelse med Forsøgene paa at udvikle en rationel kvanteelektrodynamisk Formalisme, der eftervises en principiel Begrænsning af Mulighederne for Udmaaling af elektromagnetiske Feltkomponenter, svarende til den karakteristiske komplementære Begrænsning af Maalingsmulighederne for mekaniske Størrelser, som betinger den kvantemekaniske Formalismes Modsigelsesfrihed.

Vil blive trykt i Math.-fys. Medd. Bd. XII, Nr. 8.

Dec. 2. NIELS BOHR : The limited measurability of electromagnetic fields of force. An investigation in collaboration with L. Rosenfeld proves the existence of a limitation of the measurability of electromagnetic field components, conforming with the tentative rational formulation of quantum electrodynamics, and analogous to the characteristic complementary limitation of the measurability of mechanical quantities, which secures the consistency of quantum mechanics.

[54]

II. ON THE QUESTION OF THE MEASURABILITY OF ELECTROMAGNETIC FIELD QUANTITIES

(WITH L. ROSENFELD)

ZUR FRAGE DER MESSBARKEIT
DER ELEKTROMAGNETISCHEN FELDGRÖSSEN
Mat.–Fys. Medd. Dan. Vidensk. Selsk. **12**, no. 8 (1933)

ON THE QUESTION OF THE MEASURABILITY OF
ELECTROMAGNETIC FIELD QUANTITIES
"Selected Papers of Léon Rosenfeld" (eds. R.S. Cohen and J.J. Stachel),
D. Reidel Publishing Company, Dordrecht 1979, pp. 357–400

See Introduction to Part I, sect. 1.

The English version, translated by A. Petersen with revisions by R.S. Cohen and J. Stachel for publication in "Selected Papers of Léon Rosenfeld", is not quite adequate. We have nevertheless chosen to use this translation, since it is also included in "Quantum Theory and Measurement" (ed. J.A. Wheeler and W.H. Zurek), Princeton University Press, 1983, which will continue to be the most widespread and accessible publication for the general reader.

The shortcomings of the translation should not give rise to serious misunderstandings, but in cases of doubt the German original must of course be consulted.

Det Kgl. Danske Videnskabernes Selskab.
Mathematisk-fysiske Meddelelser **XII**, 8.

ZUR FRAGE DER MESSBARKEIT DER ELEKTROMAGNETISCHEN FELDGRÖSSEN

VON

N. BOHR UND L. ROSENFELD

KØBENHAVN

LEVIN & MUNKSGAARD

1933

§ 1. Einleitung.

Die Frage nach der im Wirkungsquantum begründeten Begrenzung der Messbarkeit elektromagnetischer Feldgrössen hat durch die Diskussion der noch ungelösten Schwierigkeiten der relativistischen Atommechanik ein besonderes Interesse gewonnen. Zwar versuchte HEISENBERG[1] durch Betrachtungen orientierenden Charakters eine ähnliche Verbindung zwischen der Begrenzung der Messbarkeit von Feldgrössen und der Quantentheorie der Felder nachzuweisen, wie den durch das Unbestimmtheitsprinzip ausgedrückten Zusammenhang zwischen der komplementären Begrenzung der Messbarkeit kinematischer und dynamischer Grössen und dem nichtrelativistischen Formalismus der Quantenmechanik. Jedoch kamen LANDAU und PEIERLS[2] im Laufe einer kritischen Untersuchung der Grundlagen der relativistischen Verallgemeinerung dieses Formalismus zu dem Schluss, dass die Messbarkeit von Feldgrössen weiteren Einschränkungen unterworfen wäre, die über die Voraussetzungen der Quantentheorie der Felder wesentlich hinausgehen, und daher dieser Theorie jede physikalische Grundlage entziehen würden.

In diesem Widerspruch könnte man zunächst ein ernstliches Dilemma erblicken. Einerseits dürfte nämlich die Quantentheorie der Felder eine konsequente korrespondenz-

---

[1] W. HEISENBERG, Die physik. Prinzipien der Quantentheorie, 1930, S. 33 ff.

[2] L. LANDAU und R. PEIERLS, Zs. f. Phys., **69**, 56, 1931.

mässige Umdeutung der klassischen elektromagnetischen Theorie anzusehen sein in ähnlichem Sinne, wie die Quantenmechanik eine der Existenz des Wirkungsquantums entsprechende Umgestaltung der klassischen Mechanik darstellt. Andererseits hat eben die Quantenelektrodynamik die Schwierigkeiten der harmonischen Verschmelzung von Feldtheorie und Atomtheorie, denen wir schon in der klassischen Elektronentheorie begegnen, wesentlich vergrössert. Eine nähere Betrachtung zeigt indessen, dass die verschiedenen hier auftretenden Probleme sich weitgehend von einander trennen lassen, indem der quantenelektromagnetische Formalismus an sich von allen Vorstellungen über den atomaren Aufbau der Materie unabhängig ist. Dies erhellt schon daraus, dass in ihn als universelle Konstanten ausser der Lichtgeschwindigkeit nur das Wirkungsquantum eingeht; denn diese beiden Konstanten reichen offenbar nicht aus zur Festlegung irgendwelcher spezifischen raumzeitlichen Dimensionen. In der Quantentheorie des Atombaus wird jede solche Festlegung ja erst durch das Hinzutreten der elektrischen Elementarladung und der Ruhemassen der Elementarteilchen erreicht.

Eben die ungenügende Unterscheidung zwischen Feldtheorie und Atomtheorie bildet die Hauptursache für die Unstimmigkeiten der früheren Untersuchungen über die Messbarkeit von Feldgrössen, wo als Probekörper einzig elektrisch geladene Massenpunkte in Betracht gezogen wurden. Die der bisherigen Atommechanik zugrundegelegte korrespondenzmässige Verwertung der klassischen Elektronentheorie beruht nämlich vor allem auf der Kleinheit der Elementarladung gegenüber der Quadratwurzel des Produkts von Wirkungsquantum und Lichtgeschwindigkeit, wodurch es möglich wird, alle Strahlungsreaktionen als klein gegenüber den auf die Teilchen ausgeübten ponderomotorischen Kräften zu behandeln. Bei Messungen von Feldgrössen erweist es sich aber als wesentlich, über die Ladung der Probekörper in einem Umfang verfügen zu können, der mit letzterer Voraussetzung in Gegensatz käme, wenn man diese Körper als Punktladungen betrachten würde. Diese Schwierigkeiten verschwinden aber, wie wir sehen werden, wenn man Probekörper benutzt, deren linearen Abmessungen genügend gross gegenüber den Atomdimensionen gewählt werden, um ihre Ladungsdichte als näherungsweise konstant über den ganzen Körper betrachten zu können.

In dieser Verbindung ist es auch von wesentlicher Bedeutung, dass die übliche Beschreibung eines elektromagnetischen Feldes durch die Feldkomponenten in jedem Raumzeitpunkt, welche die klassische Feldtheorie charakterisiert, und nach welcher das Feld mittels Punktladungen im Sinne der Elektronentheorie ausmessbar wäre, eine Idealisation ist, der in der Quantentheorie nur eine beschränkte Anwendbarkeit zukommt. Dieser Umstand findet seinen sinngemässen Ausdruck gerade im quantenelektromagnetischen Formalismus, in welchem die Feldgrössen nicht mehr durch eigentliche Punktfunktionen dargestellt werden, sondern durch Funktionen von Raumzeitgebieten, die formal den Mittelwerten der idealisierten Feldkomponenten über die betreffenden Gebiete entsprechen. Nur für die Messbarkeit solcher Gebietsfunktionen lassen sich an Hand des Formalismus eindeutige Aussagen ableiten, und unsere Aufgabe wird also darin bestehen, zu untersuchen, inwieweit die in dieser Weise definierten, komplementären Begrenzungen der Messbarkeit von Feldgrössen mit den physikalischen Messungsmöglichkeiten übereinstimmen.

Soweit wir von allen in der atomistischen Struktur der Messinstrumente beruhenden Einschränkungen absehen können, lässt sich auch wirklich in dieser Hinsicht eine völlige Übereinstimmung nachweisen. Ausser einer eingehenden Untersuchung der Konstruktion und Handhabung der Probekörper verlangt aber dieser Nachweis noch die Berücksichtigung von gewissen, bei der Diskussion der Messbarkeitsfrage zutagetretenden, neuen Zügen der komplementären Beschreibungsweise, die nicht in der üblichen, der unrelativistischen Quantenmechanik entsprechenden Fassung des Unbestimmtheitsprinzips einbezogen sind. Nicht nur bedeutet es eine wesentliche Komplikation des Problems der Feldmessungen, dass wir beim Vergleich von Feldmittelwerten über verschiedene Raumzeitgebiete im allgemeinen nicht in eindeutiger Weise von einer zeitlichen Reihenfolge der Messvorgänge sprechen können, sondern schon die Deutung der einzelnen Messergebnisse erfordert bei Feldmessungen eine noch grössere Vorsicht als bei den gewöhnlichen quantenmechanischen Messproblemen.

Das Merkmal letzterer Probleme ist die Möglichkeit, jedem einzelnen Messergebnis eine im Sinne der klassischen Mechanik wohldefinierte Deutung beizulegen, indem der durch das Wirkungsquantum bedingten, prinzipiell unkontrollierbaren Wechselwirkung zwischen Messinstrument und Messobjekt völlig Rechnung getragen wird durch den Einfluss jedes Messvorgangs auf die durch nachfolgende Messungen zu prüfenden statistischen Erwartungen. Bei Messungen von Feldgrössen hingegen ist zwar jedes Messergebnis auf Grund des klassischen Feldbegriffs wohldefiniert, aber die begrenzte Anwendbarkeit der klassischen Feldtheorie auf die Beschreibung der unvermeidlichen elektromagnetischen Feldwirkungen der Probekörper bei der Messung

bringt, wie wir sehen werden, mit sich, dass diese Feldwirkungen in gewissem Umfang das Messergebnis selber
in unkompensierbarer Weise beeinflussen. Eine nähere
Untersuchung des prinzipiell statistischen Charakters der
Folgerungen des quantenelektromagnetischen Formalismus
zeigt jedoch, dass diese Beeinflussung des Messobjekts durch
den Messvorgang die Prüfbarkeit solcher Folgerungen in
keiner Weise beeinträchtigt, sondern vielmehr als einen
wesentlichen Zug der innigen Anpassung der Quantentheorie
der Felder an das Messbarkeitsproblem anzusehen ist.

Bevor wir zur näheren Ausführung der hier angedeuteten
Betrachtungen übergehen, möchten wir nochmals betonen,
dass die grundsätzlichen Schwierigkeiten, die der konse-
quenten Verwertung der Feldtheorie in der Atomtheorie
entgegenstehen, von der vorliegenden Untersuchung völlig
unberührt bleiben. Für die Beurteilung des Zusammenhangs dieser Schwierigkeiten mit den wohlbekannten Paradoxien der Messprobleme der relativistischen Quantenmechanik dürfte eben die Berücksichtigung des atomistischen
Aufbaus aller Messinstrumente wesentlich sein. Besonders
käme hier auch die durch den endlichen Wert der Elementarladung im Vergleich zur Quadratwurzel des Produkts
aus Lichtgeschwindigkeit und Wirkungsquantum bedingte
Begrenzung der korrespondenzmässigen Atommechanik in
Betracht.[1]

---

[1] Vgl. N. BOHR, Atomic Stability and Conservation Laws, Atti del
Congresso di Fisica Nucleare, 1932. Anmerkung bei der Korrektur.
In einer demnächst erscheinenden besonderen Veröffentlichung wird näher
diskutiert, welche Folgerungen die neue Entdeckung des Auftretens von
sogenannten »positiven Elektronen« unter besonderen Umständen, sowie
die Erkenntnis des Zusammenhangs dieser Erscheinung mit der relativistischen Elektronentheorie von DIRAC, für die in der zitierten Arbeit besprochenen Probleme mit sich bringt.

§ 2. Messbarkeit von Feldern nach der Quantentheorie.

Der quantenelektromagnetische Formalismus hat seinen
Ausgangspunkt gefunden in der von Dirac entwickelten
Quantentheorie der Strahlung, welche durch die Einführung
einer mit den quantenmechanischen Vertauschungsrelationen
zusammenhängenden Nichtvertauschbarkeit kanonisch kon-
jugierter Schwingungsamplituden eines Strahlungsfeldes
charakterisiert ist. Auf Grund dieser Theorie wurden
von Jordan und Pauli zunächst für ladungsfreie Felder
Vertauschungsrelationen zwischen den elektromagnetischen
Feldkomponenten aufgestellt, und der Formalismus wurde
dann von Heisenberg und Pauli zu einem gewissen
Abschluss gebracht, indem die Wechselwirkung zwischen
Feld und materiellen Ladungsträgern in korrespondenz-
mässiger Weise behandelt wurde. Die konsequente Anwendung der Theorie auf Atomprobleme ist indessen wesentlich
beeinträchtigt durch das Auftreten der wohlbekannten Paradoxien der unendlichen Selbstenergie der Elementarteilchen,
welche auch nicht durch die von Dirac vorgeschlagene
Modifikation der Darstellung des Formalismus beseitigt wurden[1]. In unserer Diskussion der Begrenzung der Messbarkeit
von Feldgrössen spielen diese Schwierigkeiten jedoch keine
Rolle, weil für diesen Zweck die atomistische Struktur der
Materie nicht wesentlich in Frage kommt. Zwar verlangt
die Messung der Felder die Benutzung materieller geladener
Probekörper, aber ihre eindeutige Anwendung als Messinstrumente wird eben durch den Umfang bedingt, in dem
wir sowohl ihre Beeinflussung durch die Felder, wie ihre
Wirkungen als Feldquellen auf Grund der klassischen
Elektrodynamik behandeln können.

Bei diesem Sachverhalt können wir uns auf die reine

[1] Vgl. L. Rosenfeld, Zs. f. Phys., **76**, 729, 1932.

Feldtheorie beschränken und also zur Untersuchung der Folgerungen des quantenelektromagnetischen Formalismus betreffend die Messbarkeit von Feldgrössen direkt von den Vertauschungsrelationen ladungsfreier Felder ausgehen. Unter Benutzung der üblichen Bezeichnung, $[p, q] = pq - qp$, haben wir sodann zwischen den Feldkomponenten in zwei Raumzeitpunkten (x_1, y_1, z_1, t_1) und (x_2, y_2, z_2, t_2) folgende typische Relationen, woraus die übrigen Vertauschungsrelationen durch zyklische Permutation hervorgehen[1]):

$$\left. \begin{aligned} \left[\mathfrak{E}_x^{(1)}, \mathfrak{E}_x^{(2)}\right] &= \left[\mathfrak{H}_x^{(1)}, \mathfrak{H}_x^{(2)}\right] = \sqrt{-1}\,\hbar\left(A_{xx}^{(12)} - A_{xx}^{(21)}\right) \\ \left[\mathfrak{E}_x^{(1)}, \mathfrak{E}_y^{(2)}\right] &= \left[\mathfrak{H}_x^{(1)}, \mathfrak{H}_y^{(2)}\right] = \sqrt{-1}\,\hbar\left(A_{xy}^{(12)} - A_{xy}^{(21)}\right) \\ \left[\mathfrak{E}_x^{(1)}, \mathfrak{H}_x^{(2)}\right] &= 0 \\ \left[\mathfrak{E}_x^{(1)}, \mathfrak{H}_y^{(2)}\right] &= -\left[\mathfrak{H}_x^{(1)}, \mathfrak{E}_y^{(2)}\right] = \sqrt{-1}\,\hbar\left(B_{xy}^{(12)} - B_{xy}^{(21)}\right). \end{aligned} \right\} \quad (1)$$

Dabei bedeuten $\mathfrak{E}_x^{(1)}, \mathfrak{E}_y^{(1)}, \mathfrak{E}_z^{(1)}, \mathfrak{H}_x^{(1)}, \mathfrak{H}_y^{(1)}, \mathfrak{H}_z^{(1)}$ die elektrischen und magnetischen Feldkomponenten im Raumzeitpunkte (x_1, y_1, z_1, t_1), während zur Abkürzung

$$\left. \begin{aligned} A_{xx}^{(12)} &= -\left(\frac{\partial^2}{\partial x_1 \partial x_2} - \frac{1}{c^2}\frac{\partial^2}{\partial t_1 \partial t_2}\right)\left\{\frac{1}{r}\delta\left(t_2 - t_1 - \frac{r}{c}\right)\right\} \\ A_{xy}^{(12)} &= -\frac{\partial^2}{\partial x_1 \partial y_2}\left\{\frac{1}{r}\delta\left(t_2 - t_1 - \frac{r}{c}\right)\right\} \\ B_{xy}^{(12)} &= -\frac{1}{c}\frac{\partial}{\partial t_1 \partial z_2}\left\{\frac{1}{r}\delta\left(t_2 - t_1 - \frac{r}{c}\right)\right\} \end{aligned} \right\} \quad (2)$$

[1] Vgl. P. JORDAN und W. PAULI, Zs. f. Phys., **47**, 151, 1928, sowie W. HEISENBERG und W. PAULI, Zs. f. Phys. **56**, 33, 1929. Abgesehen von einem unwesentlichen Vorzeichenunterschied, der von einer Unstimmigkeit in der Wahl der Zeitrichtung bei der Fourierzerlegung der Feldstärken herrührt, stimmen die angeführten Formeln mit den in den zitierten Abhandlungen abgeleiteten inhaltlich überein. Insbesondere bedeutet die hier benutzte Schreibweise, wo alle Glieder als retardierte erscheinen, eine rein formale Änderung, die auf eine möglichst anschauliche Deutung der Messprobleme hinzielt.

gesetzt ist. Weiter bedeutet $\hbar$ die durch 2π dividierte Plancksche Konstante, c die Lichtgeschwindigkeit und r den räumlichen Abstand der beiden Punkte. Schliesslich bezeichnet δ die von Dirac eingeführte symbolische Funktion, die bekanntlich durch die Eigenschaft, dass

$$\int_{t'}^{t''} \delta(t-t_0)\, dt = \begin{cases} 1 & \text{wenn } t' < t_0 < t'' \\ 0 & \text{wenn } t_0 < t' \text{ oder } t_0 > t'' \end{cases} \tag{3}$$

charakterisiert ist, und die formal wie eine gewöhnliche Funktion differenziiert wird.

Eben im Auftreten der durch (3) definierten δ-Funktion in den Vertauschungsrelationen (1) kommt die bereits erwähnte Tatsache zum Vorschein, dass die quantentheoretischen Feldgrössen nicht als eigentliche Punktfunktionen zu betrachten sind, sondern erst Raumzeitintegralen über die Feldkomponenten ein eindeutiger Sinn zukommt. Mit Rücksicht auf die einfachste Prüfungsmöglichkeit des Formalismus werden wir uns mit der Betrachtung von Mittelwerten der Feldkomponenten über einfach zusammenhängende Raumzeitgebiete begnügen, deren räumliche Ausdehnung innerhalb eines bestimmten Zeitintervalls konstant bleibt. Indem wir das Volumen eines solchen Gebiets G mit V und das zugehörige Zeitintervall mit T bezeichnen, definieren wir also zum Beispiel den Mittelwert von $\mathfrak{E}_x$ durch die Formel

$$\overline{\mathfrak{E}}_x^{(G)} = \frac{1}{VT} \int_T dt \int_V \mathfrak{E}_x\, dv. \tag{4}$$

Für die so definierten Mittelwerte zweier Feldkomponenten über zwei gegebene Raumzeitgebiete I und II gelten Vertauschungsrelationen, die sich umittelbar aus (1) durch Integration über die beiden Gebiete und Division mit dem

Produkt ihrer vierdimensionalen Ausdehnungen ergeben. Die Werte der Klammersymbole $\left[\overline{\mathfrak{E}}_x^{(I)}, \overline{\mathfrak{E}}_x^{(II)}\right], \ldots$ gehen also einfach aus (1) hervor, wenn die Grössen $A^{(12)}$, $B^{(12)}$ durch ihre Mittelwerte über die zwei Gebiete

$$
\left.\begin{aligned}
\bar{A}_{xx}^{(I, II)} &= -\frac{1}{V_I V_{II} T_I T_{II}} \int_{T_I} dt_1 \int_{T_{II}} dt_2 \int_{V_I} dv_1 \int_{V_{II}} dv_2 \\
&\quad \left(\frac{\partial^2}{\partial x_1 \partial x_2} - \frac{1}{c^2}\frac{\partial^2}{\partial t_1 \partial t_2}\right)\left\{\frac{1}{r}\,\delta\left(t_2 - t_1 - \frac{r}{c}\right)\right\} \\
\bar{A}_{xy}^{(I, II)} &= -\frac{1}{V_I V_{II} T_I T_{II}} \int_{T_I} dt_1 \int_{T_{II}} dt_2 \int_{V_I} dv_1 \int_{V_{II}} dv_2 \\
&\quad \frac{\partial^2}{\partial x_1 \partial y_2}\left\{\frac{1}{r}\,\delta\left(t_2 - t_1 - \frac{r}{c}\right)\right\} \\
\bar{B}_{xy}^{(I, II)} &= -\frac{1}{V_1 V_{II} T_I T_{II}} \int_{T_I} dt_1 \int_{T_{II}} dt_2 \int_{V_I} dv_1 \int_{V_{II}} dv_2 \\
&\quad \frac{1}{c}\frac{\partial^2}{\partial t_1 \partial z_2}\left\{\frac{1}{r}\,\delta\left(t_2 - t_1 - \frac{r}{c}\right)\right\}
\end{aligned}\right\} \quad (5)
$$

ersetzt werden.

In ganz ähnlicher Weise, wie sich die dem Unbestimmtheitsprinzip zugrundeliegende Heisenbergsche Relation

$$\Delta p \Delta q \sim \hbar \qquad (6)$$

für zwei kanonisch konjugierte mechanische Grössen aus der allgemeinen quantenmechanischen Vertauschungsrelation

$$[q,\, p] = \sqrt{-1}\,\hbar \qquad (7)$$

ableiten lässt, ergeben sich also für die Produkte der komplementären Unbestimmtheiten der betrachteten Feldmittelwerte folgende typische Formeln:

$$
\left.\begin{aligned}
\Delta \overline{\mathfrak{E}}_x^{(I)} \Delta \overline{\mathfrak{E}}_x^{(II)} &\sim \hbar \left|\bar{A}_{xx}^{(I, II)} - \bar{A}_{xx}^{(II, I)}\right|, & \Delta \overline{\mathfrak{E}}_x^{(I)} \Delta \overline{\mathfrak{E}}_y^{(II)} &\sim \hbar \left|\bar{A}_{xy}^{(I, II)} - \bar{A}_{xy}^{(II, I)}\right| \\
\Delta \overline{\mathfrak{E}}_x^{(I)} \Delta \overline{\mathfrak{H}}_x^{(II)} &= 0, & \Delta \overline{\mathfrak{E}}_x^{(I)} \Delta \overline{\mathfrak{H}}_y^{(II)} &\sim \hbar \left|\bar{B}_{xy}^{(I, II)} - \bar{B}_{xy}^{(II, I)}\right|.
\end{aligned}\right\} \quad (8)
$$

Einige für unser Problem wichtige Ergebnisse lassen sich unmittelbar aus den Ausdrücken (5) und (8) folgern. Vor allem sehen wir, dass gemäss der durch (3) ausgedrückten Eigenschaft der δ-Funktion die Grössen $\bar{A}^{(I, II)}$, $\bar{B}^{(I, II)}$ sich bei stetiger Verschiebung der Grenzen der Gebiete I und II stetig ändern, solange die Ausdehnungen dieser Gebiete, d. h. die Werte von V_I, T_I, V_{II}, T_{II} von Null verschieden bleiben. Insbesondere verschwinden die Differenzen $\bar{A}^{(I, II)} - \bar{A}^{(II, I)}$ und $\bar{B}^{(I, II)} - \bar{B}^{(II, I)}$ ohne Unstetigkeit, wenn die Grenzen der zwei Gebiete allmählich mit einander zusammenfallen. Hieraus folgt, dass die Mittelwerte aller Feldkomponenten über dasselbe Raumzeitgebiet unter einander vertauschbar sind, und somit unabhängig von einander genau messbar sein sollen. Diese Folgerung der Theorie, die über die Voraussetzung der unbeschränkten Messbarkeit jeder einzelnen Feldgrösse wesentlich hinausgeht, erscheint übrigens als ein Spezialfall zweier allgemeinerer Sätze, die aus den Symmetrieeigenschaften der Grössen $\bar{A}^{(I, II)}$ und $\bar{B}^{(I, II)}$ folgen. Aus der Tatsache, dass die Ausdrücke $A^{(12)} - A^{(21)}$ ihr Vorzeichen wechseln, wenn die Zeitpunkte t_1 und t_2 vertauscht werden, folgt nämlich, dass die Mittelwerte zweier gleichartiger (d. h. zweier elektrischer oder zweier magnetischer) Komponenten über zwei beliebige Raumgebiete immer vertauschbar sind, sobald die zugehörigen Zeitintervalle zusammenfallen. Weiter folgt aus der entsprechenden Antisymmetrie der Ausdrücke $B^{(12)}$ und $B^{(21)}$ bei Vertauschung der Raumpunkte (x_1, y_1, z_1) und (x_2, y_2, z_2), dass die Mittelwerte zweier ungleichartiger Komponenten, wie $\mathfrak{E}_x$ und $\mathfrak{H}_y$, über zwei beliebige Zeitintervalle vertauschbar sind, wenn die zugehörigen Raumgebiete zusammenfallen.

Diese Resultate könnten zunächst mit den aus der Hei-

senberg-Paulischen Darstellung des Formalismus abzulei-
tenden Vertauschungsrelationen zwischen Mittelwerten von
Feldgrössen zu einem und demselben Zeitpunkt über endliche Raumgebiete, die im zitierten Buch von HEISENBERG
diskutiert sind, unvereinbar erscheinen. Während auch dort
Mittelwerte gleichartiger Komponenten als vertauschbar an-
gegeben werden, wird auf die Nichtvertauschbarkeit von
ungleichartigen Komponenten über ein und dasselbe Raumgebiet geschlossen. Dieser Gegensatz löst sich aber einfach
dadurch, dass es sich in der Heisenbergschen Behandlung
um einen Grenzübergang handelt, bei dem zwei ursprünglich
nicht zusammenfallende Raumzeitgebiete erst dann zur
Deckung gebracht werden, nachdem ihre zeitlichen Ausdehnungen sich zu einem und demselben Zeitpunkt zu-
sammengezogen haben. Mit Rücksicht auf die Symmetrie
des durch (2) gegebenen Ausdruckes $B_{xy}^{(12)}$ in Bezug auf t_1 und
t_2, sowie auf die durch (3) definierte Eigenschaft der δ-Funk-
tion finden wir nämlich für zusammenfallende Zeitintervalle

$$\bar{B}_{xy}^{(\mathrm{I,\,II})} - \bar{B}_{xy}^{(\mathrm{II,\,I})} = \frac{2}{V_\mathrm{I} V_\mathrm{II} T^2} \cdot \frac{1}{c} \int_{V_\mathrm{I}} dv_1 \int_{V_\mathrm{II}} dv_2 \, \frac{\partial}{\partial z_1}\left(\frac{1}{r}\right), \qquad (9)$$

wo $T_\mathrm{I} = T_\mathrm{II} = T$ gesetzt ist, und die doppelte Integration
über alle Paare von Punkten je eines der betreffenden
Raumgebiete zu erstrecken ist, deren Abstand r kleiner ist
als cT. Wenn wir nun weiter die beiden Raumgebiete von
selbem Volumen $V_\mathrm{I} = V_\mathrm{II} = V$ und von selber Gestalt,
aber gegen einander in der z-Richtung verschoben annehmen,
so bekommen wir im Grenzfall, wo cT als verschwindend
klein in Vergleich mit den linearen Abmessungen der Raumgebiete betrachtet werden kann, für das in (9) auftretende
Raumintegral durch partielle Integration den Ausdruck
$\pm 2\pi c^2 T^2 F$, wo F das Areal der von der Projektion der

Schnittkurve der Begrenzungsflächen von V_I und V_{II} auf die xy-Ebene eingeschlossenen Fläche darstellt, und wo das Vorzeichen $+$ bzw. $-$ zu nehmen ist, je nachdem der Bereich II in der positiven bzw. negativen z-Richtung gegenüber dem Bereich I verschoben ist. Bei stetiger Durcheinanderschiebung der Bereiche ändert sich also die Differenz $\bar{B}_{xy}^{(I,II)} - \bar{B}_{xy}^{(II,I)}$ unstetig um den Betrag $\frac{8\pi cF}{V^2}$, indem beide Ausdrücke $\bar{B}_{xy}^{(I,II)}$ und $\bar{B}_{xy}^{(II,I)}$ ihr Vorzeichen wechseln. Im besprochenen Grenzfall weist daher die Vertauschungsrelation der instantanen Raummittelwerte von $\mathfrak{E}_x$ und $\mathfrak{H}_y$ eine wesentliche Zweideutigkeit auf, die für den erwähnten scheinbaren Widerspruch verantwortlich ist.

Der durch die früheren Untersuchungen der physikalischen Messungsmöglichkeiten vermeintlich erbrachte Nachweis einer komplementären Begrenzung der Messbarkeit ungleichartiger Feldkomponenten innerhalb eines und desselben Raumbereiches beruht auch, wie wir sehen werden, lediglich auf der Benutzung von Punktladungen als Probekörper, was keine genügend scharfe Begrenzung des Messbereichs erlaubt. Für die Prüfung des quantenelektromagnetischen Formalismus kommen, wie schon betont, nur Messungen mit Probekörpern endlich ausgedehnter Ladungsverteilung in Betracht, da jede wohldefinierte Aussage dieses Formalismus sich ja auf Mittelwerte der Feldkomponenten über endliche Raumzeitgebiete bezieht. Dieser Umstand hindert uns aber keineswegs, alle eindeutigen, aus der Heisenberg-Paulischen Darstellung zu ziehenden Schlüsse über die Zeitabhängigkeit von räumlichen Feldmittelwerten durch Feldmessungen zu prüfen. Dazu brauchen wir nur Mittelwertsbereiche heranzuziehen, deren zeitliche Ausdehnung T, mit c multipliziert, genügend klein ist gegenüber ihren linearen räumlichen Abmessungen, deren Grössenordnung wir im folgenden immer mit L bezeichnen werden.

Eben der Fall $L > cT$ eignet sich übrigens besonders zu einer eingehenden Prüfung der Folgerungen des Formalismus im eigentlichen quantentheoretischen Gebiet. Schon im Gültigkeitsbereich der klassischen Theorie bietet ja der Fall $L \leq cT$ wenig Interesse, weil alle Besonderheiten der Wellenfelder innerhalb des Volumens V sich wegen der Fortpflanzung während des Zeitintervalls T bei der Mittelwertbildung weitgehend ausgleichen. Im Quantengebiet kommen zu diesem Ausgleich noch die eigentümlichen Schwankungserscheinungen hinzu, die aus dem prinzipiell statistischen Charakter des Formalismus folgen. Während im Falle $L \leq cT$ diese Schwankungen, wie wir sogleich sehen werden, in die Lösungen gegebener Probleme wesentlich eingehen, spielen sie eben im Fall $L > cT$ eine verhältnismässig kleine Rolle.

Die erwähnten Schwankungen hängen aufs Engste zusammen mit der Unmöglichkeit, die für die Quantentheorie der Felder charakteristische Lichtquantenvorstellung mittels klassischer Begriffe zu veranschaulichen. Insbesondere geben sie Ausdruck für die gegenseitige Ausschliessung der genauen Kenntnis der Lichtquantenzusammensetzung eines elektromagnetischen Feldes und der Kenntnis des Mittelwerts irgendeiner seiner Komponenten in einem wohldefinierten Raumzeitgebiet. Denken wir uns die Dichte ω_i ($\varkappa_x$, $\varkappa_y$, $\varkappa_z$) der Lichtquanten von bestimmtem Polarisationsparameter i und gegebenen Impuls und Energie $\hbar\varkappa_x$, $\hbar\varkappa_y$, $\hbar\varkappa_z$ und $\hbar\nu = \hbar c \sqrt{\varkappa_x^2 + \varkappa_y^2 + \varkappa_z^2}$ bekannt, so sind zwar die Erwartungswerte aller Feldmittelwerte Null, aber das mittlere Schwankungsquadrat jeder Feldgrösse, wie die durch (4) definierte $\overline{\mathfrak{E}}_x^{(G)}$, wird durch die einfach abzuleitende Formel

$$S(G) = \frac{1}{V^2 T^2} \cdot \frac{\hbar}{3} \int_T dt_1 \int_T dt_2 \int_V dv_1 \int_V dv_2 \, \frac{\partial^2}{\partial t_1 \partial t_2} \int_{-\infty}^{\infty} \left(\sum_i \omega_i + 1 \right)$$
$$\cos \left[\varkappa_x (x_1 - x_2) + \varkappa_y (y_1 - y_2) + \varkappa_z (z_1 - z_2) - \nu (t_1 - t_2) \right] \qquad (10)$$
$$\frac{d\varkappa_x d\varkappa_y d\varkappa_z}{\nu}$$

gegeben. Aus der Formel (10) sehen wir, dass die betreffenden Schwankungen bei gegebener Lichtquantenzusammensetzung nie ausbleiben können, da sie schon für $\omega_i = 0$, d. h. bei völliger Abwesenheit von Lichtquanten, einen endlichen, positiven Wert annehmen, der sich durch eine leichte Rechnung auf die Form

$$S_0(G) = \frac{2}{3\pi^2} \frac{\hbar c}{V^2} \int_V dv_1 \int_V dv_2 \, \frac{1}{r^2 \left[(cT)^2 - r^2 \right]} \qquad (11)$$

bringen lässt. Für jede andere, durch eine gegebene Dichte ω_i definierte Lichtquantenverteilung wird das Schwankungsquadrat des Mittelwerts einer Feldkomponente grösser als $S_0(G)$ sein. Die nach dem Formalismus zu erwartenden Schwankungen eines Feldmittelwerts können aber beliebig klein werden, wenn eine direkte, etwa durch Messungen erreichte Kenntnis von Feldgrössen vorausgesetzt ist. In einem solchen Fall ist selbstverständlich die Lichtquantendichte ω_i nicht wohldefiniert, und wir müssen uns mit statistischen Aussagen über diese Dichte begnügen.

Für die Diskussion der Messungsmöglichkeiten ist es ferner von entscheidender Bedeutung, dass der Ausdruck (11) nicht nur für die Feldschwankungen im lichtquantenfreien Raum gilt, sondern auch das Schwankungsquadrat jedes Feldmittelwerts in dem allgemeineren Fall darstellt, wo als Feldquellen nur klassisch beschreibbare Strom- und Ladungsverteilungen vorkommen. Der Feldzustand ist hier eindeutig durch die Forderungen definiert, dass der Erwartungswert jeder Feldgrösse mit dem klassisch berechneten übereinstimmt, und dass die Anzahl der Lichtquanten von gegebenem Impuls und Polarisation um ihren korrespondenzmässig abgeschätzten Mittelwert n_0 gemäss dem für unabhängige Ereignisse gültigen Wahrscheinlichkeitsgesetz

$$w(n) = \frac{n_0^n e^{-n_0}}{n!} \qquad (12)$$

verteilt sind. Für die Feldschwankungen dieses Zustandes ergibt eine einfache Rechnung eben den Ausdruck (11). Entsprechend der Eigentümlichkeit der Hohlraumschwankungen folgt weiter, dass auch im allgemeinen Fall eines Feldes gegebener Lichtquantenzusammensetzung das Hinzukommen der Feldwirkungen irgendwelcher klassisch beschreibbaren Quellen keinerlei Einfluss auf die Schwankungserscheinungen haben wird.

Die Quadratwurzel des Ausdrucks (11) mag als eine kritische Feldgrösse $\mathfrak{S}$ angesehen werden, in dem Sinne, dass wir nur bei der Betrachtung von Feldmittelwerten, die wesentlich grösser als $\mathfrak{S}$ sind, von den betreffenden Schwankungen absehen dürfen. Für die Beurteilung der Prüfungsmöglichkeit des Formalismus im eigentlichen Quantengebiet kommt noch eine andere kritische Feldgrösse $\mathfrak{U}$ in Betracht, die gleich ist der Quadratwurzel des durch (8) gegebenen Produkts der komplementären Unbestimmtheiten zweier Feldmittelwerte über Raumzeitbereiche, die sich nur teilweise überdecken, indem sie räumlich und zeitlich um Strecken der Grössenordnung L bzw. T gegen einander verschoben sind. Für Feldstärken, die wesentlich grösser als $\mathfrak{U}$ sind, gelangen wir nämlich offenbar in den Gültigkeitsbereich der klassischen elektromagnetischen Theorie, wo alle quantentheoretischen Züge des Formalismus ihre Bedeutung verlieren. Eine einfache Abschätzung auf Grund der Formeln (8) und (11) ergibt nun, dass im Falle $L \leq cT$ beide kritische Ausdrücke $\mathfrak{U}$ und $\mathfrak{S}$ von derselben Grössenordnung

$$\mathfrak{U} \backsim \mathfrak{S} \backsim \frac{\sqrt{\hbar c}}{L \cdot cT} \tag{13}$$

sind. Im Falle $L > cT$ hingegen ist

$$\mathfrak{U} \backsim \sqrt{\frac{\hbar}{L^3 T}} \quad \text{und} \quad \mathfrak{S} \backsim \frac{\sqrt{\hbar c}}{L^2}, \tag{14}$$

so dass im Grenzfall $L \gg cT$ die kritische Feldstärke $\mathfrak{U}$ viel grösser ist als $\mathfrak{S}$, und wir daher bei der Prüfung der charakteristischen Folgerungen des Formalismus von den Feldschwankungen weitgehend absehen können.

Bevor wir zu dem Vergleich der in diesem Paragraphen besprochenen Folgerungen des quantenelektromagnetischen Formalismus mit den physikalischen Messungsmöglichkeiten von Feldgrössen übergehen, möchten wir noch hier betonen, dass die widerspruchslose Deutbarkeit dieses Formalismus in keiner Weise gefährdet ist durch solche paradoxe Züge seiner mathematischen Darstellung, wie die unendliche Nullpunktsenergie. Insbesondere hat diese letztere Paradoxie, die übrigens mittels einer die physikalische Deutung nicht störenden, formalen Änderung der Darstellung[1] beseitigt werden kann, keine direkte Verbindung zum Problem der Messbarkeit von Feldgrössen. Auf Grund der Feldtheorie würde ja eine Bestimmung der elektromagnetischen Energie in einem gegebenen Raumzeitbereich die Kenntnis der einer Messung unzugänglichen Werte der

---

[1] Vgl. L. Rosenfeld und J. Solomon, Journal de Physique **2**, 139, 1931, sowie W. Pauli, Hb. d. Physik, 2. Aufl., Bd. 24/1, S. 255, 1933.

Feldkomponenten in jedem Raumzeitpunkt des Gebiets verlangen. Eine physikalische Messung der Feldenergie liesse sich nur mittels einer geeigneten mechanischen Vorrichtung ausführen, durch welche die in einem gegebenen Raumgebiet befindlichen elektromagnetischen Felder von dem übrigen Feld getrennt werden könnten, so dass die in diesem Gebiet enthaltene Energie unter Verwendung des Erhaltungssatzes nachträglich gemessen werden könnte. Jede solche Trennung der Felder wird aber wegen der Wechselwirkung mit dem Messmechanismus eine unkontrollierbare Änderung der Feldenergie im betreffenden Gebiet mit sich bringen, deren Berücksichtigung wesentlich ist für die Aufklärung der wohlbekannten Paradoxien, die in der Diskussion der Energieschwankungen der Hohlraumstrahlung zutagetreten.[1]

§ 3. Voraussetzungen physikalischer Feldmessungen.

Definitionsgemäss beruht die Messung von elektromagnetischen Feldgrössen auf der Uebertragung von Impuls auf geeignete elektrische oder magnetische Probekörper, die sich im Feld befinden. Ganz abgesehen von der in der Quantentheorie gebotenen Vorsicht bei der Anwendung der üblichen Idealisation der in jedem Raumzeitpunkt definierten Feldkomponenten, handelt es sich dabei wesentlich immer um Mittelwerte dieser Komponenten sowohl über die für die Impulsübertragung erforderlichen endlichen Zeitintervalle, wie über die Raumbereiche, über welche die Elektrizitätsladungen bzw. magnetischen Polstärken der in Frage kommenden Probekörper verteilt sind. Selbstverständlich ist schon die Annahme einer gleichmässigen Ladungsverteilung auf einem Probekörper eine Idealisation, die wegen des

---

[1] Vgl. W. HEISENBERG, Leipziger Berichte 83, 1, 1931.

atomaren Aufbaus aller materieller Körper einer gewissen Einschränkung unterliegt, die aber für die eindeutige Definition von Feldgrössen unentbehrlich ist.

Um einen bestimmten Fall vor Augen zu haben, betrachten wir die Messung des Mittelwerts der elektrischen Feldkomponente in der x-Richtung $\mathfrak{E}_x$ über einen Raumzeitbereich von Volumen V und Zeitlänge T. Dazu benutzen wir also einen Probekörper, dessen elektrische Ladung über das Volumen V mit der Dichte ϱ gleichmässig verteilt ist und bestimmen die Werte p'_x und p''_x der Impulskomponente dieses Körpers in der x-Richtung am Anfang t' und am Ende t'' des Intervalls T. Der gesuchte Mittelwert $\overline{\mathfrak{E}}_x$ wird dann durch die Gleichung

$$p''_x - p'_x = \varrho\,\overline{\mathfrak{E}}_x VT \qquad (15)$$

bestimmt, wobei allerdings vorausgesetzt ist, dass die von den Impulsmessungen beanspruchten Zeitintervalle, die wir grössenordnungsmässig mit $\varDelta t$ bezeichnen wollen, als verschwindend klein gegenüber T betrachtet werden können, und dass wir die Verschiebung des Probekörpers sowohl infolge der Impulsmessungen, wie infolge der ihm durch das zu messende Feld innerhalb des Zeitintervalls T erteilten Beschleunigung vernachlässigen können im Verhältnis zu den linearen Abmessungen L des Raumbereichs V.

Durch die Wahl eines genügend schweren Probekörpers können wir offenbar seine Beschleunigung unter Einwirkung des Feldes beliebig herabsetzen. Bei den Impulsmessungen begegnen wir aber Verhältnissen, die von der Masse des Probekörpers unabhängig sind. Dem Unbestimmtheitsprinzip zufolge wird ja jede mit der Genauigkeit $\varDelta p_x$ ausgeführte Messung der Impulskomponente p_x mit einem Verlust $\varDelta x$ der Kenntnis der Lage des betreffenden Körpers verbunden, der grössenordnungsmässig durch die in (6) enthaltene Relation

$$\Delta p_x \Delta x \sim \hbar \tag{16}$$

gegeben wird. An und für sich bedeutet dieser Sachverhalt jedoch keine Einschränkung der durch die Feldmessung zu erzielenden Genauigkeit, da wir noch über den Wert der Ladungsdichte verfügen können. Wenn wir Δt und Δx im Verhältnis zu T und L vernachlässigen, erhalten wir in der Tat aus (15) und (16) für die Genauigkeit $\Delta\overline{\mathfrak{E}}_x$ der Feldmessung grössenordnungsmässig

$$\Delta\overline{\mathfrak{E}}_x \sim \frac{\hbar}{\varrho \Delta x \cdot VT} \tag{17}$$

welche bei jedem noch so kleinen Wert von Δx durch die Wahl eines genügend grossen Werts von ϱ beliebig klein gemacht werden kann.

Streng genommen hängt die Genauigkeit der Feldmessung noch von der absoluten Grösse des Werts von $\overline{\mathfrak{E}}_x$ selber ab, denn bei gegebenen Spielräumen für Δt und Δx wird, selbst wenn Δp_x Null wäre, der aus (15) ermittelte Wert von $\overline{\mathfrak{E}}_x$ wegen der unscharfen Begrenzung des Messbereiches mit einer Unsicherheit behaftet, die jede Grenze überschreitet, wenn $\mathfrak{E}_x$ ins Unendliche wächst. Letzterer Umstand entspricht jedoch nur der allgemeinen Begrenzung aller physikalischen Messungen, bei welchen eine Kenntnis der Grössenordnung der zu erwartenden Effekte für die Wahl der geeigneten Messinstrumente immer nötig ist. Bei unserem Problem ist eine obere Grenze der uns interessierenden Effekte dadurch gesetzt, dass wir bei wachsender Grösse der Feldkomponente allmählich ins Gültigkeitsgebiet der klassischen elektromagnetischen Theorie gelangen. In dem zur Prüfung des quantenelektromagnetischen Formalismus besonders geeigneten Fall $L > cT$ stellt, wie im vorigen Paragraphen erwähnt, der mit der rechten Seite der ersten Formel (14) äquivalente Ausdruck

$$Q = \sqrt{\frac{\hbar}{VT}} \qquad (18)$$

eine in dieser Hinsicht kritische Feldgrösse dar. Bei Einführung dieser Bezeichnung nimmt die Beziehung (17) die Form

$$\varDelta \overline{\mathfrak{E}}_x \backsim \lambda Q \qquad (19)$$

an, wo

$$\lambda = \frac{Q}{\varrho \varDelta x} \qquad (20)$$

ein dimensionsloser, für die Beurteilung der Genauigkeit der Feldmessung ausschlaggebender Faktor ist.

Die Forderung, dass λ klein gegenüber der Einheit, und gleichzeitig $\varDelta x$ klein gegenüber L sein soll, bedeutet, dass die gesamte elektrische Ladung des Körpers aus einer sehr grossen Anzahl Elementarladungen ε bestehen soll. In der Tat ist nach (20) diese Anzahl N durch

$$N = \frac{\varrho V}{\varepsilon} = \frac{QV}{\lambda \varepsilon \varDelta x} = \frac{1}{\lambda} \cdot \frac{L}{\varDelta x} \cdot \sqrt{\frac{L}{cT}} \cdot \sqrt{\frac{\hbar c}{\varepsilon^2}} \qquad (21)$$

gegeben, und ist sehr gross, wenn die erwähnten Forderungen erfüllt sind und, wie angenommen, $L > cT$ ist. Der letzte Faktor ist ja die reziproke Quadratwurzel der Feinstrukturkonstante, deren Kleinheit, wie schon in der Einleitung erwähnt, eine wesentliche Voraussetzung der korrespondenzmässigen Elektronentheorie ist. Wie dort betont, sind einer Feldmessung mit einer Elementarladung als Probekörper wesentliche Einschränkungen auferlegt, was auch direkt aus (21) ersichtlich ist, wenn man $N = 1$ setzt.[1] Die Annahme eines grossen Werts von N ist überdies eine notwendige Bedingung für die physikalische Verwirklichung einer gleichmässigen Verteilung der Ladung des Probekörpers über das Volumen V; und solange die linearen Abmessungen L des Probekörpers gross sind gegenüber den Atomdimensionen, bietet ihre Erfüllung offenbar keine prinzipielle Schwierigkeit. Es braucht auch kaum erwähnt zu werden, dass unter dieser Voraussetzung die oben benutzte Annahme über die Masse des Probekörpers, die damit gleichbedeutend ist, dass diese Masse sehr gross gegenüber derjenigen eines Lichtquants der Wellenlänge L sein muss, stets befriedigt werden kann.

Soweit haben wir indessen völlig von den elektromagnetischen Feldwirkungen abgesehen, welche die Beschleunigung jedes Probekörpers während der Impulsmessung begleiten. Diese Wirkungen überlagern sich auf das ursprüngliche Feld und müssen in die durch Gleichungen vom Typus (15) definierten Feldmittelwerte einbezogen werden. Die Hauptaufgabe der folgenden Untersuchung wird daher sein, eine Messanordnung zu finden, in welcher die Feldwirkungen der Probekörper in grösstmöglichem Umfang kontrolliert oder kompensiert werden können. An dieser Stelle müssen wir aber zunächst auf die Frage eingehen, inwieweit die Rückwirkung der durch die Beschleunigungen der Probekörper bei den Impulsmessungen erzeugten Strahlungsfelder schon die Ausführbarkeit der Messung der

[1] Vgl. V. Fock und P. Jordan, Zs. f. Phys. **66**, 206, 1930, wo auf derartige, mit der Quantentheorie der Felder nicht verbundene Einschränkungen von Feldmessungen hingewiesen ist. Vgl. auch J. Solomon, Journal de physique, **4**, 368, 1933.

in (15) auftretenden Werte der Impulskomponente des Probekörpers am Anfang und am Ende des Messintervals beeinträchtigen könnte. Eben mit Hinblick auf diese Möglichkeit haben Landau und Peierls in der anfangs zitierten Arbeit die Zuverlässigkeit der Unbestimmtheitsrelation (16) für geladene Körper angezweifelt, und geschlossen, dass sie durch eine andere, noch mehr einschränkende Relation zu ersetzen sei, in welcher die Ladung des Probekörpers wesentlich eingeht. Dabei haben sie jedoch das elektromagnetische Verhalten eines solchen Körpers mit demjenigen einer Punktladung e verglichen, und folglich zur Abschätzung der Grössenordnung der durch die Strahlungsrückwirkung hervorgerufenen Impulsänderung des Probekörpers während der Zeit $\varDelta t$ den Ausdruck

$$\delta_e p_x \sim \frac{e^2}{c^3} \frac{\varDelta x}{\varDelta t^2} \qquad (22)$$

benutzt. Wird aber $\delta_e p_x$ als eine zusätzliche Unbestimmtheit der Impulsmessung betrachtet, so bekommt man anstatt (17), wenn $\varrho V = e$ gesetzt ist und zwischen $\overline{\mathfrak{E}}_x$ und $\mathfrak{E}_x$ nicht unterschieden wird,

$$\varDelta_e \mathfrak{E}_x \sim \frac{\hbar}{eT\varDelta x} + \frac{e\varDelta x}{c^3 T\varDelta t^2}, \qquad (23)$$

dessen Minimum bei Variation von e offenbar durch

$$\varDelta_m \mathfrak{E}_x \sim \frac{\sqrt{\hbar c}}{c^2 T\varDelta t} \qquad (24)$$

gegeben ist. Wenn man weiter mit Landau und Peierls zwischen T und $\varDelta t$ nicht unterscheidet, stimmt dieser Ausdruck überein mit der von ihnen angegebenen absoluten Grenze der Messbarkeit von Feldkomponenten, auf der sie ihre Kritik der Grundlagen des quantenelektromagnetischen Formalismus begründet haben.

Die vermeintlichen Schwierigkeiten der Impulsmessung verschwinden aber sofort, wenn auf die endliche Ausdehnung der elektrischen Ladung des Probekörpers genügende Rücksicht genommen wird. Bei der unten näher zu prüfenden Idealisation einer gleichmässigen, starr verschiebbaren Ladungsverteilung, können nämlich die elektrischen Feldstärken im Gebiet V während der Beschleunigung des Probekörpers innerhalb der Zeit Δt höchstens einen Wert von der Grössenordnung $\varrho \Delta x$ erreichen. Denn ihre zeitlichen Ableitungen sind ja nach den Maxwellschen Gleichungen höchstens von derselben Grössenordnung wie die Stromdichte, die grössenordnungsmässig durch $\varrho \dfrac{\Delta x}{\Delta t}$ gegeben ist. Jede elektromagnetische Rückwirkung auf den Körper während des Messintervalls Δt kann daher nur eine Impulsübertragung von der Grössenordnung

$$\delta_{\varrho} p_x \infty \varrho^2 V \Delta x \Delta t \qquad (25)$$

mit sich bringen. Mit Rücksicht auf (18) und (20) bekommen wir also durch Vergleich von (16) und (25):

$$\delta_{\varrho} p_x \infty \Delta p_x \cdot \lambda^{-2} \dfrac{\Delta t}{T}, \qquad (26)$$

woraus folgt, das bei jeder, durch einen gegebenen Wert von λ symbolisierten, angestrebten Genauigkeit der Feldmessung der Einfluss der elektromagnetischen Rückwirkung auf die Impulsmessung des Probekörpers vernachlässigt werden kann, wenn nur Δt in Vergleich mit T genügend klein gewählt wird. Eben dieser Umstand ist für die Beurteilung der Genauigkeit der Feldmessungen ausschlaggebend;

denn es erweist sich als unmöglich, den Einfluss des Strahlungsrückstosses auf die Impuls- und Energiebilanz bei den einzelnen Impulsmessungen direkt in Betracht zu ziehen. Zum Beispiel wäre der von Pauli[1] gemachte Vorschlag, die in der Ausstrahlung enthaltenen Impuls- und Energiebeiträge durch eine besondere Vorrichtung nachträglich zu messen, schon deswegen unausführbar, weil die Strahlungsfelder, die bei den Impulsmessungen am Anfang und Ende des Intervalls T entstehen, wenigstens in dem für die Feldmessungen besonders wichtigen Fall $L > cT$ nicht in einem für diesen Zweck genügenden Mass von einander trennen lassen. Überhaupt werden wir in den folgenden Paragraphen ganz allgemein zeigen, dass jeder Versuch einer derartigen Kontrolle der Feldwirkungen der Probekörper die Verwertung der betreffenden Feldmessung wesentlich beeinträchtigen würde.

Uebrigens ist es nicht nur für die Diskussion des Verhaltens der einzelnen Probekörper während der Messungen, sondern auch für die Beurteilung der gegenseitigen Beeinflussung mehrerer Probekörper wesentlich, dass diese nicht als Punktladungen, sondern als kontinuierliche Ladungsverteilungen behandelt werden. Denn die übliche Identifizierung der Ortsunbestimmtheit eines als Punktladung betrachteten Probekörpers mit den linearen Dimensionen des Messbereichs bedeutet eine willkürliche, dem Messbarkeitsproblem fremde Annahme. Aus diesem Grunde weichen die von Heisenberg einerseits, von Landau und Peierls andererseits durch Betrachtung von Punktladungen abgeschätzten Ausdrücke für das Produkt der Unsicherheiten von $\mathfrak{E}_x$ und $\mathfrak{H}_y$ innerhalb desselben Raumzeitbereichs nicht nur, wie schon erwähnt, von den Erwartungen des quanten-

---

[1] Vgl. W. Pauli, Hb. d. Physik, 2. Aufl., Bd. 24/1, S. 257, 1933.

elektromagnetischen Formalismus ab, sondern stimmen nur im Spezialfall $L \backsim cT$ mit einander überein. In diesem Fall ergeben beide Abschätzungen den Ausdruck Q^2, welcher dem nach dem Formalismus zu erwartenden grössenordnungsmässigen Wert des Produkts der komplementären Unbestimmtheiten zweier Feldmittelwerte innerhalb Raumzeitbereiche, die gegen einander um Raumzeitstrecken derselben Grössenordnung wie L und T verschoben sind, entspricht. Für zusammenfallende Bereiche ist es indessen ein wesentlicher Zug des Formalismus, dass das betreffende Produkt identisch verschwindet. Der physikalische Sinn dieses Resultats leuchtet auch sofort ein, sobald man die gleichmässige Ladungsverteilung des zur Messung von $\overline{\mathfrak{E}}_x$ benutzten Probekörpers berücksichtigt; denn die magnetische Feldstärke, die in einem Punkt P_2 des Volumens V durch die Verschiebung der Ladung ϱdv eines im Punkt P_1 befindlichen Volumelements erzeugt wird, ist genau gleich und entgegengesetzt der magnetischen Feldstärke, die infolge der gleichen Verschiebung der im Punkt P_2 befindlichen Ladung ϱdv im Punkte P_1 entsteht, sodass der Mittelwert über das Volumen V jeder durch die Verschiebung des Probekörpers erzeugten magnetischen Feldkomponente verschwindet.

Aus dem obigen geht hervor, dass für die Diskussion der Messbarkeit von Feldgrössen die Annahme von entscheidender Bedeutung ist, dass die zu benutzenden Probekörper sich als gleichmässig geladene, starre Körper verhalten, deren Impulse innerhalb jedes gegebenen, beliebig kurzen Zeitintervalls mit der durch (16) ausgedrückten, zur begleitenden unkontrollierbaren Verschiebung komplementären Genauigkeit gemessen werden können. Natürlich dürfen wir dabei wegen der endlichen Fortpflanzung aller Kräfte nicht an die gewöhnliche mechanische Idealisation des star-

ren Körpers denken, sondern müssen uns jeden Probekörper
als ein System individueller Teilkörper von genügend kleinen
Dimensionen vorstellen, und die Messung des Gesamtimpulses dieses Systems in solcher Weise ausgeführt denken,
dass alle Teilkörper mit genügender Annäherung dieselbe
Verschiebung während der Impulsmessung erleiden. Dass
diese Forderung, jedenfalls soweit man von dem atomaren
Aufbau der Probekörper absehen kann, sich ohne prinzipielle Schwierigkeit erfüllen lässt, liegt daran, dass die erforderlichen Impulsmessungen sich vollständig auf klassischer
Grundlage beschreiben lassen, gleichgültig, ob sie auf der
Verfolgung eines Stossprozesses zwischen dem Probekörper
und einem geeigneten materiellen Stosskörper, oder etwa
auf der Untersuchung des Dopplereffekts bei Reflexion von
Strahlung am Probekörper beruhen. Wenn nur die Masse
des Stosskörpers gross genug ist, oder das zur Messung
des Dopplereffekts benutzte Strahlungsbündel genügend viele
Lichtquanten enthält, lässt sich nämlich die Wechselwirkung
zwischen Probekörper und Stosskörper mit beliebiger Annäherung klassisch verfolgen. Der die Impulsmessung be-
gleitende Verlust der Kenntnis der Lage des Probekörpers
beruht in der Tat lediglich auf der Unmöglichkeit, zugleich
den Verlauf des Stossprozesses relativ zu einem wohldefinierten raumzeitlichen Bezugssystem zu fixieren. Überhaupt
ist ja die eigentümliche Komplementarität der Beschreibungsweise letzten Endes dadurch bedingt, dass jede solche
Fixierung mit einer unvermeidlichen, prinzipiell unkontrollierbaren Impuls- und Energieübertragung an die zur
Festlegung des Koordinatensystems nötigen Massstäbe und
Uhren verknüpft ist.[1]

[1] Vgl. N. Bohr, Atomtheorie und Naturbeschreibung, Berlin, Springer,
1931. Diese Frage ist inzwischen vom Verfasser ausführlicher behandelt

Wir erinnern daran, dass der bei jeder Beschreibung offen bleibende Spielraum in der Zeit Δt nach dem Unbestimmtheitsprinzip mit der Genauigkeit ΔE der Kenntnis der beim Stossprozess zwischen Stosskörper und Probekörper ausgetauschten Energie durch die bekannte Relation

$$\Delta E \cdot \Delta t \infty \hbar \qquad (27)$$

verbunden ist. Wegen der für beide Körper gültigen Beziehung zwischen Energie und Impuls- und Geschwindigkeitskomponenten

$$dE = v_x dp_x \qquad (28)$$

folgt direkt, dass

$$\Delta p_x \left| v_x'' - v_x' \right| \Delta t \infty \hbar. \qquad (29)$$

Obwohl die hier auftretende Geschwindigkeitsänderung $\left| v_x'' - v_x' \right|$ des Probekörpers bei der Impulsmessung nach dem oben Gesagten für einen genügend schweren Probekörper als beliebig genau bekannt angesehen werden kann, bedeutet offenbar der Faktor

$$\left| v_x'' - v_x' \right| \Delta t = \Delta x \qquad (30)$$

in vollem Einklang mit der Unbestimmtheitsrelation (16) einen ganz freien Spielraum in der Lage des Körpers relativ zum festen Bezugssystem. Aus (30) folgt unmittelbar die Bedingung

$$\Delta x < c \Delta t, \qquad (31)$$

welche mit Rücksicht auf (16) bei einer Impulsmessung mit gegebener oberer Grenze Δt des Zeitspielraums der zu erzielenden Genauigkeit Δp_x eine absolute Grenze setzt. In

in einem in Wien gehaltenen, bald erscheinenden Gastvortrag, wo insbesondere auf die bei der Deutung des Unbestimmtheitsprinzips unter Berücksichtigung der Relativitätsforderung auftretenden Paradoxien näher eingegangen wird.

Anbetracht der relativistischen Invarianz der Relationen (16) und (27) und insbesondere der Beziehung (28) bedeutet dieser Umstand aber keinerlei Einschränkung in der Formulierung und Anwendbarkeit des Unbestimmtheitsprinzips. Für unser Problem ist es zumal erlaubt, von allen mechanischen Relativitätseffekten abzusehen, denn es ist immer möglich, unter Benutzung genügend schwerer Probekörper sich so einzurichten, dass die Geschwindigkeiten aller Probekörper während des ganzen Messvorgangs klein gegen die Lichtgeschwindigkeit bleiben. Infolgedessen können wir jede Verschiebung $\varDelta x$ bei den Impulsmessungen sogar als sehr klein betrachten gegenüber dem entsprechenden Wert von $c\varDelta t$, der seinerseits beliebig klein gewählt werden muss.

Eben die genaue Verfolgbarkeit des relativen raumzeitlichen Verlaufs des zur Impulsmessung dienenden Vorgangs ermöglicht die Messung des Gesamtimpulses eines ausgedehnten Körpers innerhalb jedes gegebenen Zeitintervalles mit der erforderlichen, durch (16) ausgedrückten Genauigkeit. So können wir den Gesamtimpuls des als Probekörper dienenden Systems von geladenen materiellen Teilkörpern durch einen einzigen Stossprozess bestimmen, indem wir uns eines Stosskörpers besonderer Konstruktion bedienen, der überall im Probekörpersystem eingreift, und jedem Teilkörper zur selben Zeit dieselbe Beschleunigung erteilt. Zwar stellt diese Vorrichtung der Konstruktion der Stoss- und Probekörper weitgehende Anforderungen, die jedoch keine prinzipielle Schwierigkeit bieten, soweit wir den atomaren Aufbau der Körper vernachlässigen können. Am einfachsten gestaltet sich wohl die betrachtete Messung des Gesamtimpulses des Probekörpers, wenn man sie auf optischem Wege mittels Dopplereffektbestimmung ausführen würde, wobei man etwa folgendermassen vorgehen könnte: Man denke sich jeden

Teilkörper mit einem kleinen, auf die x-Richtung senkrechten Spiegel versehen und denke sich eine Anzahl anderer Spiegel in einer solchen Weise fest angebracht, dass der Lichtweg von der Strahlungsquelle zu jedem Teilkörper derselbe ist. Wenn wir nun durch eine passende Vorrichtung ein Strahlungsbündel von der Dauer $\varDelta t$ erzeugen, das eine Anzahl von Lichtquanten enthält, die genügend gross ist gegenüber der Anzahl der Teilkörper, werden also alle diese Körper gleichzeitig einen Stoss bekommen und eine Beschleunigung erleiden, die für alle Teilkörper mit beliebiger Genauigkeit gleich gross gemacht werden kann.

Um zu zeigen, dass man mit einer solchen Anordnung tatsächlich den Gesamtimpuls des Probekörpers mit einer der Relation (16) genügenden Genauigkeit bestimmen kann, werden wir die Wechselwirkung zwischen Probekörpersystem und Strahlungsbündel etwas näher betrachten. Unter Berücksichtigung der oben erwähnten Annahme der Kleinheit der Geschwindigkeit des Probekörpers gegen die Lichtgeschwindigkeit haben wir für jeden Teilkörper

$$\left.\begin{aligned} m_\tau\left(v''_{\tau,x} - v'_{\tau,x}\right) &= \frac{\hbar}{c}\sum_{n_\tau}(\nu' + \nu''), \\ \frac{1}{2}\,m_\tau\left(v''^{\,2}_{\tau,x} - v'^{\,2}_{\tau,x}\right) &= \hbar\sum_{n_\tau}(\nu' - \nu''), \end{aligned}\right\} \tag{32}$$

wo m_τ die Masse eines Teilkörpers, $v'_{\tau,x}$, $v''_{\tau,x}$ seine Geschwindigkeit vor und nach der Reflexion bezeichnet, und die Summation sich über die am Teilkörper reflektierten n_τ Lichtquanten erstreckt, deren Frequenzen (reziproke Periode mal 2π) vor und nach der Reflexion durch ν' bzw. ν'' dargestellt werden. Für die Impulskomponente des betreffenden Teilkörpers bzw. vor und nach dem Stoss folgt aus (32)

$$p'_{\tau,x} = m_\tau v'_{\tau,x} = \left.\begin{matrix} \\ \\ \end{matrix}\right\} \; m_\tau c \; \frac{\sum\limits_{n_\tau} (v' - v'')}{\sum\limits_{n_\tau} (v' + v'')} \mp \frac{1}{2} \frac{\hbar}{c} \sum_{n_\tau} (v' + v''), \quad (33)$$

$$p''_{\tau,x} = m_\tau v''_{\tau,x} =$$

Wenn wir nun annehmen, dass die mittlere spektrale Frequenz v_0 des Strahlungsbündels sehr gross ist, sowohl gegen die mittlere Abweichung $(\varDelta t)^{-1}$ seiner Frequenzverteilung, wie gegen alle Frequenzänderungen $v' - v''$, so können wir mit genügender Annäherung für die Geschwindigkeitsänderungen der Teilkörper durch den Stoss

$$v''_{\tau,x} - v'_{\tau,x} = \frac{\hbar}{m_\tau c} \sum_{n_\tau} (v' + v'') = \frac{2 n_\tau \hbar v_0}{m_\tau c} \quad (34)$$

setzen und sie für alle Teilkörper als gleich gross annehmen. Durch den Stoss bekommen also alle Teilkörper zwar unkontrollierbare, aber beliebig genau gleiche Verschiebungen, deren Grössenordnung $\varDelta x$ der Relation (30) genügt, wo $|v''_x - v'_x|$ mit der gemeinsamen Geschwindigkeitsänderung des ganzen Probekörpersystems zu identifizieren ist. Indem wir unseren Voraussetzungen gemäss $\varDelta x$ als verschwindend klein gegenüber $c \varDelta t$ betrachten können, erhalten wir daher auf Grund von (33) und (34) für das Produkt von $\varDelta x$ mit der Unsicherheit des Gesamtimpulses des Probekörpers näherungsweise

$$\varDelta p_x \varDelta x \sim \varDelta t \cdot \varDelta \left(\sum_\tau \sum_{n_\tau} \hbar v' - \sum_\tau \sum_{n_\tau} \hbar v'' \right). \quad (35)$$

Die in der Klammer stehenden Grössen in (35) sind eben die Gesamtenergien der auf den Probekörper einfallenden und von diesem reflektierten Strahlungsbündel. Die Energie des letzteren Bündels lässt sich, etwa durch Spektralanalyse der reflektierten Strahlung, mit beliebiger Genauigkeit messen. Für das einfallende Strahlungsbündel wäre aber eine solche Analyse mit den Versuchsbedingungen offenbar unverträglich. Die Gesamtenergie dieser Strahlung lässt sich jedoch immer mit einer zu Δt komplementären, durch die Relation (27) gegebenen Genauigkeit messen. Hierfür ist nämlich eine rein mechanische Vorrichtung hinreichend, durch welche das betrachtete Bündel aus einem Strahlungsfeld abgetrennt wird, dessen Energie vor und nach der Trennung mit beliebiger Genauigkeit etwa durch Spektralmessungen ermittelt werden kann. Die Relation (35) ist also identisch mit der üblichen Unbestimmtheitsrelation (16). Man bemerke noch, dass der Nachweis dieser Identität wesentlich dadurch bedingt ist, dass wir der beschriebenen Anordnung gemäss keine Auskunft über die Impulse der einzelnen Teilkörper, sondern nur über den Gesamtimpuls des Probekörpers erhalten.

Der Umstand, dass das Probekörpersystem bei den erforderlichen Impulsmessungen eine gemeinsame Translation erleidet, ist nicht nur wichtig für die Berechnung der diese Messungen begleitenden Feldwirkungen der Probekörper, sondern gibt uns die Möglichkeit, was in diese Berechnung eine grosse Vereinfachung bringt, uns so einzurichten, dass alle bei der Feldmessung zu benutzenden Probekörper ausserhalb der von den Impulsmessungen beanspruchten kurzen Zeitintervalle als ruhend betrachtet werden können. Wir können nämlich unmittelbar nach jeder Impulsmessung, d. h. praktisch genommen noch innerhalb des Intervalls Δt, dem Probekörpersystem durch eine geeignete Vorrichtung einen zweiten, entgegengerichteten Stoss geben, durch welchen die durch den ersten Stoss erteilte Geschwindigkeitsänderung jedes Teilkörpers mit beliebiger, d. h. mit einer seiner Masse umgekehrt proportionalen Genauigkeit aufgehoben wird,

ohne dass die angestrebte Kenntnis des Gesamtimpulses des
Probekörpers verloren geht. Dabei ist es aber unmöglich,
das Zeitintervall zwischen den beiden Stossprozessen mit
einem geringeren Spielraum als Δt zu kennen, so dass der
Probekörper, wie es das Unbestimmtheitsprinzip verlangt,
durch den Gegenstoss keineswegs in seine ursprüngliche
Lage, sondern in eine unbekannte, grössenordnungsmässig
um Δx verschobene Lage, mit der betreffenden Annäherung
zur Ruhe gebracht wird.

Für die Beurteilung der im nächsten Paragraphen näher
zu untersuchenden komplementären Begrenzung der Messbarkeit von Feldgrössen ist es überhaupt erforderlich, das
Verhalten der Probekörper während des ganzen Messvorgangs möglichst genau zu verfolgen. Dabei zeigt es sich
zunächst notwendig, die Lage jedes Probekörpers zu jeder
seiner Benutzung bei der Messung vorangehenden und nachfolgenden Zeit genau zu kennen. Zweckmässigerweise wird
dieses dadurch erreicht, dass der Probekörper ausserhalb
des Zeitintervalles, während dessen der auf ihn vom Feld
übertragene Impuls zu ermitteln ist, mit einem als räum-
liches Bezugssystem dienenden, starren Gerüst fest verbunden bleibt. Am Anfang des betrachteten Intervalles muss diese
Verbindung aufgelöst und die Komponente des Impulses
des Probekörpers in der Richtung der zu bestimmenden
Feldkomponente gemessen werden, wobei wir immer annehmen werden, dass durch einen unmittelbar folgenden
Gegenstoss der oben besprochenen Art der Körper in eine
nicht genau voraussagbare Lage mit einer seiner Masse umgekehrt proportionalen Annäherung wieder zur Ruhe gebracht wird. Am Ende des Zeitintervalles wird nach erneuter
Messung der betreffenden Impulskomponente die feste Verbindung wieder hergestellt, wobei es sich als nicht unwe-

sentlich erweist, den Probekörper in genau dieselbe Lage
wie ursprünglich zu bringen. Schon diese Vorschriften stellen,
wenn die raumzeitlichen Mittelwertbereiche genügend scharf
definiert werden sollen, weitgehende Ansprüche an die verfeinerte Konstruktion der Probekörpersysteme. Denn wegen
der Retardation aller Kräfte ist es streng genommen notwendig, dass die Auflösung sowie die Wiederherstellung der
Verbindung der Probekörpersysteme mit dem festen Gerüst
für alle ihre unabhängigen Teilkörper, deren lineare Abmessungen mindestens ebenso klein wie der kleinste in Betracht
kommende Wert von $c\varDelta t$ sein müssen, gleichzeitig vorgenommen wird, d. h. innerhalb des Zeitspielraums $\varDelta t$ der
Impulsmessung, der seinerseits genügend klein gegenüber
dem Zeitintervall T gewählt werden muss.

Noch weitergehende Ansprüche an Idealisation in Bezug
auf die Konstruktion und Handhabung der Probekörpersysteme sind offenbar nötig, wenn es sich um die Messung
von Feldmittelwerten über zwei sich teilweise überdeckende
Raumzeitgebiete handelt. In diesem Fall müssen wir ja über
Probekörper verfügen, die ohne gegenseitige mechanische
Beeinflussung in einander verschoben werden können. Um
die zu messenden elektromagnetischen Felder möglichst
wenig durch die Anwesenheit der Probekörpersysteme zu
stören, werden wir überdies jedem elektrischen oder magnetischen Teilkörper einen anderen, genau entgegengesetzt
geladenen Neutralisierungskörper zur Seite gestellt denken.
Im Falle magnetischer Probekörpersysteme ist zwar zu bedenken, dass eine gleichmässige Polstärkenverteilung auf
einem streng abgegrenzten Körper nicht bestehen kann.
Man kann sich aber, wenigstens im Prinzip, vorstellen, dass
jeder Teilkörper eines solchen Systems mittels magnetisier-
barer biegsamer Fäden mit dem zugehörigen Neutralisie-

3*

rungskörper verbunden ist. Alle diese Neutralisierungskörper
sollen während des ganzen Messvorgangs mit dem festen
Gerüst verbunden bleiben, ohne die freie Beweglichkeit der
zum eigentlichen Probekörpersystem gehörigen Teilkörper
mechanisch zu beeinflussen. Die in solchen Voraussetzungen, sowie in den unten einzuführenden Annahmen über
die noch nötigen Kompensationsmechanismen, enthaltenen
Idealisationen sind natürlich nur zu verteidigen, soweit wir
den atomaren Aufbau der Probekörper vernachlässigen können. Wie schon erwähnt, bedeutet diese Vernachlässigung
jedoch keine prinzipielle Einschränkung der Prüfungsmöglichkeit des quantenelektromagnetischen Formalismus, da
in diesem Formalismus keinerlei universellen raumzeitlichen
Dimensionen auftreten. Der Zweck der vorangehenden Betrachtungen war daher auch vor allem, zu zeigen, dass es
bei den für die Feldmessungen in Frage kommenden, rein
mechanischen Problemen möglich ist, zwischen den durch
die atomistische Struktur der Materie bedingten Einschränkungen der Beschaffenheit der Probekörper und den auf
dem universellen Wirkungsquantum beruhenden, besonders
im Unbestimmtheitsprinzip formulierten Begrenzungen der
Handhabung dieser Körper streng zu unterscheiden.

§ 4. Berechnung der Feldwirkungen der Probekörper.

Nach der Untersuchung der physikalischen Voraussetzungen der Beschaffenheit der Probekörper werden wir nun übergehen zur genaueren Betrachtung der die Messung von Feldgrössen begleitenden elektromagnetischen Feldwirkungen
der Probekörper, die für die Messbarkeitsfrage von entscheidender Bedeutung sind. Gemäss den obigen Ausführungen werden wir dabei jeden Probekörper als eine das

räumliche Mittelwertgebiet gleichmässig auffüllende Ladungsverteilung behandeln, die während der Impulsmessung eine
einfache Translation erleidet. Die Berechnung der dadurch
erzeugten elektromagnetischen Felder werden wir zunächst
auf Grund der klassischen Elektrodynamik ausführen, und
erst nachher auf die durch das Wirkungsquantum bedingte
Begrenzung der Gültigkeit dieser Behandlung eingehen.

Betrachten wir zwei Raumzeitgebiete I und II, mit Volu-
mina V_I und V_{II} und Zeitlängen T_I und T_{II} und fragen wir
nach dem elektromagnetischen Feld, das in einem Punkt
(x_2, y_2, z_2, t_2) des Gebiets II durch eine Messung des Mittelwerts von $\mathfrak{E}_x$ über das Gebiet I erzeugt wird. Wir nehmen
also an, dass sich ursprünglich im Volumen V_I zwei elektrische Ladungsverteilungen mit den konstanten Dichten
$+\varrho_I$ und $-\varrho_I$ befinden. Im Intervall von t_I' bis $t_I' + \varDelta t_I$ er-
fährt die erste Ladungsverteilung eine einfache ungleich-
förmige Translation in der x-Richtung um die Strecke $D_x^{(I)}$;
im Intervall von $t_I' + \varDelta t_I$ bis t_I'' bleibt sie in Ruhe in der
verschobenen Lage; schliesslich bewegt sie sich innerhalb
des Zeitintervalls von t_I'' bis $t_I'' + \varDelta t_I$, ungleichförmig parallel
der x-Achse bis zu ihrer ursprünglichen, mit der Neutrali-
sierungsverteilung zusammenfallenden Lage zurück. Im Einklang mit der im vorigen Paragraphen besprochenen Forderung werden wir ferner annehmen, dass $\varDelta t_I$ sehr klein
gegenüber $T_I = t_I'' - t_I'$ ist, und dass $D_x^{(I)}$ nicht nur sehr klein
ist gegen die linearen Abmessungen des räumlichen Mittel-
wertgebiets von Volumen V_I, sondern auch klein ist gegenüber $c \varDelta t_I$.

Im Grenzfall verschwindend kleiner $\varDelta t_I$ lassen sich also
die Quellen des gesuchten Feldes darstellen als eine im Gebiet I während des Zeitintervalls von t_I' bis t_I'' bestehende
Polarisation in der x-Richtung von der konstanten Dichte

$P_x^{(I)} = \varrho_{\mathrm{I}} D_x^{(I)}$, sowie eine nur in unmittelbarer Nähe der Zeit-punkte t_{I}' und t_{I}'' vorhandene Stromdichte, die wir unter Benutzung des durch Formel (3) definierten Symbols

$$J_x^{(I)} = \varrho_{\mathrm{I}} D_x^{(I)} \left[\delta\left(t - t_{\mathrm{I}}'\right) - \delta\left(t - t_{\mathrm{I}}''\right) \right] \qquad (36)$$

schreiben können. Mit Hilfe desselben Symbols lässt sich ebenfalls die Polarisation $P_x^{(I)}$ zu einer beliebigen Zeit t durch die Formel

$$P_x^{(I)} = \varrho_{\mathrm{I}} D_x^{(I)} \int_{t_{\mathrm{I}}'}^{t_{\mathrm{I}}''} \delta\left(t - t_1\right) dt_1 \qquad (37)$$

ausdrücken. Die Komponenten der durch diese Quellen im Raumzeitpunkte $(x_2,\, y_2,\, z_2,\, t_2)$ erzeugten Felder berechnen sich bekanntlich aus den Formeln

$$\left.\begin{aligned}
E_x^{(I)} &= -\frac{\partial \varphi^{(I)}}{\partial x_2} - \frac{1}{c}\frac{\partial \psi_x^{(I)}}{\partial t_2}, & E_y^{(I)} &= -\frac{\partial \varphi^{(I)}}{\partial y_2}, & E_z^{(I)} &= -\frac{\partial \varphi^{(I)}}{\partial z_2}, \\[2mm]
H_x^{(I)} &= 0, & H_y^{(I)} &= \frac{\partial \psi_x^{(I)}}{\partial z_2}, & H_z^{(I)} &= -\frac{\partial \psi_x^{(I)}}{\partial y_2},
\end{aligned}\right\} \quad (38)$$

wobei wir zur Unterscheidung von den zu messenden Feld-komponenten lateinische Buchstaben gebrauchen. In (38) bedeutet $\varphi^{(I)}$ das retardierte skalare Potential

$$\varphi^{(I)} = \int_{V_{\mathrm{I}}} \frac{\partial}{\partial x_1}\left[\frac{P_x^{(I)}\left(t_2 - \dfrac{r}{c}\right)}{r} \right] dv_1 \qquad (39)$$

und $\psi_x^{(I)}$ die retardierte Vektorpotentialkomponente

$$\psi_x^{(I)} = \frac{1}{c}\int_{V_{\mathrm{I}}} \frac{J_x^{(I)}\left(t_2 - \dfrac{r}{c}\right)}{r}\, dv_1, \qquad (40)$$

wo r den Abstand der Raumpunkte (x_1, y_1, z_1) und (x_2, y_2, z_2) darstellt. Bedenkt man, dass der Ausdruck (36) auch in der Form

$$J_x^{(1)} = -\varrho_I D_x^{(I)} \int_{t_I'}^{t_I''} \frac{\partial}{\partial t_1} \delta (t - t_1) \, dt_1 \qquad (41)$$

geschrieben werden kann, so lassen sich die sich aus (38), (39) und (40) ergebenden Feldkomponenten mit Rücksicht auf (37) und (41) mittels der durch (2) definierten Abkürzungen durch die typischen Formeln

$$\left.\begin{array}{ll} E_x^{(1)} = \varrho_I D_x^{(I)} \int_{V_I} dv_1 \int_{T_I} dt_1 A_{xx}^{(12)}, & E_y^{(1)} = \varrho_I D_x^{(I)} \int_{V_I} dv_1 \int_{T_I} dt_1 A_{xy}^{(12)}, \\[2mm] H_x^{(1)} = 0, & H_y^{(1)} = \varrho_I D_x^{(I)} \int_{V_I} dv_1 \int_{T_I} dt_1 B_{xy}^{(12)} \end{array}\right\} \quad (42)$$

darstellen.

Mit Rücksicht auf die Eigenschaften der symbolischen δ-Funktion ist es leicht einzusehen, dass die durch (42) gegebenen Feldkomponenten immer endlich bleiben und sogar in keinem Raumzeitpunkt (x_2, y_2, z_2, t_2) einen Wert von der Grössenordnung $\varrho_I D_x^{(I)}$ überschreiten können. Wie schon erwähnt, sind die elektromagnetischen Kräfte, die während der Impulsmessung am Probekörper im Zeitintervall $\varDelta t$ auftreten, eben von dieser Grössenordnung (vgl. S. 25). Dass die Feldintensitäten in der nachfolgenden Zeit nicht wesentlich zunehmen, ist lediglich eine Folge des gleich nach der Impulsmessung stattfindenden Gegenstosses, wodurch der Körper zur Ruhe gebracht wird, und der in den Ansätzen (36) und (37) seinen idealisierten mathematischen Ausdruck findet.

Die uns besonders interessierenden Mittelwerte dieser Feldkomponenten über das Gebiet II gehen aus (42) durch einfache Raumzeitintegration hervor, und werden also gemäss (5) durch die Formeln

$$\left.\begin{array}{ll} \bar{E}_x^{(I, II)} = D_x^{(I)} \varrho_I V_I T_I \bar{A}_{xx}^{(I, II)}, & \bar{E}_y^{(I, II)} = D_x^{(I)} \varrho_I V_I T_I \bar{A}_{xy}^{(I, II)} \\[2mm] \bar{H}_x^{(I, II)} = 0, & \bar{H}_y^{(I, II)} = D_x^{(I)} \varrho_I V_I T_I \bar{B}_{xy}^{(I, II)} \end{array}\right\} \quad (43)$$

gegeben.

Auf Grund der schon im § 2 diskutierten Eigenschaften der Ausdrücke $\bar{A}$ und $\bar{B}$ sehen wir, dass die durch (43) gegebenen Feldmittelwerte bei gegebenem Wert von $D_x^{(I)}$ wohldefinierte, stetige Funktionen der Gebiete I und II sind. Bei abnehmenden Spielräumen $\varDelta t$ und $\varDelta x$ der Zeitdauer der Impulsmessungen und der diese begleitenden, unvoraussagbaren Verschiebungen, sind also diese Feldmittelwerte vom näheren raumzeitlichen Verlauf der Stossprozesse ganz unabhängig und einfach der im Messintervall T_I konstanten Verschiebung des Probekörpers proportional. Eben dieser Umstand erweist sich, wie wir sehen werden, als entscheidend für die Möglichkeit einer weitgehenden Kompensation der unkontrollierbaren Feldwirkungen der Probekörper.

Soweit ist die Berechnung der Feldwirkungen auf rein klassischer Grundlage ausgeführt worden. Für den genaueren Vergleich der Messungsmöglichkeiten mit den Forderungen des quantenelektromagnetischen Formalismus ist es aber notwendig, noch die Begrenzung zu berücksichtigen, die, infolge des durch die Lichtquantenvorstellung symbolisierten, quantentheoretischen Zugs jeder Feldwirkung, der klassischen Berechnungsweise auferlegt ist. Um einen Überblick über die Verhältnisse zu gewinnen, nehmen wir an, dass die betrachteten Mittelwertgebiete grössenordnungsmässig gleich sind und räumlich gegen einander verschoben sind um Strecken von derselben Grössenordnung wie ihre linearen

Abmessungen, die wir mit L bezeichnen, und dass ferner die zugehörigen Zeitintervalle von der Grössenordnung T kleiner sind als $\frac{L}{c}$. Unter diesen Bedingungen kommen in der spektralen Zerlegung der Feldwirkungen im wesentlichen nur Wellen vor, deren Länge von derselben Grössenordnung wie L ist. Da ferner im betrachteten Fall die Intensität des durch die Impulsmessung erzeugten Felds grössenordnungsmässig gleich $\varrho \varDelta x$, und folglich die im Volumen V enthaltene Feldenergie von der Grössenordnung $\varrho^2 \varDelta x^2 V$ ist, so wird die Anzahl der in Frage kommenden Lichtquanten durch den Ausdruck

$$n \backsim \varrho^2 \varDelta x^2 V \frac{L}{\hbar c} = \lambda^{-2} \frac{L}{cT} \tag{44}$$

abgeschätzt, wo λ den durch (20) definierten, für die Messgenauigkeit massgebenden Faktor bedeutet. Wir sehen also, dass in unserem Fall n immer gross gegenüber der Einheit ist, wenn eine Messgenauigkeit verlangt wird, bei der Feldstärken gemessen werden können, die kleiner sind als die kritische Feldgrösse Q.

Je grösser die bei den Feldmessungen angestrebte Genauigkeit ist, umso verhältnismässig genauer werden offenbar die klassisch berechneten Ausdrücke (42) und (43) der betrachteten Feldwirkungen. Es ist indessen wesentlich zu bemerken, dass sich die absolute Genauigkeit dieser Ausdrücke bei wachsendem Wert von n nicht ändert. Der statistische Schwankungsbereich der Feldmittelwerte wird nämlich in unserem Fall schätzungsweise durch

$$\frac{\varrho \varDelta x}{\sqrt{n}} \backsim \sqrt{\frac{\hbar c}{VL}} \backsim \frac{\sqrt{\hbar c}}{L^2}$$

gegeben. Dieser allein von den linearen Abmessungen des Messgebiets abhängige, immer endlich bleibende Ausdruck

für den Schwankungsbereich der Feldwirkungen der Probekörper stimmt tatsächlich mit dem für den Fall $L > cT$ aus
dem Formalismus abgeleiteten Ausdruck (14) für die Grössenordnung der reinen Hohlraumschwankungen überein. Überhaupt handelt es sich bei obiger Betrachtung nur um ein
Beispiel der im § 2 angeführten, allgemeinen Beziehung
zwischen Hohlraumschwankungen und den nur statistisch
beschreibbaren Abweichungen der Feldmittelwerte von den
nach der klassischen Theorie aus der Angabe der Quellen
berechneten Feldgrössen. Wie schon dort erwähnt wurde,
sind weiter in dem für die Prüfung des Formalismus besonders wichtigen Falle $L > cT$ die Hohlraumschwankungen
immer kleiner als die für die komplementäre Messbarkeit
von Feldgrössen massgebende Feldstärke Q, und zwar umso
kleiner, je grösser das Verhältnis zwischen L und cT ist.
Bei dem folgenden Vergleich zwischen Feldmessungen und
Formalismus werden wir daher immer von den klassisch
berechneten Ausdrücken (43) ausgehen, und erst nachher
die Bedeutung der Schwankungserscheinungen für die Widerspruchsfreiheit des Formalismus diskutieren.

§ 5. Messung einzelner Feldmittelwerte.

Der Untersuchung der Messungsmöglichkeiten von Feldmittelwerten legen wir definitionsgemäss die Gleichung (15)
zugrunde, welche die klassisch beschriebene Impulsbilanz
eines im Felde befindlichen Probekörpers ausdrückt. Nach
den vorangehenden Ausführungen ist dabei jede Feldkomponente, wie $\mathfrak{E}_x$, als die Überlagerung der von allen Feldquellen, einschliesslich der Probekörper selber, herrühren-
den Felder zu betrachten, und der Kern des Messproblems
ist eben die Frage, in welchem Umfang diese Felder den

verschiedenen Quellen zugeordnet werden können. Wir möchten aber gleich hier betonen, dass die strenge Anwendbarkeit des klassischen Feldbegriffs für die erwähnte Definition der Feldmittelwerte an sich durch die oben berührte begrenzte Gültigkeit der klassischen Beschreibung der Feldwirkungen der Probekörper nicht beeinträchtigt wird. Ganz abgesehen von der im § 3 diskutierten Frage der bei den Impulsmessungen der Probekörper am Anfang und Ende des Messintervalls erreichbaren Genauigkeit, dürfte die Eindeutigkeit dieser Definition lediglich verlangen, dass die Massen der Probekörper genügend gross gewählt werden, um jede von ihren Beschleunigungen im Messintervall unter Einfluss der elektromagnetischen Felder herrührende Modifikation dieser Felder vernachlässigen zu können. Würde man in dieser Vernachlässigung einen Widerspruch erblicken zum atomaren Charakter des Impulsaustausches zwischen elektromagnetischen Wellenfeldern und materiellen Körpern, so muss man bedenken, dass es sich beim betrachteten Messproblem keineswegs um die Verfolgung wohldefinierter Elementarvorgänge im Sinne der Lichtquantenvorstellung handelt. Insbesondere wird nach der beschriebenen Messanordnung ein unkontrollierbarer Impulsbeitrag vom festen Gerüst, woran jeder Probekörper vor und nach dem Messintervall gebunden ist, aufgenommen. Im Grenzfall einer klassisch beschreibbaren Wechselwirkung zwischen einem elektromagnetischen Wellenzug und einem genügend schweren, geladenen Körper würde ja die zuletzt erwähnte Impulsübertragung den im Messintervall vom Probekörper aufgenommenen Impuls offenbar genau kompensieren.

Als Vorbereitung zur allgemeinen Diskussion des Messproblems betrachten wir zunächst eine einzelne Feldmessung und fragen, wie im § 3, nach dem Mittelwert von $\mathfrak{E}_x$

in einem bestimmten Raumzeitgebiet, das wir entsprechend den Bezeichnungen des vorigen Paragraphen mit I kennzeichnen. Nach der Grundgleichung (15) bekommen wir also für die Impulsbilanz des Probekörpers

$$p_x^{(I)''} - p_x^{(I)'} = \varrho_I V_I T_I \left(\overline{\mathfrak{E}}_x^{(I)} + \bar{E}_x^{(I,I)} \right), \tag{45}$$

wo $\overline{\mathfrak{E}}_x^{(I)}$ den Anteil des Mittelwerts von $\mathfrak{E}_x$ darstellt, der im betrachteten Raumzeitgebiet I vorhanden wäre, wenn keine Impulsmessung zur Zeit t' am Probekörper vorgenommen wäre, während $\bar{E}_x^{(I,I)}$ den Anteil des Feldmittelwerts bedeutet, der von dieser Messung stammt und dessen klassisch abgeschätzter Ausdruck durch (43) gegeben wird, wenn die Gebiete I und II gleichgesetzt werden.

Nach den Ausführungen von § 3 lässt sich die in (45) auftretende Summe der Feldmittelwerte $\overline{\mathfrak{E}}_x^{(I)}$ und $\bar{E}_x^{(I,I)}$ durch die Wahl eines genügend grossen Werts von ϱ_I mit beliebiger Genauigkeit bestimmen. Je grösser aber ϱ_I gewählt wird, umso grösser wird der unkontrollierbare Wert von $\bar{E}_x^{(I,I)}$, und der durch die bisher beschriebene, einfache Messanordnung erreichbaren Genauigkeit der Bestimmung von $\overline{\mathfrak{E}}_x^{(I)}$, welche nach (45) durch

$$\varDelta \overline{\mathfrak{E}}_x^{(I)} \sim \frac{\varDelta p_x^{(I)}}{\varrho_I V_I T_I} + \varDelta \bar{E}_x^{(I,I)} \tag{46}$$

gegeben ist, wird daher eine Grenze gesetzt. Mit Rücksicht auf die Relation (16) und auf den Umstand, dass die in (43) auftretende Grösse $D_x^{(I)}$ nur mit dem Spielraum $\varDelta x_I$ voraussagbar ist, erhalten wir nämlich aus (46) für $\varDelta \overline{\mathfrak{E}}_x^{(I)}$ den Ausdruck

$$\varDelta \overline{\mathfrak{E}}_x^{(I)} \sim \frac{\hbar}{\varrho_I \varDelta x_I V_I T_I} + \varrho_I \varDelta x_I V_I T_I \left| \bar{A}_{xx}^{(I,I)} \right|, \tag{47}$$

dessen Minimalwert offenbar

$$\varDelta_m \overline{\mathfrak{E}}_x^{(I)} \sim \sqrt{\hbar \left| \bar{A}_{xx}^{(I,I)} \right|} \tag{48}$$

beträgt, und im Fall $L_I > cT_I$ eben gleich der kritischen Grösse Q_I ist. Freilich ist (48), wenn L_I gross gegenüber cT_I ist, wesentlich kleiner als der von LANDAU und PEIERLS als absolute Grenze der Messbarkeit von Feldgrössen angegebene Ausdruck (24); wäre aber (48) als eine unvermeidliche Grenze der Messgenauigkeit anzusehen, so kämen wir dennoch zu der mit der Auffassung der genannten Verfasser übereinstimmenden Schlussfolgerung, dass der quantenelektromagnetische Formalismus keine Prüfung im eigentlichen Quantengebiet zuliesse, und dass der ganzen Feldtheorie also nur im klassischen Grenzfall eine physikalische Realität zukäme.

Diese Schlussfolgerung lässt sich jedoch nicht aufrechterhalten, denn der Umstand, dass in $\bar{E}_x^{(I,I)}$ nach (43) der Faktor der unvoraussagbaren Verschiebung $D_x^{(I)}$ eine wohldefinierte, allein von den geometrischen Verhältnissen abhängige Grösse ist, erlaubt uns, bei den Messungen uns so einzurichten, dass die Wirkung des Feldes $E_x^{(I)}$, bis auf die unvermeidlichen Feldschwankungen, völlig kompensiert wird. Dies wird durch eine Messanordnung erreicht, bei welcher der Probekörper auch nicht im Messintervall T_I frei beweglich ist, sondern mit dem festen Gerüst durch einen Federmechanismus verbunden bleibt, dessen Spannung zu $D_x^{(I)}$ proportional ist. Wird die durch diesen Mechanismus in der x-Richtung auf den Probekörper ausgeübte Kraft gleich $-F_I D_x^{(I)}$ angesetzt, so wird offenbar der ganze vom Feld $E_x^{(I)}$ auf diesen Körper übertragene Impuls durch die Feder völlig aufgehoben, wenn die Spannkraft

$$F_{\mathrm{I}} = \varrho_{\mathrm{I}}^2 V_{\mathrm{I}}^2 T_{\mathrm{I}} \bar{A}_{xx}^{(\mathrm{I},\,\mathrm{I})} \tag{49}$$

gewählt wird. Dies gilt jedenfalls, wenn der Probekörper so schwer ist, dass seine Schwingungsperiode unter dem Einfluss der Feder gross gegen T_{I} und somit seine Verschiebung innerhalb der Zeit T_{I} durch die Federspannung klein gegen $D_x^{(\mathrm{I})}$ ist. Ferner lässt sich die Wirkung der Feder, die streng genommen nur im asymptotischen Grenzfall klassisch beschreibbar ist, mit umso grösserer Näherung auf Grund der klassischen Mechanik berechnen, je grösser die Masse des Probekörpers ist. Abgesehen von den Einschränkungen, die durch die atomistische Struktur aller Körper bedingt sind, dürften gegen eine solche Kompensationsvorrichtung keine prinzipiellen Einwände bestehen. Erstens werden durch die Benutzung einer mechanischen Feder alle elektromagnetischen Felder vermieden, die von den zu messenden Feldern untrennbar wären. Zweitens kann man, wenn die Länge der Feder genügend klein, d. h. klein gegenüber $c T_{\mathrm{I}}$ ist, offenbar von allen Retardationseffekten absehen. Wenn das Probekörpersystem genügend schwer ist, so ist es dabei gleichgültig, ob die Feder nur auf einen Teilkörper wirkt, oder ob man ein Federsystem verwendet, das an jedem Teilkörper gleichmässig angreift.

Wir sehen also, dass die Deutbarkeit einer einzelnen Feldmessung allein durch die Grenze beschränkt ist, die der klassischen Beschreibung der Feldwirkungen des Probekörpers gesetzt ist. Diese Begrenzung, die von umso kleinerer Bedeutung ist, je grösser L_{I} ist gegenüber $c T_{\mathrm{I}}$, hat jedoch auch im Falle $L_{\mathrm{I}} \leq c T_{\mathrm{I}}$ keinerlei Einschränkung der Prüfbarkeit der Folgerungen des quantenelektromagnetischen Formalismus zur Folge. Bei der Beurteilung dieser Frage müssen wir scharf unterscheiden zwischen der Prüfung von theoretischen Erwartungen, welche auf Feldmessungen beruhende Angaben über elektrische oder magnetische Kräfte voraussetzen, und von solchen, die sich auf eine auf anderer Grundlage gewonnene Kenntnis des Zustandes des betrachteten Felds beziehen. Was die ersteren Erwartungen betrifft, so erfordert ihre Prüfung selbstverständlich eine nähere Untersuchung der gegenseitigen Beziehungen mehrerer Feldmessungen; hier kann es sich also zunächst nur um die Prüfung von Erwartungen der letzeren Art handeln.

Es ist nun, wie im § 2 erwähnt, ein Hauptergebnis der Quantentheorie der Felder, dass alle Erwartungen über Feldmittelwerte, die nicht auf eigentlichen Feldmessungen, sondern auf der Lichtquantenzusammensetzung des zu untersuchenden Feldes oder auf der Kenntnis klassisch beschriebener Feldquellen beruhen, wesentlich statistischer Natur sein müssen. Die dortige nähere Ausführung zeigt ferner, dass die Einbeziehung der Schwankungen der Feldwirkungen der Probekörper um ihre klassisch abgeschätzten Werte keinerlei Änderung dieser statistischen Erwartungen mit sich bringt. Ohne weitere Korrektion bieten sich also die mittels der beschriebenen Versuchsanordnung erzielten Messergebnisse für die Prüfung der theoretischen Aussagen als die gesuchten Feldmittelwerte dar. Eine solche Auffassung der Messergebnisse, deren allgemeine Berechtigung wir im Folgenden näher untersuchen werden, ist auch dadurch nahegelegt, dass es sich bei allen Messungen von physikalischen Grössen definitionsmässig um die Anwendung klassischer Vorstellungen handeln muss, und dass also bei Feldmessungen jede Rücksichtnahme auf die Begrenzung der strengen Anwendbarkeit der klassischen Elektrodynamik im Widerspruch mit dem Messbegriff selber stehen würde.

Obwohl somit bei den Feldmessungen, wie schon in der Einleitung betont, der Messbegriff mit noch grösserer Vorsicht anzuwenden ist als bei den üblichen quantenmechanischen Messproblemen, weist die geschilderte Situation jedoch in bezug auf die Untrennbarkeit zwischen Phänomen und Messvorgang eine weitgehende Analogie zu diesen Problemen auf. Schon bei einer Orts- oder Impulsmessung des Elektrons eines Wasserstoffatoms von gegebenem stationären Zustand kann man ja mit gewissem Recht behaupten, dass das Messergebnis erst durch die Messung selber geschaffen wird. Wohl ist hier keine Rede von einer Begrenzung der Deutbarkeit der Messergebnisse auf Grund der klassischen Mechanik, sondern nur von einem Verzicht auf jede Kontrolle der Beeinflussung des Zustandes des Atoms durch den Messvorgang. Bei den Feldmessungen entspricht dieser für die Widerspruchsfreiheit wesentliche, komplementäre Zug der Beschreibung dem Umstand, dass die Kenntnis der Lichtquantenzusammensetzung des Feldes durch die Feldwirkungen der Probekörper verloren geht, und zwar gemäss (44) in umso grösserem Mass, je grösser die bei der Messung angestrebte Genauigkeit ist. Ausserdem ergibt sich aus der folgenden Diskussion, dass jeder Versuch, die Kenntnis der Lichtquantenzusammensetzung des Feldes durch eine nachträgliche Messung mittels irgend einer geeigneten Vorrichtung wiederherzustellen, zugleich jede weitere Verwertung der betreffenden Feldmessung verhindern würde.

Dass beim Nachweis der Übereinstimmung zwischen der Prüfbarkeit der Folgerungen des quantenelektromagnetischen Formalismus mittels einer einzelnen Feldmessung und der Deutbarkeit einer solchen Messung auf Grund der klassischen Elektrodynamik die reinen Hohlraumschwankungen als gemeinsame Begrenzung auftreten, bedeutet jedoch keineswegs, dass diese Schwankungen jeder Verwertung von Feldmessungen eine absolute Grenze setzen. In der Tat besteht eine derartige allgemeine Einschränkung weder für die Folgerungen des Formalismus betreffend Beziehungen zwischen Mittelwerten einer Feldkomponente über verschiedene Bereiche noch für die Prüfung solcher Beziehungen durch direkte Feldmessungen. Dies wird aus den Betrachtungen des folgenden Paragraphen hervorgehen, und es wird sich insbesondere zeigen, dass die für die Diskussion der Widerspruchsfreiheit der üblichen Quantenmechanik wesentliche Forderung der Wiederholbarkeit von Messungen kinematischer und dynamischer Grössen bei den Feldmessungen ihr sinngemässes Analogon besitzt.

§ 6. Messbarkeit zweier Mittelwerte einer Feldkomponente.

Bei der Untersuchung der Messbarkeit zweier Feldgrössen ist es zweckmässig, mit der Messung der Mittelwerte einer und derselben Feldkomponente über zwei verschiedene Gebiete I und II anzufangen. Indem wir, wie oben, die Feldkomponente $\mathfrak{E}_x$ betrachten, und zunächst von der Begrenzung der klassischen Beschreibbarkeit der Feldwirkungen der Probekörper absehen, haben wir also in diesem Fall für die Impulsbilanz der beiden Probekörper anstatt (45):

$$
\left.
\begin{aligned}
p_x^{(I)''} - p_x^{(I)'} &= \varrho_I V_I T_I \left(\overline{\mathfrak{E}}_x^{(I)} + \bar{E}_x^{(I,\,I)} + \bar{E}_x^{(II,\,I)} \right) \\
p_x^{(II)''} - p_x^{(II)'} &= \varrho_{II} V_{II} T_{II} \left(\overline{\mathfrak{E}}_x^{(II)} + \bar{E}_x^{(II,\,II)} + \bar{E}_x^{(I,\,II)} \right),
\end{aligned}
\right\} \quad (50)
$$

wo $\bar{E}_x^{(I,\,II)}$ durch den Ausdruck (43) definiert ist, und $\bar{E}_x^{(II,\,I)}$ sich aus diesem Ausdruck durch einfache Vertauschung der Indizes I und II ergibt.

Das Auftreten der Ausdrücke $\bar{E}_x^{(I, I)}$ und $\bar{E}_x^{(II, II)}$ in den Gleichungen (50) hat nach den Betrachtungen des vorigen Paragraphen zur Folge, dass jeder der gesuchten Feldmittelwerte $\overline{\mathfrak{E}}_x^{(I)}$ und $\overline{\mathfrak{E}}_x^{(II)}$ mittels einer einfachen Messanordnung nur mit einer durch (48) gegebenen beschränkten Genauigkeit bestimmt werden kann. Es ist also von vornherein einleuchtend, dass ein Kompensationsverfahren unvermeidlich ist, und zur vorläufigen Orientierung über das hier betrachtete, mehr verwickelte Messproblem werden wir deshalb zunächst eine Messanordnung heranziehen, in welcher die Rückwirkungen $\varrho_I V_I T_I \bar{E}_x^{(I, I)}$ und $\varrho_{II} V_{II} T_{II} \bar{E}_x^{(II, II)}$ durch zwei auf die Probekörper I und II wirkende Federn, deren Spannkräfte durch (49) und einen analogen Ausdruck gegeben sind, aufgehoben werden.

Aus den Gleichungen (50), unter Auslassung von $\bar{E}_x^{(I, I)}$ und $\bar{E}_x^{(II, II)}$, ergibt sich gemäss (16) und (43) für die Unsicherheiten der beiden Feldmessungen bei dieser Messanordnung, wenn man noch berücksichtigt, dass die in $\bar{E}_x^{(I, II)}$ und $\bar{E}_x^{(II, I)}$ auftretenden Verschiebungen $D_x^{(I)}$ und $D_x^{(II)}$ der Probekörper von einander völlig unabhängig und nur mit den Spielräumen $\varDelta x_I$ bzw. $\varDelta x_{II}$ bekannt sind:

$$\left.\begin{aligned}
\varDelta \overline{\mathfrak{E}}_x^{(I)} &\sim \frac{\hbar}{\varrho_I \varDelta x_I V_I T_I} + \varrho_{II} \varDelta x_{II} V_{II} T_{II} \left| \bar{A}_{xx}^{(II, I)} \right| \\
\varDelta \overline{\mathfrak{E}}_x^{(II)} &\sim \frac{\hbar}{\varrho_{II} \varDelta x_{II} V_{II} T_{II}} + \varrho_I \varDelta x_I V_I T_I \left| \bar{A}_{xx}^{(I, II)} \right|.
\end{aligned}\right\} \quad (51)$$

Durch passende Wahl der Werte von $\varrho_I \varDelta x_I$ und $\varrho_{II} \varDelta x_{II}$ lässt sich offenbar jede der Grössen $\varDelta \overline{\mathfrak{E}}_x^{(I)}$, $\varDelta \overline{\mathfrak{E}}_x^{(II)}$ einzeln beliebig herabsetzen, jedoch nur auf Kosten einer Zunahme der anderen. Für das Produkt der beiden Grössen bekommen wir ja nach (51) den Minimalwert

$$\varDelta\overline{\mathfrak{E}}_x^{(\mathrm{I})}\,\varDelta\overline{\mathfrak{E}}_x^{(\mathrm{II})} \sim \hbar\left[\left|\bar{A}_{xx}^{(\mathrm{I},\,\mathrm{II})}\right|+\left|\bar{A}_{xx}^{(\mathrm{II},\,\mathrm{I})}\right|\right]. \qquad (52)$$

Trotz der grossen Aehnlichkeit der Relation (52) mit den von dem Formalismus geforderten Unsicherheitsrelationen (8) besteht jedoch ein prinzipieller Unterschied darin, dass in den letzteren nicht die Summe der Beträge der Grössen $\bar{A}_{xx}^{(\mathrm{I},\mathrm{II})}$ und $\bar{A}_{xx}^{(\mathrm{II},\mathrm{I})}$, sondern ihre algebraische Differenz auftritt. Zwar stimmen (8) und (52) im allgemeinen grössenordnungsmässig mit einander überein, wenn die Gebiete I und II räumlich und zeitlich um Strecken der Grössenordnung L und T gegen einander verschoben sind, wo sie beide den Abschätzungswert Q^2 ergeben. Die in der Unbestimmtheitsrelation (8) auftretende Differenz bewirkt aber, wie im § 2 erwähnt, dass das Produkt der komplementären Unsicherheiten in wichtigen Fällen identisch verschwindet, obwohl die Grössen $\bar{A}_{xx}^{(\mathrm{I},\,\mathrm{II})}$ und $\bar{A}_{xx}^{(\mathrm{II},\,\mathrm{I})}$ einzeln von Null verschieden bleiben. Dies trifft zum Beispiel zu, wenn die zeitlichen Mittelwertsintervalle T_{I} und T_{II} zusammenfallen, und insbesondere wenn die Mittelwertbereiche I und II sich ganz überdecken. Im letzteren Fall würde sogar die durch (52) gegebene Grenze der Messbarkeit zweier Feldmittelwerte in schroffem Widerspruch stehen mit dem Resultat der obigen Diskussion der Messung eines einzelnen Feldmittelwerts. Ueberhaupt stimmen die Ausdrücke (52) und (8) nur dann genau überein, wenn mindestens eine der Grössen $\bar{A}_{xx}^{(\mathrm{I},\mathrm{II})}$ oder $\bar{A}_{xx}^{(\mathrm{II},\mathrm{I})}$ verschwindet, was im allgemeinen erfordert, dass einer der in den Integralen (5) als Argumente der δ-Funktion auftretenden Ausdrücke $t_1-t_2-\dfrac{r}{c}$ oder $t_2-t_1-\dfrac{r}{c}$ für jedes Punktepaar $(x_1,\,y_1,\,z_1,\,t_1)$ und $(x_2,\,y_2,\,z_2,\,t_2)$ der Gebiete I und II von Null verschieden bleibt.

Abgesehen vom letzterwähnten Fall, wo zwischen den beiden Feldmittelwerten keine, oder jedenfalls nur eine einseitige

Korrelation besteht, verlangt also der Nachweis der Uebereinstimmung zwischen Messbarkeit und quantenelektromagnetischem Formalismus eine verfeinerte Messanordnung, wo
die unkontrollierbaren Effekte in grösserem Masse kompensiert werden können. Zwar tritt hier, im Vergleich mit dem
schon für die Messung einer Feldgrösse nötigen Kompensationsverfahren, die weitere Komplikation auf, dass die Verschiebungen beider Probekörper nicht nur unbekannt bleiben
müssen, sondern auch von einander völlig unabhängig sind.
Dieser Umstand bedeutet aber keine prinzipielle Schwierigkeit, nur wird ein etwas komplizierteres Verfahren notwendig, um auch den Einfluss der relativen Verschiebung der
Probekörper auf die Feldmessungen möglichst zu kompensieren. Zu diesem Zweck wählen wir uns aus den Probekörper-
systemen I und II je einen Teilkörper ε_I und ε_{II}, für die der
Ausdruck $r - c(t_1 - t_2)$ für zwei innerhalb der Zeitintervalle T_I
bzw. T_{II} liegende Zeitpunkte t_I^* und t_{II}^* Null wird. Wäre eine
solche Wahl nicht möglich, so wäre ja nach dem oben
Gesagten die Uebereinstimmung zwischen Messbarkeit und
Formalismus schon ohne weitere Kompensation erreicht.
Zur Herstellung der notwendigen Korrelation zwischen den
Probekörpern könnte man zunächst an eine Feder denken,
welche die Körper ε_I und ε_{II} direkt mit einander verbinden sollte; dabei käme man jedoch wegen der Retardation
der Kräfte in Schwierigkeiten. Es ist aber möglich, mit
einer kurzen, d. h. gegen cT kleinen Feder auszukommen,
indem man zum zweiten Probekörpersystem einen neu-
tralen Teilkörper ε_{III} hinzufügt, der sich in der unmittelbaren
Nähe des zum ersten System gehörigen Teilkörpers ε_I be-
findet, und mit diesem durch eine Feder verbunden ist.

Wie alle Teilkörper der beiden Probekörpersysteme soll

zunächst der Körper $\varepsilon_{\mathrm{III}}$ an das feste Gerüst gebunden sein. Zur Zeit t'_{I} soll nun nach Auflösung dieser Verbindung sein Impuls mit derselben Genauigkeit, wie derjenige des Probekörpersystems II gemessen werden. Dadurch erleidet er eine unbekannte Verschiebung $D_x^{(\mathrm{III})}$ in der x-Richtung, die von derselben Grössenordnung wie $\varDelta x_{\mathrm{II}}$ ist. Wird nun die Spannkraft der zwischen $\varepsilon_{\mathrm{III}}$ und ε_{I} angebrachten Feder gleich $\frac{1}{2}\varrho_{\mathrm{I}}\varrho_{\mathrm{II}} V_{\mathrm{I}} V_{\mathrm{II}} T_{\mathrm{II}} \left(\bar{A}_{xx}^{(\mathrm{I,II})} + \bar{A}_{xx}^{(\mathrm{II,I})} \right)$ gewählt, so wird im Zeitintervalle T_{I} von $\varepsilon_{\mathrm{III}}$ auf ε_{I} der Impuls

$$P = \frac{1}{2}\varrho_{\mathrm{I}}\varrho_{\mathrm{II}} V_{\mathrm{I}} V_{\mathrm{II}} T_{\mathrm{I}} T_{\mathrm{II}} \left(\bar{A}_{xx}^{(\mathrm{I,II})} + \bar{A}_{xx}^{(\mathrm{II,I})} \right) \left(D_x^{(\mathrm{I})} - D_x^{(\mathrm{III})} \right) \quad (53)$$

übertragen, während $\varepsilon_{\mathrm{III}}$ im selben Zeitintervall die Impulsänderung $-P$ erleidet. Zur Zeit t''_{I} wird wieder der Impuls von $\varepsilon_{\mathrm{III}}$ mit derselben Genauigkeit gemessen. Vor dieser Messung, und zwar zur Zeit t^*_{II} soll aber ein kurzes Lichtsignal von $\varepsilon_{\mathrm{II}}$ nach $\varepsilon_{\mathrm{III}}$ gesandt werden, durch welches mittels einer geeigneten Vorrichtung die relative Verschiebung $D_x^{(\mathrm{III})} - D_x^{(\mathrm{II})}$ dieser Körper mit beliebiger Genauigkeit gemessen werden kann. Bei der Aussendung, bzw. beim Empfang des Signals erleiden beide Körper Impulsänderungen, die zwar völlig unbekannt bleiben, sich aber in der Summe der an den Körpern gemessenen Impulsdifferenzen gegenseitig genau aufheben.

Für die Impulsbilanz der beiden Probekörpersysteme während der Messung haben wir also, wenn wir den Körper $\varepsilon_{\mathrm{III}}$ zum System II mitrechnen,

$$\left. \begin{aligned} p_x^{(\mathrm{I})''} - p_x^{(\mathrm{I})'} &= \varrho_{\mathrm{I}} V_{\mathrm{I}} T_{\mathrm{I}} \left(\bar{\mathfrak{E}}_x^{(\mathrm{I})} + \bar{E}_x^{(\mathrm{II,I})} \right) + P \\ p_x^{(\mathrm{II})''} - p_x^{(\mathrm{II})'} + p_x^{(\mathrm{III})''} - p_x^{(\mathrm{III})'} &= \varrho_{\mathrm{II}} V_{\mathrm{II}} T_{\mathrm{II}} \left(\bar{\mathfrak{E}}_x^{(\mathrm{II})} + \bar{E}_x^{(\mathrm{I,II})} \right) - P. \end{aligned} \right\} (54)$$

Unter Berücksichtigung von (43) und (53) lassen sich diese
Formeln in die Gestalt

$$
\left.
\begin{aligned}
& p_x^{(\mathrm{I})''} - p_x^{(\mathrm{I})'} = \varrho_\mathrm{I} V_\mathrm{I} T_\mathrm{I} \overline{\mathfrak{E}}_x^{(\mathrm{I})} \\
& + \frac{1}{2} \varrho_\mathrm{I} \varrho_\mathrm{II} V_\mathrm{I} V_\mathrm{II} T_\mathrm{I} T_\mathrm{II} \left\{ - D_x^{(\mathrm{II})} \left(\bar{A}_{xx}^{(\mathrm{I,II})} - \bar{A}_{xx}^{(\mathrm{II,I})} \right) \right. \\
& + \left(D_x^{(\mathrm{II})} - D_x^{(\mathrm{III})} \right) \left(\bar{A}_{xx}^{(\mathrm{I,II})} + \bar{A}_{xx}^{(\mathrm{II,I})} \right) + D_x^{(\mathrm{I})} \left(\bar{A}_{xx}^{(\mathrm{I,II})} + \bar{A}_{xx}^{(\mathrm{II,I})} \right) \right\} \\
& p_x^{(\mathrm{II})''} - p_x^{(\mathrm{II})'} + p_x^{(\mathrm{III})''} - p_x^{(\mathrm{III})'} = \varrho_\mathrm{II} V_\mathrm{II} T_\mathrm{II} \overline{\mathfrak{E}}_x^{(\mathrm{II})} \\
& + \frac{1}{2} \varrho_\mathrm{I} \varrho_\mathrm{II} V_\mathrm{I} V_\mathrm{II} T_\mathrm{I} T_\mathrm{II} \left\{ D_x^{(\mathrm{I})} \left(\bar{A}_{xx}^{(\mathrm{I,II})} - \bar{A}_{xx}^{(\mathrm{II,I})} \right) \right. \\
& - \left(D_x^{(\mathrm{II})} - D_x^{(\mathrm{III})} \right) \left(\bar{A}_{xx}^{(\mathrm{I,II})} + \bar{A}_{xx}^{(\mathrm{II,I})} \right) + D_x^{(\mathrm{II})} \left(\bar{A}_{xx}^{(\mathrm{I,II})} + \bar{A}_{xx}^{(\mathrm{II,I})} \right) \right\}
\end{aligned}
\right\} \quad (55)
$$

bringen.

Die letzten Glieder in den geschweiften Klammern von
(55) sind den unbekannten Verschiebungen der Probekörper I und II proportional und können also, genau wie die
einfachen Rückwirkungen jedes Probekörpers auf sich selbst,
durch geeignete Federverbindungen mit dem festen Gerüst
aufgehoben werden. Dies läuft einfach darauf hinaus, dass
der durch (49) gegebene Ausdruck der Spannkraft der auf
den Körper I wirkenden Feder durch

$$
F_{\mathrm{I,II}} = \varrho_\mathrm{I}^2 V_\mathrm{I}^2 T_\mathrm{I} \bar{A}_{xx}^{(\mathrm{I,I})} + \frac{1}{2} \varrho_\mathrm{I} \varrho_\mathrm{II} V_\mathrm{I} V_\mathrm{II} T_\mathrm{II} \left(\bar{A}_{xx}^{(\mathrm{I,II})} + \bar{A}_{xx}^{(\mathrm{II,I})} \right) \quad (56)
$$

zu ersetzen ist, und dass die Spannkraft der Feder zwischen
dem Gerüst und dem Körper II analog zu ändern ist. Weiter
sind die zur relativen Verschiebung $D_x^{(\mathrm{III})} - D_x^{(\mathrm{II})}$ proportionalen Glieder bei der beschriebenen Messanordnung mit
beliebiger Genauigkeit bekannt und lassen sich daher bei
den Feldmessungen einfach in Rechnung ziehen. Übrigens
könnte man durch eine etwas kompliziertere Vorrichtung
sogar erreichen, dass die Differenz $D_x^{(\mathrm{III})} - D_x^{(\mathrm{II})}$ verschwin-

det, indem man, analog der im § 3 beschriebenen Anordnung für die Messung des Gesamtimpulses eines Probekörpersystems, zur Bestimmung von $p_x^{(II)} + p_x^{(III)}$ ein und dasselbe Strahlungsbündel benutzt und mittels passend angebrachter fester Spiegel die Lichtwege so regelt, dass die Reflexionen am Körper ε_{III} und an allen Teilkörpern des Systems II bei der ersten Impulsmessung zu den Zeiten t_I' bzw. t_{II}' und bei der zweiten Impulsmessung zu den Zeiten t_I'' bzw. t_{II}'' erfolgen.

Durch alle diese Vorrichtungen, deren beträchtliche Komplikation im Wesen der Sache liegt, indem sie allein durch die endliche Fortpflanzung aller Feldwirkungen bedingt ist, haben wir nun wirklich den am Anfang dieses Paragraphen beschriebenen scheinbaren Gegensatz zwischen den Bestimmungen eines und zweier Mittelwerte einer Feldkomponente zum Verschwinden gebracht. Aus (55) bekommen wir jetzt nämlich für die Unbestimmtheiten von $\overline{\mathfrak{E}}_x^{(I)}$ und $\overline{\mathfrak{E}}_x^{(II)}$ anstatt (51)

$$
\left.
\begin{aligned}
\mathit{\Delta}\overline{\mathfrak{E}}_x^{(I)} &\sim \frac{\hbar}{\varrho_I \mathit{\Delta} x_I V_I T_I} + \frac{1}{2}\varrho_{II}\, \mathit{\Delta} x_{II} V_{II} T_{II} \left| \bar{A}_{xx}^{(I,\,II)} - \bar{A}_{xx}^{(II,\,I)} \right| \\
\mathit{\Delta}\overline{\mathfrak{E}}_x^{(II)} &\sim \frac{\hbar}{\varrho_{II} \mathit{\Delta} x_{II} V_{II} T_{II}} + \frac{1}{2}\varrho_I\, \mathit{\Delta} x_I V_I T_I \left| \bar{A}_{xx}^{(I,\,II)} - \bar{A}_{xx}^{(II,\,I)} \right|,
\end{aligned}
\right\} \tag{57}
$$

woraus sich für den Minimalwert des Produkts der Unsicherheiten

$$
\mathit{\Delta}\overline{\mathfrak{E}}_x^{(I)}\, \mathit{\Delta}\overline{\mathfrak{E}}_x^{(II)} \sim \hbar \left| \bar{A}_{xx}^{(I,\,II)} - \bar{A}_{xx}^{(II,\,I)} \right| \tag{58}
$$

direkt ergibt, im Einklang mit der durch (8) ausgedrückten Folgerung der Quantentheorie der Felder.

Um die völlige Übereinstimmung zwischen der Messbarkeit der Mittelwerte einer Feldkomponente über zwei

Raumzeitgebiete und den Forderungen des quantenelektromagnetischen Formalismus nachzuweisen, müssen wir jedoch
etwas näher auf die Frage eingehen, inwiefern die benutzte Annahme der klassischen Beschreibbarkeit der Feldwirkungen der Probekörper die Prüfungsmöglichkeiten der
theoretischen Erwartungen beeinträchtigt. Eben bei der
Messung mehrerer Feldmittelwerte könnte man nämlich,
wie schon berührt, von vornherein denken, dass die Vernachlässigung der klassisch unverfolgbaren, von den reinen
Hohlraumschwankungen untrennbaren Schwankungen aller
Feldwirkungen der Probekörper in dieser Hinsicht einen
wesentlichen Verzicht bedeute. Solange es sich um Mittel-
wertsgebiete handelt, die gegen einander räumlich und zeitlich um Strecken von derselben Grössenordnung wie ihre
linearen Abmessungen L und zugehörigen Zeitintervalle T
verschoben sind, ist allerdings in dem wichtigen Fall, wo
L gross gegenüber cT ist, die in Frage stehende Vernachlässigung von geringer Bedeutung. Wenn jedoch L grössenordnungsmässig gleich oder kleiner als cT ist, so sind, wie
in § 2 erwähnt, die Hohlraumschwankungen eben von derselben Grössenordnung wie die für zwei derart verschobene
Gebiete mittels der Unbestimmtheitsrelationen definierte
kritische Feldstärke $\mathfrak{U}$, die als die Grenze der klassischen
Feldbeschreibung anzusehen ist. Besonders aber für zwei
fast zusammenfallende Gebiete, wo das durch (8) gegebene
Produkt der komplementären Unsicherheiten der Feldmittelwerte unabhängig vom Verhältnis zwischen L und
cT gegen Null strebt, und somit die kritische Feldstärke $\mathfrak{U}$
im Vergleich mit den Hohlraumschwankungen beliebig klein
sein kann, könnte die erwähnte Vernachlässigung noch
bedenklicher vorkommen und scheinbar einen völligen Verzicht auf die Wiederholbarkeit der Feldmessungen bedeuten.

Eine nähere Betrachtung zeigt indessen, dass wir eine widerspruchsfreie Deutung aller Folgerungen der Quantentheorie der Felder erreichen, wenn wir in zwangläufiger Verallgemeinerung des Messbegriffs die mittels der beschriebenen Anordnung erhaltenen Messergebnisse als die gesuchten Feldmittelwerte auffassen. Die in den Feldwirkungen sämtlicher Probekörper einbegriffenen, klassisch unbeschreibbaren Schwankungen lassen sich nämlich gar nicht trennen von den prinzipiell statistischen Zügen jeder theoretischen Aussage, deren Voraussetzungen sich nicht auf eigentliche Feldmessungen beziehen. Ohne das gestellte Messproblem in irgendwelcher Weise einzuschränken, können wir daher die betrachteten Schwankungen immer als integrierenden Bestandteil des zu messenden Feldes ansehen. Die Verhältnisse bei mehreren Feldmessungen weichen in dieser Hinsicht nur insofern von den bei der Messung eines einzigen Feldmittelwertes vorliegenden ab, als der Feldzustand, mit dem wir im allgemeinen Fall bei jeder einzelnen Messung zu tun haben, durch das Resultat der anderen Feldmessungen mitbestimmt wird.

Mit Hinblick auf die geschilderte Sachlage dürfte es jedoch nicht überflüssig sein, darauf hinzuweisen, dass wir bei der Korrelation mehrerer Feldmessungen mit einem dem üblichen Messproblem der nichtrelativistischen Quantenmechanik fremden Zug der allgemeinen Komplementarität der Beschreibung zu tun haben. Die grundsätzliche Vereinfachung, der wir in letzterer Theorie begegnen, liegt ja eben in der dort gemachten Trennung zwischen Raumkoordination und Zeitverlauf, welche es ermöglicht, alle Messvorgänge in eine einfache zeitliche Reihenfolge zu ordnen. Bei der Messung zweier Feldmittelwerte dagegen ist es nur dann möglich, von einer solchen Reihenfolge der Messvorgänge.

zu sprechen, wenn die zugehörigen Zeitintervalle ganz aus-
einanderliegen. Im allgemeinen wird auch, wie es dem
Formalismus entspricht, die Korrelation der beiden Messungen eine gegenseitige sein; und nur wenn eine der
Grössen $r - c(t_1 - t_2)$ und $r - c(t_2 - t_1)$ für alle Punktepaare der Gebiete I und II von Null verschieden bleibt,
begegnen wir ähnlichen Verhältnissen, wie beim gewöhnlichen Messproblem der Atommechanik, indem das Ergebnis
der einen Feldmessung sich dann einfach mitrechnen lässt
zu den Voraussetzungen der durch die andere Messung zu
prüfenden Erwartungen.

Ein lehrreiches Beispiel einer engen gegenseitigen Korrelation treffen wir eben bei Messungen der Mittelwerte einer
Feldkomponente über zwei beinahe zusammenfallende Raumzeitgebiete. Entsprechend der Forderung der Wiederholbarkeit der Messergebnisse verlangt hier die Theorie, dass
beide Messungen mit beliebiger Annäherung dasselbe Resultat ergeben sollen, ganz unabhängig von den durch die
Voraussetzungen bedingten statistischen Aussagen über die
Werte der zu messenden Feldgrössen. Dass diese Forderung
bei unserer Versuchsanordnung auch wirklich erfüllt ist,
folgt daraus, dass wir in diesem Fall mit zwei Probekörper-
systemen zu tun haben, die fast denselben Raumbereich
einnehmen und fast im selben Zeitintervall benutzt werden.
Definitionsgemäss werden sie also fast demselben Feld aus-
gesetzt, ganz gleichgültig, aus welchen Quellen dieses Feld
stammt, und welcher Beitrag von dem einen oder andern
Probekörper herrührt.

Aus der letzten Bemerkung folgt eigentlich, dass wir
bei zusammenfallenden Mittelwertgebieten schon ohne jede
Kompensation genau übereinstimmende Resultate der beiden Messungen bekommen würden. Wegen der Feldwir-

kungen der Probekörper würden aber die so gewonnenen Messresultate umso mehr von den zu prüfenden theoretischen Erwartungen in unvoraussagbarer Weise abweichen, je grösser die angestrebte Messgenauigkeit ist. Durch den für einzelne Feldmessungen geeigneten Kompensationsmechanismus, den wir am Anfang dieses Paragraphen unverändert beibehalten hatten, werden diese Abweichungen zwar im allgemeinen herabgesetzt, aber jede strenge Korrelation der Messergebnisse wird zugleich durch die den unabhängigen Verschiebungen der Probekörper proportionalen Federwirkungen verhindert. Bei der für zwei Feldmessungen schliesslich angenommenen Anordnung, bei welcher alle wohldefinierten Unterschiede zwischen Messergebnissen und theoretischen Erwartungen kompensiert werden, wird auch eine solche Korrelation eben für zusammenfallende Gebiete wiederhergestellt. Denn, ganz unabhängig vom Grössenverhältnis der unkontrollierbaren Verschiebungen der Probekörper, sind, wie man leicht sieht, die durch die Gesamtwirkung aller Federn auf jeden Probekörper übertragenen Impulse, durch die entsprechenden Ladungsdichten dividiert, in diesem Fall genau dieselben.

Was die Widerspruchsfreiheit der Beschreibung betrifft, möchten wir noch bemerken, dass jeder Versuch, die durch eine Feldmessung verursachte Änderung des Lichtquantenzustands des Feldes durch die Untersuchung der Strahlung des Probekörpers zu kontrollieren, wie schon mehrmals erwähnt, die Möglichkeit ausschliessen würde, das Messergebnis für einen Vergleich mit einer zweiten Feldmessung zu verwerten. Damit nämlich von einer solchen Verwertung überhaupt die Rede sein kann, muss es Punktepaare aus den Gebieten I bzw. II geben, für welche einer der Ausdrücke $r - c\,(t_1 - t_2)$ oder $r - c\,(t_2 - t_1)$ verschwindet. Dies hat aber zur Folge, dass die Strahlungsfelder, die durch die Probekörper I und II während der Feldmessungen erzeugt werden, nicht auf ihrem Wege von dem einen zum andern Probekörper aufgefangen und analysiert werden können, ohne zugleich die durch diese Körper zu messenden Felder wesentlich zu beeinflussen. Erst nach Beendigung aller Feldmessungen, wo ihre direkte Verwertung nicht mehr in Betracht kommt, kann man eine beliebig genaue Analyse des Lichtquantenzustands des Gesamtfeldes ohne Beeinträchtigung des gestellten Messproblems vornehmen.

§ 7. Messbarkeit zweier Mittelwerte verschiedener Feldkomponenten.

Was die Messungen von Mittelwerten verschiedener Feldkomponenten betrifft, so erfordert nur der Fall senkrechter, gleichartiger oder ungleichartiger, Komponenten eine nähere Untersuchung, denn die vom quantenelektromagnetischen Formalismus verlangte völlige Vertauschbarkeit und unabhängige Messbarkeit von Mittelwerten paralleler ungleichartiger Komponenten findet gemäss (42) ihre unmittelbare Deutung im identischen Verschwinden der Komponente $H_x^{(I)}$ des durch die Messung von $\overline{\mathfrak{E}}_x^{(I)}$ erzeugten Feldes. Übrigens lässt auf Grund der Gleichungen (43) die Messung von Mittelwerten senkrechter Feldkomponenten eine dem im vorigen Paragraphen beschriebenen Verfahren analoge Behandlungsweise zu.

Betrachten wir die Messung des Mittelwerts von $\mathfrak{E}_x$ über das Gebiet I und des Mittelwerts von $\mathfrak{E}_y$ oder $\mathfrak{H}_y$ über das Gebiet II. Wenn wir zunächst eine Messanordnung heranziehen, wobei die Feldwirkungen jedes Probekörpers auf sich selbst während der Messung in der im

§ 5 beschriebenen Weise kompensiert werden, so bekommen wir für die Impulsbilanz der beiden zu benutzenden Probekörper Gleichungen vom folgenden Typus:

$$\left. \begin{aligned} p_x^{(\mathrm{I})''} - p_x^{(\mathrm{I})'} &= \varrho_\mathrm{I} V_\mathrm{I} T_\mathrm{I} \left(\overline{\mathfrak{E}}_x^{(\mathrm{I})} + D_y^{(\mathrm{II})} \sigma_\mathrm{II} V_\mathrm{II} T_\mathrm{II} \bar{C}_{xy}^{(\mathrm{II,\,I})} \right) \\ p_y^{(\mathrm{II})''} - p_y^{(\mathrm{II})'} &= \sigma_\mathrm{II} V_\mathrm{II} T_\mathrm{II} \left(\overline{\mathfrak{R}}_y^{(\mathrm{II})} + D_x^{(\mathrm{I})} \varrho_\mathrm{I} V_\mathrm{I} T_\mathrm{I} \bar{C}_{xy}^{(\mathrm{I,\,II})} \right) \end{aligned} \right\} \quad (59)$$

Je nachdem es sich um die Messung gleichartiger oder ungleichartiger Komponenten handelt, vertritt dabei der Buchstabe $\mathfrak{R}$ die übliche Bezeichnung der Feldkomponenten $\mathfrak{E}$ oder $\mathfrak{H}$, während C anstatt der in (43) auftretenden Symbole A oder B geschrieben ist; weiter stellt die Bezeichnung σ_II die Ladungsdichte bzw. Polstärkenverteilung des Probekörpers II dar.

In einer zur Herleitung von (52) analogen Weise ergibt sich aus (59) die Relation

$$\varDelta \overline{\mathfrak{E}}_x^{(\mathrm{I})} \varDelta \overline{\mathfrak{R}}_y^{(\mathrm{II})} \sim \hbar \left[\left| \bar{C}_{xy}^{(\mathrm{I,\,II})} \right| + \left| \bar{C}_{xy}^{(\mathrm{II.\,I})} \right| \right], \qquad (60)$$

welche, wie (52), nicht allgemein, sondern nur in gewissen Fällen eine Übereinstimmung zwischen Messbarkeit und quantenelektromagnetischem Formalismus darstellt. Von solchen Fällen möchten wir die Messung ungleichartiger senkrechter Feldkomponenten innerhalb desselben Raumbereichs besonders erwähnen, bei welcher, wie im § 2 hervorgehoben, beide Ausdrücke $\overline{B}_{xy}^{(\mathrm{I,\,II})}$ und $\overline{B}_{xy}^{(\mathrm{II,\,I})}$ Null sind. Die Richtigkeit der Deutung dieser Tatsache als beliebig genaue unabhängige Messbarkeit der betreffenden Feldgrössen wurde bereits im § 3 für zusammenfallende Raumzeitgebiete durch eine elementare Betrachtung nahegelegt.

Zur allgemeinen Behandlung des Messbarkeitsproblems

senkrechter Feldkomponenten wählen wir, wie im vorigen Paragraphen, zwei Teilkörper ε_I und ε_{II} der Probekörpersysteme I bzw. II, deren Abstand $r = c(t_I^* - t_{II}^*)$ ist, wobei t_I^* und t_{II}^* innerhalb der Zeitintervalle T_I bzw. T_{II} liegen. Weiter bringen wir in unmittelbare Nähe von ε_I einen dritten Körper ε_{III}, dessen Impuls in der y-Richtung zu den Zeiten t_I' und t_I'' gemessen wird; die relative Verschiebung $D_y^{(III)} - D_y^{(II)}$ der Körper ε_{III} und ε_{II} wird wieder durch ein Lichtsignal bestimmt, wodurch beide Körper entgegengesetzt gleiche Impulsänderungen erleiden. Anstatt ε_{III} mit ε_I direkt durch eine Feder zu verbinden, müssen wir aber, um die Kraftübertragung durch den Federmechanismus proportional zu $D_y^{(III)} - D_x^{(I)}$ zu machen, eine Vorrichtung gebrauchen, die aus zwei Federn und einem Winkelhebel mit zwei gleich langen, zu einander senkrecht stehenden Armen besteht, der um ein am festen Gerüst angebrachtes Gelenk drehbar ist, und dessen Arme anfänglich parallel zur x-, bzw. y-Richtung stehen. Zwischen dem ersten Arm und dem Körper ε_{III} wird eine zur y-Achse parallele Feder angebracht, und zwischen dem zweiten Arm und dem Körper ε_I wirkt eine zur x-Achse parallele Feder. Die Spannkraft der Feder sei so gewählt, dass die Kraft, die während des Zeitintervalls T_I den Körper ε_{III} in der y-Richtung und den Körper ε_I in der x-Richtung angreift, durch

$$\frac{1}{2} \varrho_I \sigma_{II} V_I V_{II} T_{II} \left(\bar{C}_{xy}^{(I,\,II)} + \bar{C}_{xy}^{(II,\,I)} \right) \left(D_x^{(I)} - D_y^{(III)} \right)$$

gegeben wird.

Die Impulsbilanz der beiden Probekörpersysteme schreibt sich also nach zweckmässiger Umformung in folgender, zu (55) analoger Gestalt:

$$\left.\begin{aligned}
p_x^{(\mathrm{I})''} - p_x^{(\mathrm{I})'} &= \varrho_{\mathrm{I}} V_{\mathrm{I}} T_{\mathrm{I}} \overline{\mathfrak{E}}_x^{(\mathrm{I})} \\
&+ \frac{1}{2} \varrho_{\mathrm{I}} \sigma_{\mathrm{II}} V_{\mathrm{I}} V_{\mathrm{II}} T_{\mathrm{I}} T_{\mathrm{II}} \left\{ - D_y^{(\mathrm{II})} \left(\bar{C}_{xy}^{(\mathrm{I},\,\mathrm{II})} - \bar{C}_{xy}^{(\mathrm{II},\,\mathrm{I})} \right) \right. \\
&\left. + \left(D_y^{(\mathrm{II})} - D_y^{(\mathrm{III})} \right) \left(\bar{C}_{xy}^{(\mathrm{I},\,\mathrm{II})} + \bar{C}_{xy}^{(\mathrm{II},\,\mathrm{I})} \right) + D_x^{(\mathrm{I})} \left(\bar{C}_{xy}^{(\mathrm{I},\,\mathrm{II})} + \bar{C}_{xy}^{(\mathrm{II},\,\mathrm{I})} \right) \right\} \\
p_y^{(\mathrm{II})''} - p_y^{(\mathrm{II})'} + p_y^{(\mathrm{III})''} - p_y^{(\mathrm{III})'} &= \sigma_{\mathrm{II}} V_{\mathrm{II}} T_{\mathrm{II}} \mathfrak{R}_y^{(\mathrm{II})} \\
&+ \frac{1}{2} \varrho_{\mathrm{I}} \sigma_{\mathrm{II}} V_{\mathrm{I}} V_{\mathrm{II}} T_{\mathrm{I}} T_{\mathrm{II}} \left\{ D_x^{(\mathrm{I})} \left(\bar{C}_{xy}^{(\mathrm{I},\,\mathrm{II})} - \bar{C}_{xy}^{(\mathrm{II},\,\mathrm{I})} \right) \right. \\
&\left. + \left(D_y^{(\mathrm{II})} - D_y^{(\mathrm{III})} \right) \left(\bar{C}_{xy}^{(\mathrm{I},\,\mathrm{II})} + \bar{C}_{xy}^{(\mathrm{II},\,\mathrm{I})} \right) - D_y^{(\mathrm{II})} \left(\bar{C}_{xy}^{(\mathrm{I},\,\mathrm{II})} + \bar{C}_{xy}^{(\mathrm{II},\,\mathrm{I})} \right) \right\}.
\end{aligned}\right\} \quad (61)$$

Nach Kompensation der letzten Glieder erhalten wir somit für die Unsicherheiten der Feldmittelwerte:

$$\left.\begin{aligned}
\varDelta \overline{\mathfrak{E}}_x^{(\mathrm{I})} &\sim \frac{\hbar}{\varrho_{\mathrm{I}} \varDelta x_{\mathrm{I}} V_{\mathrm{I}} T_{\mathrm{I}}} + \frac{1}{2} \sigma_{\mathrm{II}} \varDelta y_{\mathrm{II}} V_{\mathrm{II}} T_{\mathrm{II}} \left| \bar{C}_{xy}^{(\mathrm{I},\,\mathrm{II})} - \bar{C}_{xy}^{(\mathrm{II},\,\mathrm{I})} \right| \\
\varDelta \mathfrak{R}_y^{(\mathrm{II})} &\sim \frac{\hbar}{\sigma_{\mathrm{II}} \varDelta y_{\mathrm{II}} V_{\mathrm{II}} T_{\mathrm{II}}} + \frac{1}{2} \varrho_{\mathrm{I}} \varDelta x_{\mathrm{I}} V_{\mathrm{I}} T_{\mathrm{I}} \left| \bar{C}_{xy}^{(\mathrm{I},\,\mathrm{II})} - \bar{C}_{xy}^{(\mathrm{II},\,\mathrm{I})} \right|,
\end{aligned}\right\} \quad (62)$$

woraus sich für den Minimalwert ihres Produktes

$$\varDelta \overline{\mathfrak{E}}_x^{(\mathrm{I})} \varDelta \mathfrak{R}_y^{(\mathrm{II})} \sim \hbar \left| \bar{C}_{xy}^{(\mathrm{I},\,\mathrm{II})} - \bar{C}_{xy}^{(\mathrm{II},\,\mathrm{I})} \right| \qquad (63)$$

ergibt, wiederum in vollem Einklang mit dem quanten-elektromagnetischen Formalismus.

Aus den allgemeinen Ausführungen am Ende des vorigen Paragraphen folgt weiter, dass auch in dem hier betrachteten Fall die Verwertung der Feldmessungen für die Prüfung der Aussagen des Formalismus in keiner Weise durch die klassische Abschätzung der Feldwirkungen beeinträchtigt wird. Übrigens kommt bei Messungen von Mittelwerten

verschiedener Feldkomponenten die Frage der Wiederholbarkeit gar nicht in Betracht, und die reinen Hohlraumschwankungen sind als unvermeidlicher statistischer Zug in
allen theoretischen Aussagen einbegriffen.

§ 8. Schlussbemerkungen.

Wir kommen also zu der bereits anfangs erwähnten
Schlussfolgerung, dass die Quantentheorie der Felder in
bezug auf die Messbarkeitsfrage eine widerspruchsfreie
Idealisation darstellt in dem Umfang, in dem wir von
allen Einschränkungen, die auf der atomistischen Struktur der Feldquellen und der Messinstrumente beruhen, absehen können. Eigentlich dürfte dieses Ergebnis, wie schon
in der Einleitung betont, anzusehen sein als eine unmittelbare Konsequenz der gemeinsamen korrespondenzmässigen
Grundlage des quantenelektromagnetischen Formalismus
und der Gesichtspunkte, von welchen die Prüfungsmöglichkeiten dieses Formalismus zu beurteilen sind. Nichtsdesto-
weniger dürfte der etwas komplizierte Charakter der zum
Nachweis der völligen Übereinstimmung zwischen Formalismus und Messbarkeit herangezogenen Betrachtungen kaum
zu vermeiden sein. Erstens sind ja die an der Messanordnung zu stellenden physikalischen Forderungen bedingt
durch die in Integralform gekleideten Aussagen des quantenelektromagnetischen Formalismus, wodurch die besondere
Einfachheit der klassischen Feldtheorie als reiner Diffe-
rentialtheorie verloren geht. Weiter erfordert die Deutung
der Messergebnisse und ihre Verwertung an Hand des
Formalismus, wie wir gesehen haben, die Berücksichtigung

von gewissen in den Messproblemen der unrelativistischen Quantenmechanik nicht auftretenden Zügen der komplementären Beschreibungsweise.

––––––––––

Bei der Abschliessung dieser Arbeit möchten wir nicht unerwähnt lassen, dass wir in vielen Diskussionen über die behandelten Fragen mit früheren und jetzigen Mitarbeitern des Instituts, worunter sowohl HEISENBERG und PAULI wie LANDAU und PEIERLS, manche Anregung und Hilfe gefunden haben.

Universitetets Institut for teoretisk Fysik.

København, April 1933.

––––––––––

Meddelt paa Mødet den 2. December 1932.
Færdig fra Trykkeriet den 19. December 1933.

CHAPTER 1

ON THE QUESTION OF THE MEASURABILITY OF ELECTROMAGNETIC FIELD QUANTITIES*

[1933b]

1. INTRODUCTION

The question of limitations on the measurability of electromagnetic field quantities, rooted in the quantum of action, has acquired particular interest in the course of the discussion of the still unsolved difficulties in relativistic atomic mechanics. On the basis of exploratory considerations, Heisenberg[1] attempted to demonstrate that the connection between the limitation on the measurability of field quantities and the quantum theory of fields is similar to the relationship between the complementary limitations on the measurability of kinematical and dynamical quantities expressed in the indeterminacy principle and the non-relativistic formalism of quantum mechanics. However, in the course of a critical investigation of the foundations of the relativistic generalization of this formalism, Landau and Peierls[2] came to the conclusion that the measurability of field quantities is subjected to further restrictions which go essentially beyond the presuppositions of quantum field theory, and which therefore deprive this theory of any physical basis.

At first glance, one might see in this contradiction a serious dilemma. On the one hand, the quantum theory of fields surely ought to be considered a consistent re-interpretation of classical electromagnetic theory in the sense of the correspondence argument, just as quantum mechanics represents a reformulation of classical mechanics adapted to the existence of the quantum of action. On the other hand, quantum electrodynamics has essentially increased the difficulties, already encountered in classical electron theory, of a harmonious blending of field theory and atomic theory. However, closer consideration shows that the various problems involved here can be to a large extent separated from each other, since the quantum-electromagnetic formalism in itself is independent of all ideas concerning the atomic constitution of matter. This is evident from the fact that in addition to the velocity of light only the quantum of action enters the formalism as a universal constant; for these two constants obviously do not suffice to determine any specific space-time dimensions. In the quantum theory of atomic structure such a determination will be obtained only by the inclusion of the electric charge and the rest masses of the elementary particles.

Just the insufficient distinction between field theory and atomic theory is the principal reason for the conflicting results of previous investigations of the measurability of field quantities in which single electrically charged mass points were used as test bodies. The utilization of classical electron theory, in the sense of the correspondence argument, which underlies current atomic mechanics, rests above all on the smallness of the elementary charge in comparison with the square root of the product of the quantum of action and the velocity of light, which makes it possible to treat all radiation reactions as small compared to the ponderomotive forces exerted on the particles. However, in measurements of field quantities it turns out to be essential to be able to adjust the charge of the test bodies to an extent which would be in conflict with the latter presupposition if one were to consider these bodies as point charges. As we shall see, however, these difficulties disappear if one uses test bodies whose linear extensions are chosen sufficiently large, compared to atomic dimensions, that their charge density can be considered approximately constant over the whole body.

In this connection, it is also of essential importance that the customary description of an electric field in terms of the field components at each space-time point, which characterizes classical field theory and according to which the field should be measurable by means of point charges in the sense of the electron theory, is an idealization which has only a restricted applicability in quantum theory. This circumstance finds its proper expression in the quantum-electromagnetic formalism, in which the field quantities are no longer represented by true point functions but by functions of space-time regions, which formally correspond to the average values of the idealized field components over the regions in question. The formalism only allows the derivation of unambiguous predictions about the measurability of such region-functions, and our task will thus consist in investigating whether the complementary limitations on the measurability of field quantities, defined in this way, are in accordance with the physical possibilities of measurement.

Insofar as we can disregard all restrictions arising from the atomistic structure of the measuring instruments, it is actually possible to demonstrate a complete accord in this respect. Besides a thorough investigation of the construction and handling of the test bodies, this demonstration requires, however, consideration of certain new features of the complementary mode of description, which come to light in the discussion of the measurability question, but which were not included in the customary formulation of the indeterminacy principle in connection with non-relativistic quantum mechanics. Not only is it an essential complication of the problem of field measurements that, when comparing field averages over different space-time regions, we cannot in an unambiguous way speak about a temporal sequence of the measurement processes; but even the interpretation of individual measurement results requires a still greater caution in the case of field measurements than in the usual quantum-mechanical measurement problem.

Characteristic of the latter problem is the possibility of attributing to each individual measurement result a well-defined meaning in the sense of classical mechanics, while the quantum-imposed interaction, uncontrollable in principle, between instrument and object is fully taken into account through the influence of each measuring process on the statistical expectations testable in succeeding measurements. In contrast, in measurements of field quantities, indeed, every measuring result is well defined on the basis of the classical field concept; but, the limited applicability of classical field theory to the description of the unavoidable electromagnetic field effects of the test bodies during the measurement implies, as we shall see, that these field effects to a certain degree influence the measurement result itself in a way which cannot be compensated for. However, a closer investigation of the fundamentally statistical character of the consequences of the quantum-electromagnetic formalism shows that this influence on the object of measurement by the measuring process in no way impairs the possibility of testing such consequences, but rather is to be regarded as an essential feature of the intimate adaptation of quantum field theory to the measurability problem.

Before we turn to a detailed exposition of the considerations indicated above, we want to stress once more that the fundamental difficulties which confront the consistent utilization of field theory in atomic theory remain entirely untouched by the present investigation. Indeed, consideration of

360

the atomistic constitution of all measuring instruments would be essential for an assessment of the connection between these difficulties and the well-known paradoxes of the measurement problem in relativistic quantum mechanics. Also particularly relevant in this context is the limitation which the finite value of the elementary charge compared to the square root of the product of the velocity of light and the quantum of action places on atomic mechanics based on the correspondence argument.[3]

2. MEASURABILITY OF FIELDS ACCORDING TO QUANTUM THEORY

The quantum electromagnetic formalism found its starting point in the quantum theory of radiation developed by Dirac, which is characterized by the introduction of a non-commutativity, consistent with the quantum-mechanical commutation relations, of canonically conjugate amplitudes of vibration of a radiation field. On the basis of this theory, Jordan and Pauli set up commutation relations between the electromagnetic field components for the case of charge-free fields, and the formalism was then brought to a certain completion by Heisenberg and Pauli who treated the interaction between field and material charges on correspondence lines. However, the consistent application of the theory to atomic problems is essentially impaired by the occurrence of the well-known paradoxes of the infinite self-energy of the elementary particles, which were not removed by Dirac's proposed modifications in the representation of the formalism.[4] Yet, in our discussion of the limitations of the measurability of field quantities these difficulties play no role, since for this purpose the atomistic structure of matter is not an essential issue. It is true that the measurement of fields requires the use of material charged test bodies, but their unambiguous application as measuring instruments depends exactly on the extent to which we can treat their response to the fields as well as their influence as field sources on the basis of classical electrodynamics.

In these circumstances we may restrict ourselves to the pure field theory, and thus take the commutation relations for charge-free fields as our starting point for the investigation of the consequences of the quantum electromagnetic formalism with respect to the measurability of field quantities. Using the usual notation, $[p, q] = pq - qp$, we thus have the following typical relations between the field components in two space-time points (x_1, y_1, z_1, t_1) and (x_2, y_2, z_2, t_2), from which the remaining commutation relations are obtained by cyclic permutation:[5]

$$(1) \quad \begin{cases} [\mathfrak{E}_x^{(1)}, \mathfrak{E}_x^{(2)}] = [\mathfrak{H}_x^{(1)}, \mathfrak{H}_x^{(2)}] = \sqrt{-1}\, \hbar (A_{xx}^{(12)} - A_{xx}^{(21)}) \\[2mm] [\mathfrak{E}_x^{(1)}, \mathfrak{E}_y^{(2)}] = [\mathfrak{H}_x^{(1)}, \mathfrak{H}_y^{(2)}] = \sqrt{-1}\, \hbar (A_{xy}^{(12)} - A_{xy}^{(21)}) \\[2mm] [\mathfrak{E}_x^{(1)}, \mathfrak{H}_x^{(2)}] = 0 \\[2mm] [\mathfrak{E}_x^{(1)}, \mathfrak{H}_y^{(2)}] = -[\mathfrak{H}_x^{(1)}, \mathfrak{E}_y^{(2)}] = \sqrt{-1}\, \hbar (B_{xy}^{(12)} - B_{xy}^{(21)}). \end{cases}$$

Here, $\mathfrak{E}_x^{(1)}, \mathfrak{E}_y^{(1)}, \mathfrak{E}_z^{(1)}, \mathfrak{H}_x^{(1)}, \mathfrak{H}_y^{(1)}, \mathfrak{H}_z^{(1)}$ are the electric and magnetic field components at the space-time point (x_1, y_1, z_1, t_1) while the following abbreviations have been used:

$$(2) \quad \begin{cases} A_{xx}^{(12)} = -\left(\dfrac{\partial^2}{\partial x_1 \partial x_2} - \dfrac{1}{c^2} \dfrac{\partial^2}{\partial t_1 \partial t_2} \right) \left\{ \dfrac{1}{r} \delta \left(t_2 - t_1 - \dfrac{r}{c} \right) \right\} \\[4mm] A_{xy}^{(12)} = -\dfrac{\partial^2}{\partial x_1 \partial y_2} \left\{ \dfrac{1}{r} \delta \left(t_2 - t_1 - \dfrac{r}{c} \right) \right\} \\[4mm] B_{xy}^{(12)} = -\dfrac{1}{c} \dfrac{\partial}{\partial t_1 \partial z_2} \left\{ \dfrac{1}{r} \delta \left(t_2 - t_1 - \dfrac{r}{c} \right) \right\} \end{cases}$$

Further, $\hbar$ is Planck's constant divided by 2π, c the velocity of light, and r the spatial separation between the two points. Finally, δ denotes the symbolic function introduced by Dirac which, as is well known, is characterized by the property

$$(3) \quad \int_{t'}^{t''} \delta(t - t_0)\, dt = \begin{cases} 1 \text{ for } t' < t_0 < t'' \\ 0 \text{ for } t_0 < t' \text{ or } t_0 > t'' \end{cases}$$

and which is formally differentiated like an ordinary function.

The occurrence of the δ-function, defined in (3), in the commutation relations (1) brings to the fore the fact mentioned above that the quantum theoretical field quantities are not to be considered as true point functions but that unambiguous meaning can be attached only to space-time integrals of the field components. With a view to the simplest possibility of testing the formalism we shall consider only averages of field components over simply connected space-time regions whose spatial extension remains constant during a given time interval. Thus, for example, if the volume of such a region G is denoted by V and the corresponding time interval is T, we define the average value of $\mathfrak{E}_x$ by the formula

(4) $\qquad \mathfrak{E}_x^{(G)} = \frac{1}{VT} \int_T dt \int_V \mathfrak{E}_x \, dv.$

For the average values of two field components over two given space-time regions I and II there exist commutation relations which follow immediately from (1) by integration over the two regions and division by the product of their four-dimensional extensions. Thus, the value of the bracket symbols $[\mathfrak{E}_x^{(I)}, \mathfrak{E}_x^{(II)}], \ldots$ are obtained from (1) simply by replacing the quantities $A^{(12)}$, $B^{(12)}$ by their average values over the two regions

(5)

$$
\begin{cases}
\bar{A}_{xx}^{(I,\,II)} = -\frac{1}{V_I V_{II} T_I T_{II}} \int_{T_I} dt_1 \int_{T_{II}} dt_2 \int_{V_I} dv_1 \int_{V_{II}} dv_2 \\
\qquad\qquad \left(\frac{\partial^2}{\partial x_1 \partial x_2} - \frac{1}{c^2} \frac{\partial^2}{\partial t_1 \partial t_2} \right) \left\{ \frac{1}{r} \delta \left(t_2 - t_1 - \frac{r}{c} \right) \right\} \\[2ex]
\bar{A}_{xy}^{(I,\,II)} = -\frac{1}{V_I V_{II} T_I T_{II}} \int_{T_I} dt_1 \int_{T_{II}} dt_2 \int_{V_I} dv_1 \int_{V_{II}} dv_2 \\
\qquad\qquad\qquad \frac{\partial^2}{\partial x_1 \partial y_2} \left\{ \frac{1}{r} \delta \left(t_2 - t_1 - \frac{r}{c} \right) \right\} \\[2ex]
\bar{B}_{xy}^{(I,\,II)} = -\frac{1}{V_I V_{II} T_I T_{II}} \int_{T_I} dt_1 \int_{T_{II}} dt_2 \int_{V_I} dv_1 \int_{V_{II}} dv_2 \\
\qquad\qquad\qquad \frac{1}{c} \frac{\partial^2}{\partial t_1 \partial z_2} \left\{ \frac{1}{r} \delta \left(t_2 - t_1 - \frac{r}{c} \right) \right\}
\end{cases}
$$

In exactly the same way as the Heisenberg relation for two canonically conjugate mechanical quantities

(6) $\qquad \Delta p \, \Delta q \sim \hbar,$

which is the basis for the uncertainty principle, can be derived from the general quantum mechanical commutation relation

(7) $\qquad [q, p] = \sqrt{-1} \, \hbar,$

one obtains for the products of the complementary uncertainties of the field averages in question the following typical formulae:

$$\Delta\bar{\mathfrak{E}}_x^{(I)} \, \Delta\bar{\mathfrak{E}}_x^{(II)} \sim \hbar \left| \bar{A}_{xx}^{(I,\,II)} - \bar{A}_{xx}^{(II,\,I)} \right|, \quad \Delta\bar{\mathfrak{E}}_x^{(I)} \, \Delta\bar{\mathfrak{E}}_y^{(II)} \sim \hbar \left| \bar{A}_{xy}^{(I,\,II)} - \bar{A}_{xy}^{(II,\,I)} \right|$$

$$\Delta\bar{\mathfrak{E}}_x^{(I)} \, \Delta\bar{\mathfrak{H}}_x^{(II)} = 0, \qquad\qquad \Delta\bar{\mathfrak{E}}_x^{(I)} \, \Delta\bar{\mathfrak{H}}_y^{(II)} \sim \hbar \left| \bar{B}_{xy}^{(I,\,II)} - \bar{B}_{xy}^{(II,\,I)} \right|.$$

(8)

Several results of importance for our problem follow immediately from the expressions (5) and (8). Above all, we see that, in accord with the property of the δ-function expressed in (3), the quantities $\bar{A}^{(I,\,II)}$, $\bar{B}^{(I,\,II)}$ change continuously as a result of a continuous displacement of the boundaries of regions I and II, as long as the extension of these regions, i.e. the values of V_I, T_I, V_{II}, T_{II}, remain different from zero. In particular, the differences $\bar{A}^{(I,\,II)} - \bar{A}^{(II,\,I)}$ and $\bar{B}^{(I,\,II)} - \bar{B}^{(II,\,I)}$ vanish without discontinuity when the boundaries of the two regions gradually are made to coincide. From this it follows that the averages of all field components over the same space-time region commute, and thus should be exactly measurable, independently of one another. In fact, this consequence of the theory, which goes essentially beyond the presupposition of unrestricted measurability of each single field quantity, appears as a special case of two more general theorems which follow from the symmetry properties of the quantities $\bar{A}^{(I,\,II)}$ and $\bar{B}^{(I,\,II)}$. For the fact that the expressions $A^{(12)} - A^{(21)}$ change their sign when the times t_1 and t_2 are interchanged, implies that the averages of two components of the same kind (i.e. two electric or two magnetic components) over two arbitrary space-time regions always commute if the associated time intervals coincide. Further, the corresponding antisymmetry of the expressions $B^{(12)}$ and $B^{(21)}$ with respect to interchange of the spatial points (x_1, y_1, z_1) and (x_2, y_2, z_2) implies that the averages of two components of different kind, e.g. $\mathfrak{E}_x$ and $\mathfrak{H}_y$, over two arbitrary time intervals commute when the corresponding spatial regions coincide.

At first sight these results might seem incompatible with the commutation relations between averages of field quantities at one and the same time and over finite space regions, which can be derived from the Heisenberg–Pauli representation of the formalism and are discussed in the book by Heisenberg previously cited. While it is stated there too that averages of components of the same kind commute, it is deduced that components of different kind over one and the same region of space do not commute. However, this contradiction is solved easily by noting that Heisenberg's treatment involves a limiting process in which two originally non-coinciding space-time regions are brought into coincidence only after their temporal extension has been contracted to one and the same instant of time. For from the symmetry of the expression (2) for $B_{xy}^{(1,2)}$ with respect to t_1 and t_2, and from the property of the δ-function stated in (3), we find in the case of coinciding time intervals

$$(9) \qquad \bar{B}_{xy}^{(I, II)} - \bar{B}_{xy}^{(II, I)} = \frac{2}{V_I V_{II} T^2} \cdot \frac{1}{c} \int\limits_{V_I} dv_1 \int\limits_{V_{II}} dv_2 \frac{\partial}{\partial z_1} \left(\frac{1}{r}\right),$$

where we have put $T_I = T_{II} = T$ and where the double integration is over all pairs of points, one in each of the spatial regions, whose separation r is smaller than cT. If we now further assume that the two spatial regions have the same volume, $V_I = V_{II} = V$, and the same shape, but are displaced relative to each other in the z-direction, then in the limit where cT can be considered negligible compared with the linear extensions of the space regions, the space integral in (9) gives by partial integration $\pm 2\pi c^2 T^2 F$, where F is the area enclosed by the projection on the xy-plane of the curve in which the surfaces of V_I and V_{II} intersect, and where the sign is + or − depending on whether region II is displaced relative to region I in the positive or the negative z-direction. Thus, if the regions are displaced continuously through each other the difference $\bar{B}_{xy}^{(I, II)} - \bar{B}_{xy}^{(II, I)}$ undergoes a discontinuous change of $8\pi cF/V^2$, while both expressions $\bar{B}_{xy}^{(I, II)}$ and $\bar{B}_{xy}^{(II, I)}$ change their sign. Therefore, the commutation relation between the instantaneous spatial averages of $\mathfrak{E}_x$ and $\mathfrak{H}_y$ in the limiting case under discussion displays an essential ambiguity, which is responsible for the apparent contradiction mentioned above.

Furthermore, as we shall see, the supposed demonstration, on the basis of previous investigations of the physical possibilities of measurement, that there is a complementary limitation on the measurability of field components of different kinds in one and same region of space, also depends entirely on the use of point charges as test bodies, which does not permit a sufficiently sharp delimitation of the region of measurement. As we have already emphasized, only measurements employing test bodies with a charge distribution of finite extension should be considered for the testing of the quantum-electromagnetic formalism, since every well-defined statement of this formalism refers to averages of field components over finite space-time regions. Yet this circumstance in no way prevents us from testing by field measurements all unambiguous conclusions that can be drawn from the Heisenberg–Pauli representation concerning the time dependence of spatial field averages. For this purpose we need only introduce averaging regions whose temporal extension T, multiplied by c, is sufficiently small compared to their linear spatial dimensions, the order of magnitude of which we shall in the following always denote by L.

In fact, the case $L > cT$ is particularly suited for a thorough testing of the consequences of the formalism in the properly quantum-theoretical domain. Of course, already in the domain of validity of classical theory the case $L \leqslant cT$ is of little interest, since all the peculiarities of the wave fields inside the volume V are smoothed out to a large extent by the averaging procedure, because of the propagation during the time interval T. In addition to this smoothing out, there are, in the quantum domain, the peculiar fluctuation phenomena which derive from the basically statistical character of the formalism. As we shall see shortly, these fluctuations, while essentially entering the solutions of problems in the case $L \leqslant ct$, play a comparatively small role in the case $L > cT$.

The fluctuations in question are intimately related to the impossibility, which is characteristic of the quantum theory of fields, of visualizing the concept of light quanta in terms of classical concepts. In particular, they give expression to the mutual exclusiveness of an accurate knowledge of the light quantum composition of an electromagnetic field and of knowledge of the average value of any of its components in a well-defined space-time region. Let the density $\omega_i(\kappa_x, \kappa_y, \kappa_z)$ of light quanta with definite polarization parameter i and given momentum and energy, $\hbar\kappa_x, \hbar\kappa_y, \hbar\kappa_z$ and $\hbar\nu = \hbar c \sqrt{\kappa_x^2 + \kappa_y^2 + \kappa_z^2}$ be known; then the expectation value of all field averages are indeed zero, but the mean square fluctuation of each field quantity, such as $\mathfrak{E}_x^{(G)}$ defined in (4), is given by the easily derivable formula

(10)
$$
\begin{cases}
S(G) = \dfrac{1}{V^2 T^2} \cdot \dfrac{\hbar}{3} \int_T dt_1 \int_T dt_2 \int_V dv_1 \int_V dv_2 \dfrac{\partial^2}{\partial t_1 \, \partial t_2} \int_{-\infty}^{\infty} \left(\sum_i \omega_i + 1 \right) \\[2ex]
\cos\left[\kappa_x(x_1 - x_2) + \kappa_y(y_1 - y_2) + \kappa_z(z_1 - z_2) - \nu(t_1 - t_2) \right] \\[2ex]
\dfrac{d\kappa_x \, d\kappa_y \, d\kappa_z}{\nu}
\end{cases}
$$

366

From formula (10) we see that for a given light quantum composition the fluctuations in question can never vanish, since even when $\omega_i = 0$, i.e., in the complete absence of light quanta, they assume a finite positive value, which by an easy calculation can be put in the form

$$(11) \qquad S_0(G) = \frac{2}{3\pi^2} \frac{\hbar c}{V^2} \int\limits_V dv_1 \int\limits_V dv_2 \frac{1}{r^2[(cT)^2 - r^2]} .$$

For every other light quantum distribution defined by a given density ω_i, the mean square fluctuation of the average of a field component becomes larger than $S_0(G)$. However, the fluctuations of a field average expected according to the formalism can be arbitrarily small when a direct know-ledge of field quantities, obtained by measurement for example, is assumed. In such a case the light quantum density ω_i is obviously not well defined, and we must content ourselves with statistical statements about this density.

Furthermore, it is of decisive importance for the discussion of the measurement possibilities that the expression (11) holds not only for the field fluctuations in a region free of light quanta, but also represents the mean square fluctuation of each field average in the more general case where only classically describable current and charge distributions occur as sources of the field. The state of the field is then uniquely defined by the requirement that the expectation value of every field quantity coincide with the classically computed value, and that the numbers of light quanta of given momentum and polarization be distributed around their mean value n_0, estimated by means of the correspondence argument, according to the probability law valid for independent events

$$(12) \qquad w(n) = \frac{n_0^n e^{-n_0}}{n!}$$

An easy calculation shows that the field fluctuations in this state are given just by the expression (11). Moreover, in correspondence with the characteristics of black-body fluctuations, it follows that also in the general case of a field of given light quantum composition, the inclusion of field effects of any classically describable sources will have no influence on the fluctuation phenomena.

The square root of the expression (11) may be regarded as a critical field strength, $\mathfrak{S}$, in the sense that only when considering field averages essentially larger than $\mathfrak{S}$ are we allowed to neglect the corresponding fluctuations. To assess the possibilities of testing the formalism in the properly quantum domain, still another critical field size, $\mathfrak{U}$, is relevant, which equals the square root of the products, given by (8), of the complementary uncertainties of two field averages over space-time regions that only partially coincide, being displaced relative to each other by spatial and temporal distances of order of magnitude L and T, respectively. For when the field strengths are essentially larger than $\mathfrak{U}$ we obviously enter the domain of validity of classical electromagnetic theory, where all quantum mechanical features of the formalism lose their significance. A simple estimate based on formulae (8) and (11), shows that in the case $L \lesssim cT$ both critical expressions $\mathfrak{U}$ and $\mathfrak{S}$ are of the same order of magnitude

$$(13) \qquad \mathfrak{U} \sim \mathfrak{S} \sim \frac{\sqrt{\hbar c}}{L \cdot cT}$$

On the other hand, in the case $L > cT$ one has

$$(14) \qquad \mathfrak{U} \sim \sqrt{\frac{\hbar}{L^3 T}} \text{ and } \mathfrak{S} \sim \frac{\sqrt{\hbar c}}{L^2},$$

so that in the limiting case $L \gg cT$ the critical field strength $\mathfrak{U}$ is much larger than $\mathfrak{S}$ and, therefore, in testing the characteristic consequences of the formalism we can to a large extent disregard the field fluctuations.

Before we turn to the comparison of the consequences of the quantum electromagnetic formalism discussed in this section, with the physical measurement possibilities for field quantities, we want to emphasize here once again that the consistent interpretability of this formalism is in no way endangered by such paradoxical features of its mathematical representation as the infinite zero-point energy. In particular, this latter paradox, which moreover can be removed by a formal change in the representation[6] that does not influence the physical interpretation, has no direct connection with the problem of measurability of field quantities. In fact, a field-theoretic determination of the electromagnetic energy in a given space-time domain would require knowledge of the values of the field components at each space-time point of a region, which are inaccessible to measurement. A physical measurement of the field energy can be carried out only by means of a suitable mechanical device that would make it possible to separate the electromagnetic fields in a given region from the rest of the field, so that the energy contained in the region could be measured subsequently by application of the conservation law. However, because of the interaction with the measuring mechanism, any such separation of the fields would be accompanied by an uncontrollable change in the field energy in the region in question, the consideration of which is essential for clarification of the well-known paradoxes that arise in the discussion of energy fluctuations in black-body radiation.[7]

3. Presuppositions for physical field measurements

The measurement of electromagnetic field quantities rests by definition on the transfer of momentum to suitable electric or magnetic test bodies situated in the field. Quite apart from the caution required by quantum theory in applying the customary idealization of field components defined in each space-time point, we are here always concerned with averages of these components over the finite time intervals necessary for the momentum transfer as well as over the spatial domains in which the electric charges or magnetic pole strengths of the test bodies in question are distributed. Obviously, even the assumption of a uniform charge distribution on a test body is an idealization, subject to a certain restriction because of the atomic constitution of all material bodies, but indispensable for the unambiguous definition of field quantities.

In order to have a definite case in mind we consider the measurement of the average of the electrical field component in the x-direction, $\mathfrak{E}_x$, over a space-time domain of volume V and duration T. For this purpose we therefore use a test body whose electric charge is uniformly distributed over the volume V with a density ρ, and determine the values p'_x and p''_x of this body's momentum components in the x-direction at the beginning t' and at the end t'' of the interval T. The average $\mathfrak{E}_x$ we are looking for is then determined by the equation

$$(15) \qquad p''_x - p'_x = \rho \mathfrak{E}_x VT,$$

where it is assumed that the time intervals required for the momentum measurements, whose order of magnitude we shall denote by Δt, can be regarded as negligibly small compared to T, and that we can disregard the displacements suffered by the test body due to the momentum measurements as well as the acceleration given to it during the time interval T by the field that is being measured, in comparison with the linear dimensions L of the spatial domain V.

By choosing a sufficiently heavy test body we can obviously make its acceleration due to the field arbitrarily small. In the momentum measurements, however, we encounter conditions which are independent of the mass of the test body. As a consequence of the indeterminacy principle, any measurement of the momentum component p_x carried out with the accuracy Δp_x is accompanied by a loss Δx in the knowledge of the position of the body in question, the order of magnitude of which is given by the relation contained in (6)

(16) $\Delta p_x \, \Delta x \sim \hbar.$

Nevertheless, in itself this state of affairs does not imply any restriction on the accuracy to be achieved by the field measurement, because we still have at our disposal the value of the charge density. In fact, if we neglect Δt and Δx in comparison with T and L, we get from (15) and (16) for the order of magnitude of the accuracy $\Delta \bar{\mathfrak{C}}_x$ of the field measurement

.(17) $$\Delta \bar{\mathfrak{C}}_x \sim \frac{\hbar}{\rho \Delta x \cdot VT},$$

which for any value of Δx, however small, can be made arbitrarily small by choosing a sufficiently large value of ρ.

Strictly speaking, the accuracy of the field measurement is also dependent on the absolute magnitude of the value of $\mathfrak{C}_x$ itself, for with given ranges of Δt and Δx the value of $\mathfrak{C}_x$ ascertained from (15), even if Δp_x were zero, would be affected with an uncertainty arising from the latitude in the delineation of the domain of measurement which would surpass any limit as $\mathfrak{C}_x$ increases indefinitely. Yet, the latter circumstance merely reflects the general limitation on all physical measurements, for which a knowledge of the order of magnitude of the expected effects is always required in order to choose the appropriate measuring instruments. In our problem an upper limit to the effects that we are interested in is set by the fact that as the magnitude of the field components increases we gradually reach the domain of validity of classical electromagnetic theory. As mentioned in the previous section, in the case $L > cT$, which is particularly suited for testing the quantum electromagnetic formalism, the expression

$$(18) \qquad Q=\sqrt{\frac{\hbar}{VT}},$$

which is equivalent to the right-hand side of the first formula (14), represents a critical field size in this respect. Substituting this into (17), the latter relation assumes the form

$$(19) \qquad \Delta\mathfrak{E}_x \sim \lambda Q,$$

where

$$(20) \qquad \lambda=\frac{Q}{\rho\Delta x}$$

is a dimensionless factor determining the accuracy of the field measurement.

The requirement that λ be small compared to unity and simultaneously Δx be small compared to L means that the total electric charge of the test body must consist of a large number of elementary charges ε. In fact, according to (20) this number is given by

$$(21) \qquad N=\frac{\rho V}{\varepsilon}=\frac{QV}{\lambda\varepsilon\Delta x}=\frac{1}{\lambda}\cdot\frac{L}{\Delta x}\cdot\sqrt{\frac{L}{cT}}\cdot\sqrt{\frac{\hbar c}{\varepsilon^2}}$$

and is very large when the above requirements are fulfilled and when $L>cT$ as assumed. The last factor is of course the reciprocal square root of the fine-structure constant whose smallness, as we already mentioned in the Introduction, is an essential presupposition for the correspondence approach to electron theory. As emphasized there, essential restrictions are imposed on a field measurement with an elementary charge as a test body, a fact which is also directly visible from (21) if one puts $N=1$.[8] Moreover, the assumption of a large value of N is a necessary condition for the physical realization of a uniform distribution of the charge of the test body over the volume V; and as long as the linear dimensions of the test body are large compared to the atomic dimensions, its fulfillment obviously presents no difficulties in principle. It need hardly be mentioned that with this presupposition the assumption used above about the mass of the test body, equivalent to the requirement that this mass be very large compared to that of a light quantum of wave length L, always can be satisfied.

Thus far we have completely disregarded the electromagnetic field effects which accompany the acceleration of any test body during the momentum measurement. These effects superpose themselves on the original field and must be included in the field averages defined by equations of type (15). Hence, the main task of the following investigation will be to find a measuring arrangement in which the field effects of the test bodies can be controlled or compensated to the largest possible extent. Yet here we must first of all discuss the question of whether the reaction of the radiation fields produced by the acceleration of the test body in the momentum measurements could impair even the practicability of measuring the values occurring in (15) of the test body's momentum components at the beginning and end of the measuring interval. It was just this possibility that led Landau and Peierls, in the work cited at the beginning, to doubt the reliability of the indeterminacy relation (16) for charged bodies and to conclude that it should be replaced by another even more restrictive relation in which the charge of the test body enters in an essential way. However, they likened the electromagnetic behavior of such a body to that of a point charge e, and consequently used the following expression for estimating the order of magnitude of the test body's momentum change, brought about by radiation recoil, during the time Δt

$$(22) \qquad \delta_e p_x \sim \frac{e^2}{c^3} \frac{\Delta x}{\Delta t^2}.$$

If, however, $\delta_e p_x$ is considered an additional indeterminacy of the momentum measurement, then if one puts $\rho V = e$ and does not distinguish between $\mathfrak{E}_x$ and $\mathfrak{E}_x$, one gets instead of (17)

$$(23) \qquad \Delta_e \mathfrak{E}_x \sim \frac{\hbar}{eT\Delta x} + \frac{e\Delta x}{c^3 T\Delta t^2},$$

whose minimum under variation of e is obviously given by

$$(24) \qquad \Delta_m \mathfrak{E}_x \sim \frac{\sqrt{\hbar c}}{c^2 T\Delta t}.$$

If, still following Landau and Peierls, one does not distinguish between T and Δt, this expression agrees with the absolute limit on the measurability of field components that they gave, on which they based their criticism of the foundations of the quantum electromagnetic formalism.

However, the supposed difficulties of the momentum measurement disappear as soon as sufficient account is taken of the finite extension of the test body's electric charge. Using the idealization of a uniform, rigidly displaceable charge distribution, to be more closely examined below, the electric field strengths in the region V during the acceleration of the test body within the time Δt can at most reach a value of the order of magnitude $\rho\Delta x$; since, according to Maxwell's equations, their time derivatives are at most of the same order of magnitude as that of the current density, given by $\rho(\Delta x/\Delta t)$. Hence, any electromagnetic reaction on the body during the measuring interval Δt can only contribute a momentum transfer of the order of magnitude

$$(25) \qquad \delta_\rho p_x \sim \rho^2 V \Delta x \Delta t.$$

Thus, in view of (18) and (20), we get by comparison of (16) and (25)

$$(26) \qquad \delta_\rho p_x \sim \Delta p_x \cdot \lambda^{-2} \frac{\Delta t}{T},$$

which implies that for any desired accuracy of the field measurement, symbolized by a given value of λ, the influence of the electromagnetic reaction on the momentum measurement of the test body can be neglected if only Δt is chosen sufficiently small in comparison with T. It is precisely this circumstance that is decisive for assessing the accuracy of the field measurements; for it turns out to be impossible to directly take into account the influence of the radiation reaction on the momentum and energy balance in the individual momentum measurements. For example, Pauli's proposal[9] to measure subsequently by means of a special device the momentum and energy contained in the emitted radiation would already be impracticable because of the fact that, at least in the case $L > cT$ which is of particular importance for field measurements, the radiation fields that are produced in the momentum measurements at the beginning and at the end of the interval T cannot be separated from each other to the degree sufficient for this purpose. In fact, in the following sections we shall show quite generally that any attempt at such a control of the test body's field effects would essentially impair the realization of the field measurement in question.

Besides, not only for discussing the behavior of an individual test body during the measurement but also for assessing the mutual influence of several test bodies, it is essential to treat these not as point charges but as continuous charge distributions. This is because the customary identification of the position indeterminacy of a test body, considered as a point charge, with the linear dimensions of the domain of measurement is an arbitrary assumption that is foreign to the measurability problem. For this reason, not only do the estimates of the product of the uncertainties of $\mathfrak{E}_x$ and $\mathfrak{H}_y$ inside the same space-time domain, obtained by Heisenberg and by Landau and Peierls by considering point charges, deviate from the predictions of the quantum-electromagnetic formalism as already mentioned; they are in agreement with each other only in the special case $L \sim cT$. In this case, both estimates give the expression Q^2, which corresponds to the order of magnitude, to be expected from the formalism, of the value of the product of the complementary indeterminacies of two field averages in space-time regions that are displaced relative to each other through space-time distances of the same order of magnitude as L and T. Moreover, it is an essential feature of the formalism that the product in question vanishes identically for coinciding regions. The physical meaning of this result becomes obvious as soon as one takes into account the uniform charge distribution of the test body used to measure $\bar{\mathfrak{E}}_x$; since the magnetic field strength which is produced at a point P_2 of the volume V by the displacement of the charge ρdv contained in a volume element situated at the point P_1 is exactly equal and opposite to the magnetic field strength produced at the point P_1 by the same displacement of the charge ρdv at the point P_2, so that the average over the volume V of every magnetic field component produced by the displacement of the test body disappears.

From the foregoing it emerges that in the investigation of the measureability of field quantities it is decisively important to assume that the test bodies to be used behave like uniformly charged rigid bodies whose momenta can be measured, in any given arbitrarily small time interval, with an accuracy, expressed by (16), complementary to the accompanying uncontrollable displacement. Of course, in view of the finite propagation of all forces, we should not think here of the usual mechanical idealization of rigid bodies, but must think of every test body as a system of individual components of sufficiently small dimensions; and think of the measurement of the total momentum of this system as carried out in such a way that, to a sufficient approximation, all the components undergo the same displacement during the momentum measurement. That this requirement can be fulfilled without difficulties of principle, at least insofar as one can disregard the atomic constitution of the test body, is due to the fact that the required momentum measurements can be fully described on a classical basis, irrespective of whether they depend on looking at a collision process between the test body and a suitable material colliding body; or, say, on the study of the Doppler effect involved in the reflection of radiation from the test body. For, if only the mass of the colliding body is sufficiently large or if the packet of radiation that is used to measure the Doppler effect contains a sufficient number of light quanta, then the interaction between the test body and the colliding body can be described classically to any approximation. In fact, the loss of knowledge of the position of the test body, which accompanies the momentum measurement, is due solely to the impossibility of simultaneously fixing the course of the collision process relative to a well-defined space-time reference frame. Indeed, the peculiar complementarity of the mode of description ultimately derives from the fact that any such fixation is bound up with an unavoidable, and in principle uncontrollable, transfer of energy and momentum to the scales and clocks needed to establish the co-ordinate system.[10]

We recall that, according to the indeterminacy principle, the latitude in the time Δt that is left open in any description and the accuracy with which the energy exchanged in the collision process between colliding body and test body is known are connected by the well-known relation

$$(27) \qquad \Delta E \cdot \Delta t \sim \hbar.$$

Because of the relation between energy and momentum and velocity components

$$(28) \qquad dE = v_x \, dp_x$$

which is valid for both bodies, it follows directly that

$$(29) \qquad \Delta p_x |v''_x - v'_x| \Delta t \sim \hbar.$$

Even though, as noted above, the change of velocity $|v''_x - v'_x|$ of the test body in the momentum measurement can be considered as arbitrarily well known for a sufficiently heavy test body, the factor

$$(30) \qquad |v''_x - v'_x| \Delta t = \Delta x$$

obviously implies a completely free latitude in the position of the body relative to the fixed frame of reference, in complete agreement with the indeterminacy relation (16). From (30) the condition

(31) $\Delta x < c \Delta t,$

follows immediately, which, because of (16), imposes an absolute limit on the accuracy Δp_x that can be achieved in a momentum measurement with a given upper limit for the time latitude Δt. However, in view of the relativistic invariance of the relations (16) and (27) and in particular of formula (28), this circumstance implies no restriction in the formulation and applicability of the indeterminacy principle. In our problem it is even permissible to disregard all mechanical relativistic effects, for by using sufficiently heavy test bodies we can always arrange that the velocities of all test bodies during the whole measurement process remain small compared to the velocity of light. Consequently, we can even consider any displacement Δx in the momentum measurements very small compared to the corresponding value of $c \Delta t$ which itself must be chosen arbitrarily small.

It is exactly the possibility of accurately tracing the relative space-time course of the process serving as momentum measurement that enables us to measure the total momentum of an extended body within any given time interval with the required accuracy expressed by (16). Thus, we can determine the total momentum of the system of charged material component bodies serving as test body by a single collision process if we make use of a colliding body of special construction which intervenes everywhere in the test body system and gives every component the same acceleration at the same time. It is true that this device imposes severe demands on the construction of the colliding and test bodies, but these demands do not present any difficulties of principle as long as we can disregard the atomic constitution of the bodies. The measurement of the total momentum of the test body would presumably be performed most simply by optical means, i.e. by determination of the Doppler effect; for this purpose one might proceed as follows: imagine that every component body is equipped with a small mirror at right angles to the x-direction, and that a number of other mirrors are placed in fixed positions in such a way that the light path from the radiation source to each component body is the same. If now by means of a suitable device we produce a packet of radiation of duration Δt and containing a number of light quanta sufficiently large compared to the number of component bodies, then all these bodies will suffer a collision simultaneously and undergo an acceleration which for all component bodies can be made equal with arbitrary accuracy.

In order to show that one can in fact measure the total momentum of the test body with an accuracy satisfying relation (16) by means of such an arrangement, we shall consider somewhat more closely the interaction between the test body system and the packet of radiation. In view of the assumed smallness of the velocity of the test body compared to the velocity of light, we have for each component body

$$(32) \quad \begin{cases} m_\tau(v''_{\tau,\,x} - v'_{\tau,\,x}) = \dfrac{\hbar}{c} \sum_{n_\tau} (v' + v''), \\[2mm] \dfrac{1}{2} m_\tau(v''^{\,2}_{\tau,\,x} - v'^{\,2}_{\tau,\,x}) = \hbar \sum_{n_\tau} (v' - v''), \end{cases}$$

where m_τ denotes the mass of a component body, $v'_{\tau,\,x}, v''_{\tau,\,x}$ are its velocity before and after the reflection, and where the summation extends over the n_τ light quanta reflected from the component body whose frequency (reciprocal period times 2π) before and after the reflection are denoted by v' and v'', respectively. It follows from (32) that the momentum of the component body in question before and after the collision is

$$(33) \quad \left.\begin{aligned} p'_{\tau,\,x} &= m_\tau v'_{\tau,\,x} = \\[2mm] p''_{\tau,\,x} &= m_\tau v''_{\tau,\,x} = \end{aligned}\right\} m_\tau c \,\frac{\displaystyle\sum_{n_\tau}(v' - v'')}{\displaystyle\sum_{n_\tau}(v' + v'')} \mp \frac{1}{2}\frac{\hbar}{c}\sum_{n_\tau}(v' + v''),$$

If we now assume that the mean spectral frequency v_0 of the radiation packet is very large compared to the mean width $(\Delta t)^{-1}$ of its frequency distribution as well as to all frequency changes $v' - v''$, then we can take the change of velocity of the component body in the collision, to a sufficient approximation, as

$$(34) \quad v''_{\tau,\,x} - v'_{\tau,\,x} = \frac{\hbar}{m_\tau c}\sum_{n_\tau}(v' + v'') = \frac{2n_\tau \hbar v_0}{m_\tau c}$$

and we can assume that it is the same for all component bodies. Thus, in the collision all component bodies suffer displacements which, although

[142]

uncontrollable, are arbitrarily close to being identical and whose order of magnitude Δx satisfies relation (30) where $|v''_x - v'_x|$ is to be identified with the common velocity change of the whole test body system. Consequently, since according to our assumptions Δx can be considered negligibly small compared to $c\Delta t$, we get from (33) and (34), for the product of Δx and the uncertainty of the total momentum of the test body, approximately

$$(35) \qquad \Delta p_x \Delta x \sim \Delta t \cdot \Delta \left(\sum_\tau \sum_{n_\tau} \hbar v' - \sum_\tau \sum_{n_\tau} \hbar v'' \right).$$

The quantities in the bracket in (35) are precisely the total energies of the radiation packets impinging on and reflected from the test body. The energy of the latter packet can be measured with arbitrary accuracy, e.g. by spectral analysis of the reflected radiation. For the incoming radiation packet, however, such an analysis would obviously be incompatible with the experimental conditions. Yet the total energy of this radiation can be measured with an accuracy that is complementary to Δt, as given by (27). To do that it suffices to use a purely mechanical device by means of which the packet in question is separated from a radiation field, whose energy before and after the separation can be determined with arbitrary accuracy, e.g. by spectral measurements. Thus, the relation (35) is identical with the usual indeterminacy relation (16). Note further that the demonstration of this identity is essentially dependent on the fact that in accordance with the described arrangement we obtain no information about the momentum of the single component bodies but only about the total momentum of the test body.

The fact that the test body system suffers a common translation during the required momentum measurements is not only important for the calculation of the field effects of the test body which accompany these measurements, but also gives us the possibility to arrange things in such a way that outside the short time intervals occupied by the momentum measurements all the test bodies employed in the field measurement can be considered as being at rest, which greatly simplifies the calculation. For immediately after each momentum measurement, i.e. practically speaking still inside the interval Δt, we can give the test body system a second push in the opposite direction by means of a suitable device, such that the velocity change which every component body suffered in the first collision is cancelled out; this can be done with an arbitrary accuracy, i.e.

an accuracy that is inversely proportional to the mass of the component body, and without losing the desired knowledge of the total momentum of the test body. However, with this arrangement it is impossible to know the time interval between the two collision processes with a latitude smaller than Δt, and so, as required by the indeterminacy principle, the test body is not returned to its original position by the counter-collision, but rather is brought to rest to the required approximation at an unknown position, displaced by a distance of the order Δx.

To assess the complementary limitations on the measurability of field quantities, which will be more closely investigated in the following sections, we must follow the behavior of the test body as accurately as possible during the whole measuring process. It turns out that for this purpose it is necessary first of all to know accurately the position of each test body at all times before and after its use in the measurement. This is achieved most expediently by having the test body firmly attached to a rigid frame serving as a spatial reference system, except in the time interval during which the momentum transfer to the test body from the field is to be determined. At the beginning of this interval the attachment must be disconnected, and the momentum component of the test body in the direction of the field component that is to be determined must be measured. We always assume that by an immediately following counter-impulse, as discussed above, the body is brought back to rest with an accuracy inversely proportional to its mass, at a position which is not accurately predictable. At the end of the time interval and after renewed measurement of the momentum component in question, the firm attachment is re-established; here it turns out to be not unessential that the test body be brought back into exactly the same position as it had originally. If the space-time averaging domains are to be sufficiently sharply defined, these prescriptions alone impose far-reaching demands on the detailed construction of the test body system. For due to the retardation of all forces it is strictly speaking necessary that the severance as well as the re-establishment of the attachment of the test body system to the fixed frame are performed in such a way that all its independent component bodies, whose linear dimensions must be at least as small as the smallest relevant value of $c\Delta t$, are unfastened and fastened simultaneously, i.e. within the time latitude Δt of the momentum measurement, which itself must be chosen sufficiently small compared to the time interval T.

Still more far-reaching demands on idealization with respect to the construction and handling of the test body system are obviously needed to measure field averages over two partially overlapping space-time regions. For in this case we must have test bodies at our disposal which can be displaced inside each other without mutual mechanical influence. In order that the electromagnetic field to be measured be disturbed as little as possible by the presence of the test body system, we shall imagine, moreover, that every electric or magnetic component body is placed adjacent to a neutralization body with exactly the opposite charge. In the case of a magnetic test body system it is to be noted that a uniform pole strength distribution cannot exist on a strictly delimited body. However, one can imagine, at least in principle, that every component body of such a system is connected with the corresponding neutralization body by magnetizable flexible threads. All these neutralization bodies are to remain connected with the fixed frame during the whole measurement process without mechanically influencing the free mobility of the component bodies belonging to the test body system proper. Of course, the idealizations entailed in such presuppositions, as well as in the still needed compensation mechanisms to be introduced below, are justifiable only as long as we can neglect the atomic constitution of the test body. However, as already mentioned, this neglect does not imply any restrictions in principle on the possibility of testing the quantum electromagnetic formalism, since no universal space-time dimensions appear in this formalism. Accordingly, the purpose of the preceding considerations was above all to show that in the purely mechanical problems which are relevant to the field measurements, it is possible to distinguish strictly between the restrictions on the constitution of the test body stemming from the atomic structure of matter and the restrictions on the handling of these bodies that are due to the quantum of action, formulated in particular in the principle of indeterminacy.

4. CALCULATION OF THE FIELD EFFECTS OF THE TEST BODY

After having investigated the physical presuppositions for the constitution of the test body we now turn to a closer consideration of the electromagnetic field effects of the test body which accompany the measurement of field quantities and which are of decisive importance for the measurability question. In accordance with the discussion above we shall treat each test body as a charge distribution which uniformly fills the spatial averaging domain and which undergoes a simple translation during the momentum measurement. We shall first carry out the calculation of the electromagnetic fields thereby produced on the basis of classical electrodynamics, and only afterwards discuss the restriction on the validity of this treatment due to the quantum of action.

Let us consider two space-time regions, I and II, with volumes V_I and V_{II} and durations T_I and T_{II}, and let us ask for the electromagnetic field 'which is produced at a point (x_2, y_2, z_2, t_2) of region II by a measurement of the average of $\mathfrak{E}_x$ over the region I. Thus, we assume that in volume V_I there are originally two electric charge distributions with the constant densities $+\rho_I$ and $-\rho_I$. In the interval from t'_I to $t'_I + \Delta t_I$ the first charge distribution experiences a simple non-uniform translation in the x-direction through a distance $D_x^{(I)}$; in the interval from $t'_I + \Delta t_I$ to t''_I it remains at rest at the displaced position; finally, in the interval from t''_I to $t''_I + \Delta t_I$ it moves non-uniformly parallel to the x-axis back to its original position, which coincides with that of the neutralization distribution. In accordance with the requirement discussed in the preceding sections, we assume further that Δt_I is very small compared to $T_I = t''_I - t'_I$ and that $D_x^{(I)}$ is very small not only compared to the linear dimensions of the spatial averaging region of volume V_I, but also small compared with $c\Delta t_I$.

Hence, in the limiting case of vanishingly small Δt_I, the sources of the field that we are looking for may be represented as a polarization in the x-direction of constant density, $P_x^{(I)} = \rho_I D_x^{(I)}$, existing in the region I during the time interval from t'_I to t''_I, as well as a current density present only in the immediate vicinity of the times t'_I and t''_I, which we can write as

$$(36) \qquad J_x^{(I)} = \rho_I D_x^{(I)} [\delta(t - t'_I) - \delta(t - t''_I)],$$

using δ symbol defined by formula (3). By means of the same symbol we can similarly express the polarization at an arbitrary time t by the formula

$$(37) \qquad P_x^{(I)} = \rho_I D_x^{(I)} \int_{t'_I}^{t''_I} \delta(t - t_1)\, dt_1.$$

As is well-known, the components of the fields at the space time point

(x_2, y_2, z_2, t_2) produced by these sources may be calculated from the formulae

(38)
$$
\begin{cases}
E_x^{(l)} = -\dfrac{\partial \varphi^{(l)}}{\partial x_2} - \dfrac{1}{c}\dfrac{\partial \psi_x^{(l)}}{\partial t_2}, \quad E_y^{(l)} = -\dfrac{\partial \varphi^{(l)}}{\partial y_2}, \quad E_z^{(l)} = -\dfrac{\partial \varphi^{(l)}}{\partial z_2}, \\[2ex]
H_x^{(l)} = 0, \qquad\qquad\qquad H_y^{(l)} = \dfrac{\partial \psi_x^{(l)}}{\partial z_2}, \quad H_z^{(l)} = -\dfrac{\partial \psi_x^{(l)}}{\partial y_2},
\end{cases}
$$

where we have used latin letters to distinguish these fields from the field components that are to be measured. In (38), $\varphi^{(l)}$ signifies the retarded scalar potential

(39)
$$
\varphi^{(l)} = \int_{V_1} \frac{\partial}{\partial x_1} \left[\frac{P_x^{(l)}(t_2 - r/c)}{r} \right] dv_1
$$

and $\psi_x^{(l)}$ the retarded vector potential component

(40)
$$
\psi_x^{(l)} = \frac{1}{c} \int_{V_1} \frac{J_x^{(l)}(t_2 - r/c)}{r} \, dv_1,
$$

where r is the distance between the space points (x_1, y_1, z_1) and (x_2, y_2, z_2). Noting that the expression (36) can also be written in the form

(41)
$$
J_x^{(l)} = -\rho_1 D_x^{(l)} \int_{t_1'}^{t_1''} \frac{\partial}{\partial t_1} \delta(t - t_1) \, dt_1
$$

and taking into account (37) and (41) one sees that the field components given by (38), (39) and (40) can be expressed by the typical formulae

(42)
$$
\begin{cases}
E_x^{(l)} = \rho_1 D_x^{(l)} \int_{V_1} dv_1 \int_{T_1} dt_1 A_{xx}^{(12)}, \quad E_y^{(l)} = \rho_1 D_x^{(l)} \int_{V_1} dv_1 \int_{T_1} dt_1 A_{xy}^{(12)}, \\[2ex]
H_x^{(l)} = 0, \qquad\qquad\qquad H_y^{(l)} = \rho_1 D_x^{(l)} \int_{V_1} dv_1 \int_{T_1} dt_1 B_{xy}^{(12)},
\end{cases}
$$

where the abbreviations defined in (2) have been used.

[147]

In view of the properties of the symbolic δ-function it is easy to see that the field components given by (42) always remain finite and even cannot surpass a value of the order of magnitude $\rho_1 D_x^{(1)}$ at any space-time point (x_2, y_2, z_2, t_2). As already mentioned, the electromagnetic forces which occur during the momentum measurement of the test body in the time interval Δt are just of this order of magnitude (cf. p. 372). The fact that the field intensities do not subsequently increase essentially is solely a consequence of the counter collision taking place right after the momentum measurement which brings the body back to rest, and finds its idealized mathematical expressions in (36) and (37).

The averages of these field components over the region II, which are of particular interest to us, are obtained from (42) by a simple space-time integration and, in accordance with (5), are given by the formulae

$$(43) \qquad \begin{cases} \bar{E}_x^{(\mathrm{I,\,II})} = D_x^{(1)}\rho_1 V_1 T_1 \bar{A}_{xx}^{(\mathrm{I,\,II})}, & \bar{E}_y^{(\mathrm{I,\,II})} = D_x^{(1)}\rho_1 V_1 T_1 \bar{A}_{xy}^{(\mathrm{I,\,II})} \\ \bar{H}_x^{(\mathrm{I,\,II})} = 0, & \bar{H}_y^{(\mathrm{I,\,II})} = D_x^{(1)}\rho_1 V_1 T_1 \bar{B}_{xy}^{(\mathrm{I,\,II})}. \end{cases}$$

As a result of the properties of the expressions $\bar{A}$ and $\bar{B}$, already discussed in Section 2, we see that for a given value of $D_x^{(1)}$ the field averages given by (43) are well-defined continuous functions of the regions I and II. Hence, for decreasing latitudes, Δt and Δx, of the duration of the momentum measurements and of the accompanying unpredictable displacements, these field averages are completely independent of the detailed space-time course of the collision process, and simply proportional to the constant displacement of the test body in the measuring interval T_1. As we shall see, this very fact turns out to be decisive for the possibility of an extensive compensation of the uncontrollable field effects of the test bodies.

Thus far, the calculation of the field effects has been carried out on a purely classical basis. Yet for a more detailed comparison of the measurement possibilities and the requirements of the quantum-electromagnetic formalism one must also take into account the limitation imposed on the classical mode of calculation by the quantum-theoretical features of any field effect, symbolized by the concept of light quanta. In order to get a general view of the situation we assume that the averaging regions in question are of the same order of magnitude and spatially displaced relative to each other through distances of the same order of magnitude as their linear dimensions, which we denote by L; and that further the corresponding time intervals of order of magnitude T are smaller than L/c. Under these conditions, the spectral decomposition of the field effects contains essentially only waves of wave length of the same order of magnitude as L. Since, furthermore, in the case under consideration the intensity of the field produced in the momentum measurement is of the order $\rho \Delta x$, and consequently the field energy contained in the volume V is of the order $\rho^2 \Delta x^2 V$, then the number of light quanta in question is approximated by the expression

$$(44) \qquad n \sim \rho^2 \Delta x^2 V \frac{L}{\hbar c} = \lambda^{-2} \frac{L}{cT},$$

where λ is the factor, defined in (20), that provides the measure for the accuracy of the measurement. Thus we see that in our case n is always large compared to unity if an accuracy of measurement is required that permits field strengths to be measured which are smaller than the critical field quantity Q.

Evidently, the classically calculated expressions (42) and (43) for the field effects become relatively more exact, the greater the accuracy aimed at for the field measurements. However, it is essential to note that the absolute accuracy of these expressions does not change for increasing values of n. For in our case the statistical range of fluctuation of the field averages is approximately given by

$$\frac{\rho \Delta x}{\sqrt{n}} \sim \sqrt{\frac{\hbar c}{VL}} \sim \frac{\sqrt{\hbar c}}{L^2}.$$

This expression for the range of fluctuations of the field effects of the test body, which depends only on the linear dimensions of the measurement domain and which always remains finite, agrees in fact with the expression (14) for the order of magnitude of the pure black-body fluctuations which was derived from the formalism in the case $L > cT$. Actually, in the above consideration we are dealing merely with an example of the general relation, mentioned in Section 2, between black-body fluctuations and the deviations, only describable statistically, of field averages from field quantities that are calculated according to classical theory from specification of the sources. Furthermore, as was already there pointed out, in the case $L > cT$, especially important for testing the formalism, the black-body fluctuations are always smaller than the field strength Q which is a measure of the complementary measurability of field quantities; and, indeed, so much the smaller, the larger the ratio between L and cT. Thus, in the following comparison between field measurements and formalism we shall always start from the classically calculated expressions (43), and only afterwards discuss the significance of the fluctuation phenomena for the consistency of the formalism.

5. Measurement of single field averages

By definition we base the investigation of the measurement possibilities for field averages on equation (15) which expresses the classically described momentum balance for a test body situated in the field. According to the preceding arguments, each field component, such as $\mathfrak{E}_x$, is in this context to be regarded as the superposition of the fields originating from all field sources, including the test body itself, and the core of the measurement problem is precisely the question of the extent to which these fields can be associated with the various sources. However, we must emphasize here immediately that the strict applicability of the classical field concept in defining field averages is not in itself impaired by the previously discussed limited validity of the classical description of the field effects of the test body. Quite apart from the question of the accuracy attainable in the momentum measurements of the test body at the beginning and end of the measuring interval, which was discussed in Section 3, the unambiguous character of this definition would itself require that the masses of the test bodies be chosen sufficiently large that any modifications of the electromagnetic fields, stemming from their accelerations under the influence of these fields during the measuring interval, may be neglected. If one were inclined to regard this neglect as in contradiction with the atomic character of the momentum transfer between electromagnetic wave fields and material bodies, then it must be recalled that in the measurement problem under consideration there is no question of tracing well-defined elementary processes in the sense of the light quantum concept. In particular, in the measuring arrangement described, an uncontrollable amount of momentum is absorbed by the rigid frame to which every test body is attached before and after the measuring interval. In the limiting case of a classically describable interaction between an electromagnetic wave train and a sufficiently heavy charged body, the momentum transfer just mentioned would obviously exactly compensate the momentum absorbed by the test body in the measuring interval.

As a preparation for the general discussion of the measurement problem, we first consider a single field measurement and, as in Section 3, ask for the average value of $\mathfrak{E}_x$ in a certain space-time domain which we ·denote, as in the previous sections, by I. Thus, from the fundamental equation (15), we get for the momentum balance of the test body

$$(45) \qquad p_x^{(I)''} - p_x^{(I)'} = \rho_I V_I T_I (\mathfrak{E}_x^{(I)} + \bar{E}_x^{(I,\,I)}),$$

where $\mathfrak{E}_x^{(I)}$ represents the part of the average of $\mathfrak{E}_x$ which would be present in the space-time domain I under consideration if no momentum measurement were made on the test body at time t', while $\bar{E}_x^{(I,\,I)}$ is the part of the field average that arises from this measurement, whose classically estimated expression is given by (43), if regions I and II are set equal.

According to the arguments of Section 3, the sum of the field averages $\mathfrak{E}_x^{(I)}$ and $\bar{E}_x^{(I,\,I)}$ appearing in (45) can be determined with arbitrary accuracy by choosing a sufficiently large value of ρ_I. However, the larger ρ_I is chosen, the larger will be the uncontrollable value of $\bar{E}_x^{(I,\,I)}$; and therefore, the attainable accuracy in the determination of $\mathfrak{E}_I^{(I)}$ by means of the simple measuring arrangement previously described, which according to (45) is given by

$$(46) \qquad \Delta\mathfrak{E}_x^{(I)} \sim \frac{\Delta p_x^{(I)}}{\rho_I V_I T_I} + \Delta\bar{E}_x^{(I,\,I)},$$

has a limit imposed upon it. Indeed, due to relation (16) and to the fact that the quantity $D_x^{(I)}$ appearing in (43) is predictable only with a latitude Δx_I, we get from (46) the expression

$$(47) \qquad \Delta\mathfrak{E}_x^{(I)} \sim \frac{\hbar}{\rho_I \Delta x_I V_I T_I} + \rho_I \Delta x_I V_I T_I |\bar{A}_{xx}^{(I,\,I)}|,$$

for $\Delta\mathfrak{E}_x^{(I)}$, whose minimal value obviously is

$$(48) \qquad \Delta_m\mathfrak{E}_x^{(I)} \sim \sqrt{\hbar|\bar{A}_{xx}^{(I,\,I)}|}$$

and in the case $L_I > cT_I$ precisely equals the critical quantity Q_I. It is true that when L_I is large compared to cT_I, (48) is essentially smaller than the expression (24) which was given by Landau and Peierls as the absolute limit on the measurability of field quantities; yet if (48) were to be considered as an unavoidable limit on the accuracy of measurement, we should still arrive at the conclusion, in agreement with the view of these authors, that the quantum-electromagnetic formalism admits of no test in the properly quantum domain, and that therefore physical reality can be ascribed to the entire field theory only in the classical limit.

However, this conclusion cannot be maintained, for the fact that according to (43) the coefficient of the unpredictable displacement $D_x^{(I)}$ in $\bar{E}_x^{(I,\,I)}$ is a well-defined quantity depending solely on geometrical relations, allows us to so arrange things in the measurements that the effect of the field $E_x^{(I)}$ is completely compensated except for the unavoidable field fluctuations. This is achieved by a measuring arrangement in which the test body is not freely movable, even during the measuring interval T_I, but remains connected with the rigid frame through a spring mechanism whose tension is proportional to $D_x^{(I)}$. If the force in the x-direction exerted by this mechanism on the test body is $-F_x D_x^{(I)}$, then the total momentum transferred from the field $E_x^{(I)}$ to this body will obviously be completely cancelled by the spring if the spring constant is chosen to be

$$(49) \qquad F_I = \rho_I^2 V_I^2 T_I \bar{A}_{xx}^{(I,\,I)}.$$

At any rate, this holds when the test body is so heavy that its oscillation period under the influence of the spring is large compared to T_I and thus its displacement due to the spring tension during the time T_I is small compared to $D_x^{(I)}$. Furthermore, the action of the spring, which strictly speaking is classically describable only in the asymptotic limiting case, may be calculated on the basis of classical mechanics with an accuracy which is the greater the larger the mass of the test body. Apart from the limitations due to the atomic structure of all bodies, no objection of principle could exist against such a compensation device. In the first place, by using a mechanical spring all electromagnetic fields are avoided, which would be inseparable from the fields to be measured. Secondly, if the length of the spring is sufficiently small, i.e. small compared to cT_I, one may obviously disregard all retardation effects. In doing so, if the test body system is sufficiently heavy, it is clearly immaterial whether the spring acts only on a component body or whether one uses a system of springs that affects each component body uniformly.

Thus we see that the sharpness of a single field measurement is restricted solely by the limit set for the classical description of the field effects of the test body. However, even in the case $L_1 \leqslant cT_1$ this limitation, which is the more insignificant the larger L_1 is compared to cT_1, implies no restriction at all on the possibility of testing the consequences of the quantum-electromagnetic formalism. In assessing this question, we must distinguish sharply between the testing of theoretical predictions which presuppose data concerning electric or magnetic forces obtained by field measurements, and of those which depend on knowledge of the state of the field in question obtained on some other basis. As for the former predictions, their testing obviously requires a closer investigation of the mutual relations between several field measurements; thus, to begin with, here it can only be a matter of testing predictions of the latter kind.

Now, as mentioned in Section 2, it is a major result of the quantum theory of fields that all predictions concerning field averages which do not rest on true field measurements, but on the light quantum composition of the field to be investigated or on the knowledge of classically described field sources, must be of an essentially statistical nature. Further, the more detailed argument presented there shows that inclusion of the fluctuations of the test body's field effects around their classically estimated value brings about no change whatsoever in these statistical predictions. Without further correction, the measurement results obtained by means of the experimental arrangement described thus appear as the desired field averages for testing the theoretical statements. Such a view of the measuring results, whose general justification we shall investigate more closely in the following, is also suggested by the fact that all measurements of physical quantities, by definition, must be a matter of the application of classical concepts; and that, therefore, in field measurements any consideration of limitations on the strict applicability of classical electrodynamics would be in contradiction with the measurement concept itself.

Even though, consequently, as already stressed in the Introduction, the measurement concept is to be applied with even greater caution in field measurements than in the usual quantum mechanical measurement problems; nevertheless, as regards the inseparability of phenomenon and measuring process the situation described exhibits a far-reaching analogy to these problems. Indeed, even in a position or momentum measurement on the electron in a hydrogen atom in a given stationary state one can assert with a certain right that the measuring result is produced only by the measurement itself. It is here not a question of a limitation on the sharpness of the measuring result on the basis of classical mechanics, indeed; but only of abandonment of any control over the influence of the measuring process on the state of the atom. In field measurements, this complementary feature of the description, essential for consistency, corresponds to the fact that the knowledge of the light quantum composition of the field is lost through the field effects of the test body; and in fact, according to (44) the more so, the greater the desired accuracy of the measurement. Moreover, it will appear from the following discussion that any attempt to re-establish the knowledge of the light quantum composition of the field through a subsequent measurement by means of any suitable device would at the same time prevent any further utilization of the field measurement in question.

However, the fact that pure black-body fluctuations appear as the common limitation in the demonstration of the correspondence between the testability of the consequences of the quantum-electromagnetic formalism by means of a single field measurement and the interpretability of such a measurement on the basis of classical electrodynamics in no way implies that these fluctuations set an absolute limit for any utilization of field measurements. Indeed, such a general limitation exists neither for the consequences of the formalism regarding relations between averages of a field component over different regions, nor for the testing of such relations through direct field measurements. This will become clear from the considerations in the following sections; and in particular it will be shown that the requirement of repeatability of measurements of kinematical and dynamical quantities, essential for the discussion of the consistency of the usual quantum mechanics, possesses its natural analog in field measurements.

6. Measurability of two-fold averages of a field component

In investigating the measurability of two field quantities it is convenient to start with the measurement of the average of one and the same field component over two different regions, I and II. Thus, considering as above the field component $\mathfrak{E}_x$, and disregarding to begin with the limitations of the classical describability of the test bodies' field effects, we have in this case for the momentum balance of the two test bodies, instead of (45):

$$(50) \quad \begin{cases} p_x^{(I)''} - p_x^{(I)'} = \rho_I V_I T_I (\mathfrak{C}_x^{(I)} + \bar{E}_x^{(I,\,I)} + \bar{E}_x^{(II,\,I)}), \\ p_x^{(II)''} - p_x^{(II)'} = \rho_{II} V_{II} T_{II} (\mathfrak{C}_x^{(II)} + \bar{E}_x^{(II,\,II)} + \bar{E}_x^{(I,\,II)}), \end{cases}$$

where $\bar{E}_x^{(I,\,II)}$ is defined by expression (43), and $\bar{E}_x^{(II,\,I)}$ is obtained from this expression by simple interchange of the indices I and II.

According to the considerations in the previous sections, the appearance of the expressions $\bar{E}_x^{(I,\,I)}$ and $\bar{E}_x^{(II,\,II)}$ in equations (50) implies that each of the desired field averages, $\mathfrak{C}_x^{(I)}$ and $\mathfrak{C}_x^{(II)}$, can only be determined with a limited accuracy, given by (48), by means of a simple measuring arrangement. Thus, it is evident from the beginning that a compensation procedure is unavoidable, and for preliminary orientation about the more complicated measuring problem considered here we therefore first use a measuring arrangement in which the reactions, $\rho_I V_I T_I \bar{E}_x^{(I,\,I)}$ and $\rho_{II} V_{II} T_{II} \bar{E}_x^{(II,\,II)}$, are cancelled by means of two springs acting on the test bodies I and II, the spring constants being given by (49) and an analogous expression.

From equations (50), with $\bar{E}_x^{(I,\,I)}$ and $\bar{E}_x^{(II,\,II)}$ omitted, it follows, using (16) and (43), that in this arrangement the uncertainties of the two field measurements, taking into account that the displacements of the test bodies, $D_x^{(I)}$ and $D_x^{(II)}$, appearing in $\bar{E}_x^{(I,\,II)}$ and $\bar{E}_x^{(II,\,I)}$ are completely independent of each other and known only with the latitudes Δx_I and Δx_{II}, are given by

$$(51) \quad \begin{cases} \Delta \mathfrak{C}_x^{(I)} \sim \dfrac{\hbar}{\rho_I \Delta x_I V_I T_I} + \rho_{II} \Delta x_{II} V_{II} T_{II} |\bar{A}_{xx}^{(II,\,I)}| \\[2ex] \Delta \mathfrak{C}_x^{(II)} \sim \dfrac{\hbar}{\rho_{II} \Delta x_{II} V_{II} T_{II}} + \rho_I \Delta x_I V_I T_I |\bar{A}_{xx}^{(I,\,II)}|. \end{cases}$$

By suitable choice of the values of $\rho_I \Delta x_I$ and $\rho_{II} \Delta x_{II}$ either one of the quantities $\Delta \mathfrak{C}_x^{(I)}$, $\Delta \mathfrak{C}_x^{(II)}$ can obviously be arbitrarily diminished, but only at the expense of an increase of the other. For according to (51) we get for the product of the two quantities the minimum value

$$(52) \quad \Delta \mathfrak{C}_x^{(I)} \Delta \mathfrak{C}_x^{(II)} \sim \hbar [|\bar{A}_{xx}^{(I,\,II)}| + |\bar{A}_{xx}^{(II,\,I)}|].$$

In spite of the great similarity of relation (52) to the uncertainty relations (8) required by the formalism, there is, nevertheless, a fundamental difference in that the latter contains not the sum of the magnitudes of the quantities $\bar{A}_{xx}^{(\mathrm{I,II})}$ and $\bar{A}_{xx}^{(\mathrm{II,I})}$ but their algebraic difference. It is true that (8) and (52) are in general in agreement as to order of magnitude when the regions I and II are spatially and temporally displaced relative to each other by distances of order of magnitude L and T, in which case they both have the approximate value Q^2. However, as mentioned in Section 2, the difference sign appearing in the indeterminacy relation (8) has the effect that in important cases the product of the complementary uncertainties vanishes identically, even though the quantities $\bar{A}_{xx}^{(\mathrm{I,II})}$ and $\bar{A}_{xx}^{(\mathrm{II,I})}$ each remain different from zero. This happens, for example, when the temporal averaging intervals T_I and T_II coincide, and in particular when the two averaging regions I and II completely overlap. In the latter case even the limit on the measurability of two field averages given by (52) would be in glaring contradiction to the result of the previous discussion of measurement of a single field average. In general, the two expressions (52) and (8) agree exactly only when at least one of the quantities $\bar{A}_{xx}^{(\mathrm{I,II})}$ or $\bar{A}_{xx}^{(\mathrm{II,I})}$ vanishes which in general requires that one of the expressions $t_1 - t_2 - r/c$ or $t_2 - t_1 - r/c$, appearing as arguments of the δ-function in the integrals (5), remain different from zero for every pair of points (x_1, y_1, z_1, t_1) and (x_2, y_2, z_2, t_2) of regions I and II.

Thus, apart from the last mentioned case in which there exists no correlation, or at any rate only a one-way correlation, between the two field averages, the demonstration of the agreement between measurability and quantum electromagnetic formalism requires a more refined measuring arrangement in which the uncontrollable effects can be compensated to a larger extent. It is true that there appears here, in comparison with the compensation procedure needed already for measuring a single field quantity, the further complication that the displacements of the two test bodies not only must remain unknown but are also completely independent of each other. However, this circumstance implies no fundamental difficulty; only a somewhat more complicated procedure is necessary in order to compensate as much as possible the influence of the relative displacement of the test bodies on the field measurements. For this purpose we select two component bodies ε_I and ε_II, one from each test body system I and II, for which the expression $r - c(t_1 - t_2)$ vanishes for two times t_I^* and t_II^* lying in the time intervals T_I and T_II, respectively. If such a choice were not possible, then as said above the agreement between measurability and formalism would already be attained without further compensation. To establish the necessary correlation between the test bodies one might at first think of a spring which should connect the bodies ε_I and ε_{II} directly with each other; however, due to the retardation of the forces one would thereby run into difficulties. But one can manage with a short spring, i.e. small compared to cT, if one adds to the second test body system a neutral component body ε_{III} which is situated in the immediate neighborhood of component body ε_I, belonging to the first system, and connected with it by a spring.

Like all component bodies of the two test body systems, the body ε_{III} is initially to be bound to the rigid frame. Then at time t_I', its momentum is to be measured, after severing of this link, with the same accuracy as that of test body system II. It thereby undergoes an unknown displacement $D_x^{(III)}$ in the x-direction which is of the same order of magnitude as Δx_{II}. If now the tension of the spring mounted between ε_{III} and ε_I is chosen equal to $\frac{1}{2}\rho_I\rho_{II}V_IV_{II}T_{II}(\bar{A}_{xx}^{(I,II)}+\bar{A}_{xx}^{(II,I)})$, then in the time interval T_I the momentum

$$(53) \qquad P=\tfrac{1}{2}\rho_I\rho_{II}V_IV_{II}T_IT_{II}(\bar{A}_{xx}^{(I,II)}+\bar{A}_{xx}^{(II,I)})(D_x^{(I)}-D_x^{(III)})$$

will be transferred from ε_{III} to ε_I, while ε_{III} undergoes the momentum change $-P$ during the same time interval. At time t'' the momentum of ε_{III} is measured again with the same accuracy. However, before this measurement, and in fact at time t_{II}^*, a short light signal is to be sent from ε_{II} to ε_{III}, by which the relative displacement $D_x^{(III)}-D_x^{(II)}$ of these bodies can be measured with arbitrary accuracy by means of a suitable device. At the emission and absorption of the signal the two bodies undergo momentum changes which indeed remain completely unknown, but cancel each other exactly in the sum of the momentum changes measured on the bodies.

Thus, for the momentum balance of the two test body systems during the measurement we have, if we include the body ε_{III} in system II,

$$(54) \qquad \begin{cases} p_x^{(I)''}-p_x^{(I)'}=\rho_IV_IT_I(\mathfrak{E}_x^{(I)}+\bar{E}_x^{(II,I)})+P \\ p_x^{(II)''}-p_x^{(II)'}+p_x^{(III)''}-p_x^{(III)'}=\rho_{II}V_{II}T_{II}(\bar{\bar{\mathfrak{E}}}_x^{(II)}+\bar{E}_x^{(I,II)})-P. \end{cases}$$

Taking into account (43) and (53), these formulae can be put into the form

$$
(55) \quad
\begin{cases}
\begin{aligned}
& p_x^{(I)''} - p_x^{(I)'} = \rho_I V_I T_I \mathfrak{C}_x^{(I)} \\
& + \tfrac{1}{2}\rho_I \rho_{II} V_I V_{II} T_I T_{II} \{ -D_x^{(II)}(\bar{A}_{xx}^{(I,\,II)} - \bar{A}_{xx}^{(II,\,I)}) \\
& + (D_x^{(II)} - D_x^{(III)})(\bar{A}_{xx}^{(I,\,II)} + \bar{A}_{xx}^{(II,\,I)}) + D_x^{(I)}(\bar{A}_{xx}^{(I,\,II)} + \bar{A}_{xx}^{(II,\,I)}) \} \\
& p_x^{(II)''} - p_x^{(II)'} + p_x^{(III)''} - p_x^{(III)'} = \rho_{II} V_{II} T_{II} \; \mathfrak{C}_x^{(II)} \\
& + \tfrac{1}{2}\rho_I \rho_{II} V_I V_{II} T_I T_{II} \{ D_x^{(I)}(\bar{A}_{xx}^{(I,\,II)} - \bar{A}_{xx}^{(II,\,I)}) \\
& - (D_x^{(II)} - D_x^{(III)})(\bar{A}_{xx}^{(I,\,II)} + \bar{A}_{xx}^{(II,\,I)}) + D_x^{(II)}(\bar{A}_{xx}^{(I,\,II)} + \bar{A}_{xx}^{(II,\,I)}) \}
\end{aligned}
\end{cases}
$$

The last terms in the curly brackets in (55) are proportional to the unknown displacements of the test bodies I and II and can therefore, exactly as the simple reactions of each test body on itself, be cancelled by means of suitable spring connections with the rigid frame. This simply amounts to replacing the expression (49) for the tension of the spring acting on body I by

$$
(56) \qquad F_{I,\,II} = \rho_I^2 V_I^2 T_I \bar{A}_{xx}^{(I,\,I)} + \tfrac{1}{2}\rho_I \rho_{II} V_I V_{II} T_{II}(\bar{A}_{xx}^{(I,\,II)} + \bar{A}_{xx}^{(II,\,I)})
$$

and similarly changing the tension of the spring between the frame and body II. Furthermore, in the measuring arrangement described, the terms proportional to the relative displacement $D_x^{(III)} - D_x^{(II)}$ are known with arbitrary accuracy and can therefore easily be taken into account in the field measurements. In fact, by means of a somewhat more complicated device one could even obtain the vanishing of the difference $D_x^{(III)} - D_x^{(II)}$ by using $P_x^{(II)} + P_x^{(III)}$ (in analogy to the arrangement for measuring the total momentum of a test body system described in Section 3) to determine one and the same radiation packet and, by means of suitably placed fixed mirrors, by regulating the light path in such a way that in the first momentum measurement the reflections at body ε_{III} and at all component bodies of system II occur at the times t_I' and t_{II}', respectively, and in the second momentum measurement occur at the times t_I'' and t_{II}''.

By means of all these contrivances, whose considerable complexity lies in the nature of the problem, being due solely to the finite propagation of all field effects, we now have actually removed the apparent conflict between the determination of single and two-fold averages of a field component described at the beginning of this section. For from (55) we now obtain for the indeterminacies of $\mathfrak{C}_x^{(I)}$ and $\mathfrak{C}_x^{(II)}$, instead of (51),

$$
(57) \quad
\begin{cases}
\Delta\mathfrak{E}_x^{(\mathrm{I})} \sim \dfrac{\hbar}{\rho_{\mathrm{I}}\Delta x_{\mathrm{I}} V_{\mathrm{I}} T_{\mathrm{I}}} + \dfrac{1}{2}\rho_{\mathrm{II}}\Delta x_{\mathrm{II}} V_{\mathrm{II}} T_{\mathrm{II}} \big|\bar{A}_{xx}^{(\mathrm{I},\,\mathrm{II})} - \bar{A}_{xx}^{(\mathrm{II},\,\mathrm{I})}\big| \\[2ex]
\Delta\mathfrak{E}_x^{(\mathrm{II})} \sim \dfrac{\hbar}{\rho_{\mathrm{II}}\Delta x_{\mathrm{II}} V_{\mathrm{II}} T_{\mathrm{II}}} + \dfrac{1}{2}\rho_{\mathrm{I}}\Delta x_{\mathrm{I}} V_{\mathrm{I}} T_{\mathrm{I}} \big|\bar{A}_{xx}^{(\mathrm{I},\,\mathrm{II})} - \bar{A}_{xx}^{(\mathrm{II},\,\mathrm{I})}\big|,
\end{cases}
$$

which immediately yields for the minimal value of the product of the uncertainties

$$
(58) \quad \Delta\mathfrak{E}_x^{(\mathrm{I})}\Delta\mathfrak{E}_x^{(\mathrm{II})} \sim \hbar\big|\bar{A}_{xx}^{(\mathrm{I},\,\mathrm{II})} - \bar{A}_{xx}^{(\mathrm{II},\,\mathrm{I})}\big|
$$

in agreement with the consequence of the quantum theory of fields expressed by (8).

However, in order to demonstrate the complete accord between the measurability of the averages of a field component over two space-time regions and the requirements of the quantum electromagnetic formalism, we must go somewhat further into the question of the extent to which the assumption of the classical describability of the test body's field effects impairs the possibilities of testing theoretical predictions. For, as already indicated, exactly in the case of measurement of several field averages one might think beforehand that the neglect of the fluctuations of all field effects of the test body, which cannot be followed classically, and are inseparable from the pure black-body fluctuations, in this respect signifies an essential renunciation. At any rate, as long as we are dealing with averaging regions which are displaced spatially and temporally relative to each other by distances of the same order of magnitude as their linear dimensions L and corresponding time intervals T, this neglect is of little significance in the important case where L is large compared to cT. However, if L is of the same order of magnitude or smaller than cT, then the black-body fluctuations, as mentioned in Section 2, are of just the same order of magnitude as the critical field strength $\mathfrak{U}$, which is defined for such displaced regions by means of the indeterminacy relations and is to be regarded as the limit of the classical field description. The neglect in question might appear even more doubtful and seem to imply a complete renunciation of the repeatability of field measurements in the case of two almost coinciding domains, in which the product of the complementary uncertainties of the field averages, given by (8), tends to zero independently of the ratio between L and cT, and where thus the critical field strength $\mathfrak{U}$ can be arbitrarily small compared to the black-body fluctuations.

Nevertheless, a closer consideration shows that we obtain a consistent interpretation of all consequences of the quantum theory of fields, if, in a necessary generalization of the measurement concept, we interpret the measuring results obtained by the arrangement described as the desired field averages. For the classically undescribable fluctuations included in the field effects of all test bodies cannot be separated in any way from the fundamentally statistical features of every theoretical assertion whose conditions do not refer to actual field measurements. Without in any way limiting the given measurement problem we can therefore always regard the fluctuations in question as an integral part of the field to be measured. The situation in multiple field measurements differs in this respect from that obtaining in measurement of a single field average only insofar as the state of the field, with which we are concerned in every single measurement in the general case, is codetermined by the result of the other field measurements.

However, with regard to this state of affairs it may not be superfluous to point out that in the correlation of several field measurements we have to do with a feature of the general complementarity of description which is alien to the usual measuring problem of non-relativistic quantum mechanics. Indeed, the fundamental simplification which we meet in the latter theory lies precisely in the separation made there between spatial coordination and temporal evolution, which makes it possible to order all measuring processes in a simple temporal sequence. On the other hand, it is possible to speak of such a sequence of measuring processes during the measurement of two field averages only when the corresponding time intervals do not overlap at all. In general, in accordance with the formalism, the correlation of the two measurements is also a reciprocal one; and only when one of the quantities $r - c(t_1 - t_2)$ and $r - c(t_2 - t_1)$ remains different from zero for all pairs of points of the regions I and II, do we encounter conditions similar to those of the usual measurement problem of atomic mechanics, since the result of the one field measurement may then be simply included in the preconditions used in the predictions to be tested by the other measurement.

We meet an instructive example of an intimate reciprocal correlation in measurements of the averages of a field component over two almost coinciding space-time domains. In conformity with the requirement of the repeatability of measurement results, the theory demands in this case that both measurements yield the same result to an arbitrary degree of approximation quite independently of the statistical assertions about the values of the field quantities to be measured which are implied by the preconditions. That this requirement is actually fulfilled in our experimental arrangement follows from the fact that in this case we have to do with two test body systems which occupy almost the same spatial region and are used during almost the same interval of time. Thus, by definition, they are exposed to almost the same field, quite irrespective of the sources producing this field, and of which contribution comes from one or the other test body.

Actually, it follows from the last remark that in the case of coinciding averaging regions we would get exactly identical results of the two measurements even without any compensation. However, on account of the field effects of the test bodies the measurement results so obtained would differ in an unpredictable way from the theoretical predictions to be tested, the more so, the greater the desired measuring accuracy. It is true that by means of the compensation mechanism suitable for single field measurements, which we had retained unaltered at the beginning of this section, these deviations are in general diminished; but at the same time any strict correlation of the measuring results is prevented by the effects of the springs which are proportional to the independent displacements of the test bodies. In the arrangement for two field measurements finally adopted, in which all well-defined differences between measurement results and theoretical predictions are compensated, such a correlation is also re-established just in the case of coinciding regions. For, as one easily sees, quite independently of the relation between the magnitudes of the uncontrollable displacements of the test bodies, the momenta transferred to each test body through the combined effect of all the springs, divided by the corresponding charge densities, are exactly identical in this case.

As far as the consistency of the description is concerned, we might still remark that any attempt to control the change in the light quantum composition of the field caused by the field measurement by investigation of the test body's radiation, as already mentioned several times, would exclude the possibility of utilizing the measurement result for a comparison with a second field measurement. For in order even to be able to speak of such a utilization, there must exist pairs of points from the regions I and

II, respectively, for which one of the expressions $r-c(t_1-t_2)$ or $r-c(t_2-t_1)$ vanishes. But this implies that the radiation fields produced by the test bodies I and II during the measurement cannot be intercepted and analyzed on their way from one test body to the other without at the same time essentially influencing the fields to be measured by these bodies. Only after the completion of all field measurements, when their direct utilization is no longer of concern, is it possible to perform an arbitrarily accurate analysis of the light quantum composition of the entire field without adversely affecting the given measurement problem.

7. MEASURABILITY OF TWO AVERAGES OF DIFFERENT FIELD COMPONENTS

As far as measurements of averages of different field components are concerned, only the case of perpendicular, similar or dissimilar components needs closer investigation; for the complete commutativity and independent measurability of averages of parallel dissimilar components required by the quantum electromagnetic formalism finds its direct interpretation in the identical vanishing of the component $H_x^{(I)}$ of the field produced by the measurement of $\mathfrak{E}_x^{(I)}$, as shown by (42). Besides, on the basis of equations (43), the measurement of averages of perpendicular field components allows a method of treatment analogous to the procedure described in the previous section.

Let us consider the measurement of the average of $\mathfrak{E}_x$ over the region I and the average of $\mathfrak{E}_y$ or $\mathfrak{H}_y$ over the region II. If to begin with we use a measuring arrangement in which the field effects of each test body on itself during the measurement are compensated in the manner described in Section 5, we get equations of the following type for the momentum balance of the two test bodies to be used:

$$(59) \quad \begin{cases} p_x^{(I)''} - p_x^{(I)'} = \rho_I V_I T_I (\bar{\mathfrak{E}}_x^{(I)} + D_y^{(II)} \sigma_{II} V_{II} T_{II} \bar{C}_{xy}^{(II,\,I)}) \\ p_y^{(II)''} - p_y^{(II)'} = \sigma_{II} V_{II} T_{II} \bar{\mathfrak{R}}_y^{(II)} + D_x^{(I)} \rho_I V_I T_I \bar{C}_{xy}^{(I,\,II)}) \end{cases}$$

Depending on whether we are dealing with a measurement of similar or dissimilar components, the letter $\mathfrak{R}$ here represents the usual designation of the field components $\mathfrak{E}$ or $\mathfrak{H}$, while C is written instead of the symbols A or B appearing in (43); further, the designation σ_{II} represents the charge density or the pole strength distribution of the test body II accordingly.

In a manner similar to the derivation of (52) one obtains from (59) the relation

$$(60) \qquad \Delta \mathfrak{E}_x^{(I)} \Delta \mathfrak{R}_y^{(II)} \sim \hbar [|\bar{C}_{xy}^{(I,\,II)}| + |\bar{C}_{xy}^{(II,\,I)}|],$$

which, like (52), does not generally, but only in certain cases, represent an agreement between measurability and quantum electromagnetic formalism. Of such cases we might mention in particular the measurement of dissimilar perpendicular field components inside the same spatial region, for which, as stressed in Section 2, both expressions $\bar{B}_{xy}^{(I,\,II)}$ and $\bar{B}_{xy}^{(II,\,I)}$ vanish. The correctness of interpreting this fact as an arbitrarily accurate independent measurability of the field quantities in question was suggested already in Section 3 by elementary considerations in the case of coinciding space-time regions.

For the general treatment of the measurability problem of perpendicular field components we choose, as in the previous section, two component bodies ε_I and ε_{II} of the test body systems I and II, respectively, whose separation is $r = c(t_I^* - t_{II}^*)$, where t_I^* and t_{II}^* are within the time intervals T_I and T_{II}, respectively. Furthermore, in the immediate neighborhood of ε_I we introduce a third body ε_{III}, whose momentum in the y-direction is measured at the times t_I' and t_I''; the relative displacements $D_y^{(III)} - D_y^{(II)}$ of the bodies ε_{III} and ε_{II} are again determined by means of a light signal, as a result of which both bodies undergo equal and opposite momentum changes. Instead of connecting ε_{III} directly with ε_I by a spring we must, however, in order to make the force transfer through the spring mechanism proportional to $D_y^{(III)} - D_x^{(I)}$, use a device which consists of two springs and an angular level with two equally long mutually perpendicular arms which can rotate on a hinge mounted on the rigid frame, and the arms of which are initially parallel to the x- and y-directions, respectively. A spring parallel to the y-axis is fastened between the first arm and the body ε_{III}, and a spring parallel to the x-axis acts between the second arm and the body ε_I. Let the tension of the springs be so chosen that the force which acts on the body ε_{III} in the y-direction and on the body ε_I in the x-direction during the time interval T_I is given by

$$\tfrac{1}{2} \rho_I \sigma_{II} V_I V_{II} T_{II} (\bar{C}_{xy}^{(I,\,II)} + \bar{C}_{xy}^{(II,\,I)})(D_x^{(I)} - D_y^{(III)}).$$

Thus, the momentum balance of the two test body systems, after suitable rearrangement, may be expressed in the following form, analogous to (55)

$$
(61) \quad
\begin{cases}
p_x^{(\mathrm{I})''} - p_x^{(\mathrm{I})'} = \rho_\mathrm{I} V_\mathrm{I} T_\mathrm{I} \mathfrak{E}_x^{(\mathrm{I})} \\
\quad + \tfrac{1}{2}\rho_\mathrm{I}\sigma_\mathrm{II} V_\mathrm{I} V_\mathrm{II} T_\mathrm{I} T_\mathrm{II}\{ -D_y^{(\mathrm{II})}(\bar{C}_{xy}^{(\mathrm{I,\,II})} - \bar{C}_{xy}^{(\mathrm{II,\,I})}) \\
\quad + (D_y^{(\mathrm{II})} - D_y^{(\mathrm{III})})(\bar{C}_{xy}^{(\mathrm{I,\,II})} + \bar{C}_{xy}^{(\mathrm{II,\,I})}) + D_x^{(\mathrm{I})}(\bar{C}_{xy}^{(\mathrm{I,\,II})} + \bar{C}_{xy}^{(\mathrm{II,\,I})})\} \\
p_y^{(\mathrm{II})''} - p_y^{(\mathrm{II})'} + p_y^{(\mathrm{III})''} - p_y^{(\mathrm{III})'} = \sigma_\mathrm{II} V_\mathrm{II} T_\mathrm{II} \mathfrak{R}_y^{(\mathrm{II})} \\
\quad + \tfrac{1}{2}\rho_\mathrm{I}\sigma_\mathrm{II} V_\mathrm{I} V_\mathrm{II} T_\mathrm{I} T_\mathrm{II}\{ D_x^{(\mathrm{I})}(\bar{C}_{xy}^{(\mathrm{I,\,II})} - \bar{C}_{xy}^{(\mathrm{II,\,I})}) \\
\quad + (D_y^{(\mathrm{II})} - D_y^{(\mathrm{III})})(\bar{C}_{xy}^{(\mathrm{I,\,II})} + \bar{C}_{xy}^{(\mathrm{II,\,I})}) - D_y^{(\mathrm{II})}(\bar{C}_{xy}^{(\mathrm{II,\,I})} + \bar{C}_{xy}^{(\mathrm{II,\,I})})\}.
\end{cases}
$$

After compensation of the last terms we thus obtain for the uncertainties of the field averages:

$$
(62) \quad
\begin{cases}
\Delta\mathfrak{E}_x^{(\mathrm{I})} \sim \dfrac{\hbar}{\rho_\mathrm{I}\Delta x_\mathrm{I} V_\mathrm{I} T_\mathrm{I}} + \dfrac{1}{2}\sigma_\mathrm{II}\Delta y_\mathrm{II} V_\mathrm{II} T_\mathrm{II} |\bar{C}_{xy}^{(\mathrm{I,\,II})} - \bar{C}_{xy}^{(\mathrm{II,\,I})}| \\
\Delta\mathfrak{R}_y^{(\mathrm{II})} \sim \dfrac{\hbar}{\sigma_\mathrm{II}\Delta y_\mathrm{II} V_\mathrm{II} T_\mathrm{II}} + \dfrac{1}{2}\rho_\mathrm{I}\Delta x_\mathrm{I} V_\mathrm{I} T_\mathrm{I} |\bar{C}_{xy}^{(\mathrm{I,\,II})} - \bar{C}_{xy}^{(\mathrm{II,\,I})}|,
\end{cases}
$$

from which the minimal value of their product is obtained as

$$
(63) \qquad \Delta\mathfrak{E}_x^{(\mathrm{I})}\Delta\mathfrak{R}_y^{(\mathrm{II})} \sim \hbar |\bar{C}_{xy}^{(\mathrm{I,\,II})} - \bar{C}_{xy}^{(\mathrm{II,\,I})}|,
$$

again in complete agreement with the quantum electromagnetic formalism.

Furthermore, from the general arguments at the end of the previous section it follows that also in the case considered here the utilization of field measurements for testing the formalism's assertions in no way is impaired by the classical evaluation of the field effects. Besides, in measurements of averages of dissimilar field components the question of repeatability does not arise at all, and the pure black-body fluctuations are included in all theoretical assertions as an unavoidable statistical feature.

8. CONCLUDING REMARKS

We thus arrive at the conclusion already stated at the beginning, that with respect to the measurability question the quantum theory of fields represents a consistent idealization to the extent that we can disregard all limitations due to the atomic structure of the field sources and the measuring instruments. As already emphasized in the Introduction, this result should properly be regarded as an immediate consequence of the fact that both the quantum electromagnetic formalism and the viewpoints on which the possibilities of testing this formalism are to be assessed have as their common foundation the correspondence argument. Nevertheless, it would seem that the somewhat complicated character of the considerations used to demonstrate the agreement between formalism and measurability are hardly avoidable. For in the first place the physical requirements to be imposed on the measuring arrangement are conditioned by the integral form in which the assertions of the quantum-electromagnetic formalism are expressed, whereby the peculiar simplicity of the classical field theory as a purely differential theory is lost. Furthermore, as we have seen, the interpretation of the measuring results and their utilization by means of the formalism require consideration of certain features of the complementary mode of description which do not appear in the measurement problems of non-relativistic quantum mechanics.

At the completion of this work we should not like to leave unmentioned that we have found much stimulation and help in many discussions of the questions considered with past and present colleagues at the Institute, among them Heisenberg and Pauli as well as Landau and Peierls.

Universitetets Institut for teoretisk Fysik
Copenhagen, April 1933

NOTES

* Translated by Prof. Aage Petersen; revised by RSC and JS.
[1] W. Heisenberg, *The Physical Principles of the Quantum Theory*, transl. by C. Eckart and F. Hoyt (Dover, New York, 1930), pp. 42 ff.
[2] L. Landau and R. Peierls, *Zs. f. Phys.* **69** (1931), 56.
[3] Cf. N. Bohr, 'Atomic Stability and Conservation Laws', *Atti del Congresso di Fisica Nucleare* (1932). Added in the proof: A separate publication to appear shortly will contain a discussion of the consequences for the problems discussed in the cited reference implied by the recent discovery of the occurrence, under special circumstances, of so-called 'positive electrons'; and by the recognition of the connection of this discovery with Dirac's relativistic electron theory.

[4] Cf. L. Rosenfeld, *Zs. f. Phys.* **76** (1932), 729.

[5] Cf. P. Jordan and W. Pauli, *Zs. f. Phys.* **47** (1928), 151 and also W. Heisenberg and W. Pauli, *Zs. f. Phys.* **56** (1929), 33. Apart from an unessential difference in sign resulting from a difference in the choice of time direction in the Fourier decomposition of the field strengths, the formulae above are equivalent in content with those derived in the papers quoted. In particular, the notation used here, where all terms appear as retarded, is a purely formal change which aims at an interpretation of the measurement problems that is as intuitive as possible.

[6] Cf. L. Rosenfeld and J. Solomon, *J. de Physique* **2** (1931), 139 and also W. Pauli, *Handbuch d. Physik*, 2nd edition, Vol. 24/1 (1933), p. 255.

[7] Cf. W. Heisenberg, *Leipziger Berichte* **83** (1931), 1.

[8] Cf. V. Fock and P. Jordan, *Zs. f. Phys.* **66** (1930), 206, where reference is made to such restrictions on field measurements, which are unrelated to the quantum theory of fields. Cf. also J. Solomon, *J. de Physique* **4** (1933), 368.

[9] Cf. W. Pauli, *Handbuch d. Physik*, 2nd edition, Vol. 24/1 (1933), p. 257.

[10] See N. Bohr, *Atomic Theory and the Description of Nature* (Cambridge University Press, Cambridge, England, 1934). In the meantime, this question has been treated in more detail by the author in a guest lecture in Vienna, to appear shortly, in which in particular the paradoxes arising in the interpretation of the indeterminacy principle when account is taken of the requirements of relativity are further discussed.

III. GENERAL DISCUSSION
AT THE SEVENTH SOLVAY CONFERENCE

ON THE CORRESPONDENCE METHOD
IN ELECTRON THEORY

SUR LA MÉTHODE DE CORRESPONDANCE
DANS LA THÉORIE DE L'ÉLECTRON

Structure et propriétés de noyeaux atomiques, Rapports et discussions du septième Conseil de physique tenu à Bruxelles du 22 au 29 octobre 1933, Gauthier-Villars, Paris 1934, pp. 216–228

TEXT AND TRANSLATION

DISCUSSION REMARKS
on P.A.M. Dirac, *Théorie du positron* (*ibid.*, pp. 203–212), *ibid.*, pp. 214–215

TEXT AND TRANSLATION

See Introduction to Part I, sect. 2.

[167]

The French text printed in the Proceedings of the Solvay Meeting was written by Rosenfeld mainly on the basis of a Danish manuscript entitled "Bemærkninger om Korrespondensargumentet i Elektronteorien" which is preserved in the Niels Bohr Archive (Bohr MSS, microfilm no. 13). Since his French was shaky, Bohr could not have revised Rosenfeld's manuscript in his usual painstaking way. Therefore, the editor of this volume contemplated using the English translation of the Danish manuscript as a basis. However, Erik Rüdinger, the former director of the Niels Bohr Archive, pointed out that a couple of paragraphs in the printed French version differed considerably from the Danish draft, to which no amendments have been found. Consequently, Erik Rüdinger undertook the preparation of a verbatim translation of the published French report. This procedure left of course Gallicisms that were foreign to Bohr's style, so the editor has revised the English translation assuming the *Danish* manuscript as the original, *except* where it differed substantially from the printed version. Furthermore, when in doubt the editor has tried to collate the text with Bohr's correspondence.

For the sake of completeness an English translation of a comment by Bohr in connection with Dirac's report at the meeting is appended.

M. N. Bohr. — *Sur la méthode de correspondance dans la théorie de l'électron* (¹). — La merveilleuse confirmation apportée à la théorie de l'électron de Dirac par la découverte du positron a éclairé d'un jour nouveau les paradoxes qui avaient semblé un moment limiter d'une manière décisive l'application de la méthode de correspondance dans la mécanique quantique relativiste.

Le point de départ de cette méthode est la théorie classique de l'électron, qui constitue une application directe de la méca-

---

(¹) Exposé, sous forme remaniée, des remarques générales présentées par l'auteur lors de la discussion.

nique et de l'électrodynamique classiques à des systèmes de points matériels chargés. On sait toutefois qu'une telle idéalisation n'est applicable que pour autant qu'on puisse faire abstraction de toute variation des forces agissant sur chaque particule à l'intérieur de domaines spatio-temporels dont les dimensions sont données, dans un système où l'électron est en repos instantané, par le « diamètre de l'électron »

$$\delta = \frac{e^2}{mc^2},$$

et par l'intervalle de temps propre correspondant

$$\tau = \frac{\delta}{c} = \frac{e^2}{mc^3}.$$

Je rappellerai notamment que ces conditions sont nécessaires pour que la réaction de rayonnement due à l'accélération de l'électron soit suffisamment petite par rapport aux forces extérieures agissant sur lui, de telle sorte qu'on puisse faire un usage univoque du concept de masse. Le fait que δ et τ sont très petits par rapport au domaine où l'électron est lié, même dans les atomes les plus lourds, est également essentiel pour permettre une application étendue des concepts classiques dans la théorie quantique de la constitution des atomes et des spectres. La petitesse du couplage entre atome et rayonnement par rapport au couplage entre les particules de l'atome permet notamment de négliger complètement, en première approximation, la réaction de rayonnement. Toutes ces circonstances sont, comme on sait, conditionnées exclusivement par la petitesse de la quantité sans dimension

$$\epsilon = \frac{e^2}{hc},$$

qui est la condition fondamentale permettant l'interprétation nouvelle de la théorie classique de l'électron par le principe de correspondance.

Les paradoxes imprévus, concernant l'intervention d'énergies négatives et d'un couplage infini entre un électron et son propre champ, qu'ont fait apparaître les essais d'édification d'une mécanique quantique relativiste et d'une électrodynamique quantique générale, ont conduit cependant à soupçonner que l'applicabilité des concepts spatio-temporels ainsi que du concept de champ dans la méthode de correspondance devait être soumise à des limitations plus restrictives que dans la théorie classique de l'électron. C'est ainsi que Landau et Peierls, en étudiant la possibilité de localiser un électron à l'aide de la diffusion de faisceaux de rayonnement, sont arrivés, comme on sait, à la conclusion que la longueur

$$\lambda = \frac{h}{mc},$$

rencontrée dans la théorie de l'effet Compton, et le temps correspondant

$$\theta = \frac{h}{mc^2}$$

désigné déjà précédemment par Richardson et Flint comme l'intervalle ultime de temps propre, devaient représenter, dans un système, où l'électron est approximativement au repos, des limites absolues pour tout usage rationnel des concepts spatio-temporels dans la théorie de l'électron. En outre, les mêmes auteurs ont conclu de l'étude de la mensurabilité des grandeurs caractéristiques de champ que le concept classique de champ ne pouvait aucunement être employé dans des problèmes où le quantum d'action joue un rôle essentiel. Ceci voudrait évidemment dire que tous les essais faits jusqu'ici pour obtenir une extension quantique strictement relativiste de la théorie de l'électron ainsi que de la théorie des champs renferment de telles contradictions qu'on ne pourrait leur attribuer aucune signification physique. Comme nous le verrons, ces conclusions ne peuvent toutefois pas être maintenues, et je chercherai notamment à montrer que la situation est essentiellement différente dans le domaine de la théorie de l'électron et dans celui de la théorie des champs.

Pour ce qui concerne la théorie de l'électron, il est vrai qu'on ne peut pas effectuer de mesures spatio-temporelles dans des intervalles plus petits que λ et θ d'une manière aussi simple qu'on le suppose dans la déduction habituelle des relations d'incertitude de Heisenberg. A ce sujet, il ne faut cependant

pas oublier que, conformément au caractère complémentaire de
la description des phénomènes quantiques, la question de l'interprétation des mesures se pose d'une manière essentiellement
différente, suivant que le but de la mesure est de fixer les conditions initiales du problème considéré, ou bien qu'il s'agit de
vérifier les conséquences essentiellement statistiques de la théorie
pour un problème dont les conditions initiales sont données.
Dans le premier cas, nous ne rencontrons évidemment aucune
restriction de principe dans la mécanique quantique relativiste,
comme cela résulte déjà de la forme invariante des relations
d'incertitude. Après une mesure ayant pour but une localisation
spatio-temporelle plus exacte que λ et θ, il ne peut non plus être
question, en raison de l'indétermination complémentaire de l'impulsion, de la connaissance d'un système de référence où l'électron
a une vitesse petite par rapport à c. Dans l'autre cas, où le but
des mesures est de vérifier des conséquences statistiques de la
théorie, on ne peut aucunement exiger que chaque mesure ait
une interprétation univoque, mais seulement qu'il soit possible
de montrer à l'aide de mesures répétées un nombre suffisant de
fois, dans les mêmes conditions initiales du problème, l'exactitude
de toutes les propriétés bien définies de sa solution, comme par
exemple la dépendance spatio-temporelle de la densité caractéristique du symbolisme de la mécanique quantique. C'est
ainsi que dans des problèmes tels que ceux qu'ont envisagés
Landau et Peierls, et dans lesquels intervient un système de référence où l'électron est approximativement au repos, on ne peut
sans doute pas interpréter l'observation d'un seul effet Compton
comme une mesure, d'exactitude supérieure à λ, de la position
qu'occupait l'électron avant l'observation; mais une telle interprétation ne correspondrait pas non plus à la vérification d'aucune
prédiction bien définie de la théorie. Par contre, en effectuant
un nombre suffisamment grand de mesures de la diffusion de
faisceaux de rayonnement suffisamment variés dans l'espace et
dans le temps, on peut en principe vérifier toutes les propriétés
bien définies de la densité, pour des systèmes comme ceux que
nous considérons, où toutes les dimensions qui interviennent
explicitement dans l'énoncé du problème sont grandes par rapport
à λ et θ, et où les modifications relativistes n'entraînent que

des écarts relativement faibles à partir des formules non relativistes.

Dès que nous passons à des problèmes dans lesquels interviennent explicitement des grandeurs d'espace-temps égales ou inférieures à λ et ϑ, et que, par suite, nous sortons tout à fait du domaine où le formalisme non relativiste est approximativement exact, nous devons tenir compte, pour tous les dispositifs de mesure éventuels, de la création de couples de positrons et d'électrons, ce qui entraînerait naturellement dans l'interprétation de telles mesures de grandes complications pour ce qui concerne non seulement les réactions des instruments de mesure sur l'objet, mais également la constitution même des instruments de mesure. Étant donné que λ et ϑ représentent les dimensions minima des domaines à l'intérieur desquels l'électron dans un champ de forces quelconque peut se trouver dans des états stationnaires, ainsi que les valeurs minima des périodes des photons qui peuvent être émis lors des transitions entre ces états, il ne peut tout d'abord être question d'utiliser comme instruments de mesure des corps solides édifiés, comme les substances ordinaires, au moyen d'électrons; de plus, d'après la théorie de Dirac, tout dispositif capable d'agir sur des électrons dans des domaines inférieurs à λ et ϑ provoquera une création incessante de couples d'électrons et de positrons, qui, non seulement apporteront une perturbation aux mesures, mais rendront même impossible l'existence permanente du dispositif. A ce propos, il faut toutefois observer que dans toutes les questions de mesures proprement relativistes, où nous avons affaire essentiellement à des valeurs moyennes des grandeurs physiques dans des domaines finis tant spatiaux que temporels, nous n'avons nullement besoin d'instruments de mesure rigoureusement stationnaires, mais seulement de dispositifs dont la durée est suffisamment grande par rapport aux intervalles de temps intervenant dans la mesure. Or, même des dispositifs de mesure dont les dimensions linéaires sont du même ordre de grandeur que λ ont certainement, au moins en principe, une vie moyenne qui, à cause précisément de la petitesse de la quantité ε, est beaucoup plus longue que ϑ; c'est ce que montre par exemple la finesse considérable des rayons X caractéristiques même les plus pénétrants. La probabilité de création de couples de positrons et d'électrons sous l'influence de champs de forces, qui provoque entre autres cette curieuse polarisation du vide qui est une conséquence si intéressante de la théorie de Dirac, est justement aussi, comme nous l'avons vu, proportionnelle à ε et n'est par suite, tant que les dimensions critiques sont grandes par rapport à δ et τ, que d'une importance relativement secondaire pour la réaction entre instrument et objet qui détermine l'exactitude des mesures.

Lorsque l'on considère que la formule déduite par Klein et Nishina de la théorie de Dirac pour la dispersion du rayonnement par les électrons a été assez bien vérifiée pour des rayons γ de longueurs d'onde considérablement moindres que λ, et que les conclusions de la théorie concernant l'apparition des positrons ont été confirmées au moins qualitativement, il n'y a vraiment pas lieu de douter qu'une étude suffisamment approfondie du problème des mesures dans la théorie relativiste de l'électron ne révèle entre les possibilités de mesures et les axiomes et conséquences de la théorie une harmonie qui soit aussi parfaite que celle qui pour l'idéalisation de la mécanique quantique non relativiste est exprimée par le principe d'indétermination, et qui puisse tout aussi naturellement être englobée dans le concept général de complémentarité. Il me semble par conséquent que du point de vue de la méthode de correspondance, il est permis de considérer la théorie de Dirac comme une application logique des concepts fondamentaux de la théorie classique de l'électron, dont les limites ultimes sont données par les quantités δ et τ, et dont on peut attendre une bonne approximation dans la solution de problèmes dont les dimensions caractéristiques sont sensiblement inférieures à λ et θ.

Cette attitude ne signifie nullement que je sous-estime la remarquable différence qui existe entre la théorie classique de l'électron et la théorie de Dirac, et que révèlent les effets de spin et surtout l'apparition de positrons; je veux simplement dire qu'après la découverte de Dirac tous ces phénomènes nous apparaissent comme des conséquences nécessaires des propriétés élémentaires de l'électron et de l'existence du quantum d'action. A ce propos, il est intéressant de se rappeler qu'il y a une grande différence entre les effets de spin et le phénomène du positron au point de vue de leur interprétation univoque à l'aide de concepts classiques. Tandis que les concepts de moment cinétique et de moment magnétique propres de l'électron et du positron, de même que le principe d'exclusion de Pauli, font partie intégrante du formalisme de la mécanique quantique et ne sont susceptibles d'aucune définition classique, la charge et la masse du positron peuvent être mesurées d'une manière purement classique, comme les propriétés correspondantes de l'électron. C'est justement cette circonstance qui rend possible, comme l'a surtout signalé Pauli, une présentation entièrement symétrique de la théorie des lacunes, dans laquelle électrons et positrons sont traités dès le début comme des éléments également essentiels des définitions classiques qui sont à la base de la théorie.

Jusqu'ici nous avons complètement négligé les paradoxes relatifs au couplage entre l'électron et son propre champ, auxquels conduit le formalisme de l'électrodynamique quantique, et qui sont en flagrante contradiction non seulement avec le postulat de l'existence d'états stationnaires, fondamental pour la théorie de la constitution des atomes et des spectres, mais même avec les expériences qui sont à la base de la théorie classique de l'électron. Par conséquent, il est clair que ces paradoxes ne touchent pas à proprement parler la théorie de correspondance de l'électron; celle-ci étant, comme je l'ai déjà dit, un procédé approximatif pour traiter les problèmes de rayonnement, qui repose sur l'hypothèse que l'on peut en première approximation négliger complètement la réaction de rayonnement dans le calcul des états stationnaires des atomes, de la même manière que dans tous les problèmes de la théorie classique de l'électron dans lesquels interviennent des dimensions caractéristiques spatio-temporelles grandes par rapport à δ et τ. La justification théorique de cette hypothèse, dont la validité est confirmée par toutes les expériences, doit être cherchée exclusivement, comme je l'ai dit souvent, dans la petitesse effective de la constante ε. Une toute autre question est de savoir comment il faut procéder d'une manière logique pour obtenir non seulement des valeurs approchées des fréquences des raies spectrales, mais aussi les écarts à partir de ces fréquences, qui déterminent la largeur naturelle des raies, et qui sont très importants pour les détails des phénomènes de dispersion optique.

A mon avis, ce problème très discuté ne présente aucune difficulté de principe, puisque la détermination, effectuée déjà dans la première approximation à l'aide de la méthode de correspondance, de la probabilité d'émission d'un photon lors de la transition d'un état stationnaire à un autre suffit entièrement à déterminer aussi la largeur de la raie à l'approximation même où ce problème est défini. En effet, il me semble possible et naturel de regarder toute la question de la largeur des raies comme une partie intégrante de la manière simple de traiter les problèmes de dispersion, basée sur le principe de correspondance. Comme on sait, la forme des lignes déduite de la théorie classique de la dispersion à partir de l'amortissement des vibrations par rayonnement correspond à une loi de répartition pour les valeurs de l'énergie d'un état stationnaire, qui a justement la propriété de fournir pour la différence entre des couples de valeurs quelconques des énergies de deux états distincts une loi de répartition du même type, avec la seule différence que la largeur de la raie considérée est la somme des largeurs correspondant aux deux lois de répartition. Ce procédé, comme d'ailleurs toute la théorie de la dispersion dont il est question ici, ne se justifie que par la petitesse extrême des probabilités des transitions de rayonnement, qui permet de considérer ces probabilités comme des quantités additives indépendantes, en harmonie avec le principe général de superposition de la théorie classique du rayonnement. Il est vrai que le symbolisme non commutatif introduit par Dirac pour décrire les champs de rayonnement a été très utile dans l'étude du problème de la largeur des raies, mais en raison des paradoxes qu'entraîne l'application rigoureuse de ce symbolisme, on ne peut, pour les problèmes proprement atomiques, le considérer que comme un artifice permettant de condenser d'une manière commode des raisonnements de correspondance essentiellement approximatifs.

Si l'on adopte ce point de vue, la seule question qui se pose est d'examiner en quel point précis l'électrodynamique quantique générale édifiée à partir du symbolisme de Dirac dépasse le domaine d'application légitime de la méthode de correspondance. Comme je l'ai dit, Landau et Peierls ont essayé de répondre à cette question en disant que le concept de champ n'est susceptible d'aucune application logique dans le domaine où la théorie quantique joue un rôle essentiel; en effet, ils ont cru pouvoir montrer qu'aucune mesure de grandeur de champ n'est possible dans ce domaine. Toutefois dans leurs raisonnements ils ont utilisé exclusivement comme corps d'épreuve des points matériels chargés, dont l'emploi apparaît de prime abord comme particulièrement désavantageux. En effet, la condition, nécessaire pour les mesures, que l'interaction entre le corps d'épreuve et le champ soit suffisamment grande par rapport à l'interaction incontrôlable entre ce corps et les instruments de mesure nécessaires pour établir sa coordination spatio-temporelle, exige que la charge du corps d'épreuve soit grande par rapport à $\sqrt{hc}$, et donne par suite lieu pour une charge ponctuelle, contrairement à ce qui se passe dans la théorie de correspondance de l'électron basée sur la petitesse de la constante ε, à une réaction de rayonnement qui perturbe d'une manière essentielle et incontrôlable le comportement mécanique du corps d'épreuve.

Mais un examen plus approfondi montre que des corps d'épreuve ponctuels ne conviennent aucunement à des mesures de champs dans le domaine de l'électrodynamique quantique. L'idéalisation des composantes de champ définies en chaque point de l'espace-temps, qui est caractéristique de la théorie de l'électrodynamique classique, n'est en effet pas applicable dans la théorie quantique, où l'on a essentiellement affaire aux valeurs moyennes des grandeurs de champs dans des domaines spatio-temporels finis. Pour mesurer de telles valeurs moyennes, on ne peut naturellement utiliser que des corps d'épreuve d'étendue finie et dont la charge est répartie d'une manière continue. Bien que ceci implique que dans toutes ces mesures il faille faire abstraction de la constitution atomique des corps d'épreuve, il n'en résulte aucune restriction essentielle pour la vérification des conséquences de la théorie quantique des champs, puisque celle-ci ne renferme pas d'autres constantes universelles que c et h, qui à elles seules ne suffisent pas à fixer de dimensions spatio-temporelles absolues. Aussi bien, dans un travail (¹) qui paraîtra bientôt, Rosenfeld et moi avons pu montrer que pour autant qu'on puisse faire abstraction de la constitution atomique des instruments de mesure,

---

(¹) N. Bohr et L. Rosenfeld, *Acad. Copenhague, Math. Phys. Com.*, t. XII, 1933, p. 8.

on obtenait effectivement une concordance complète entre les possibilités de mesure des grandeurs de champ électromagnétiques et les axiomes et conséquences de la théorie quantique du rayonnement. On constate en particulier que ces curieuses fluctuations des grandeurs de champ dans le vide, qui sont une conséquence caractéristique de cette théorie, et dans lesquelles on avait cru voir un argument en faveur des conclusions de Landau et Peierls, constituent un élément essentiel pour la non-contradiction du mode de description complémentaire, parce qu'elles sont essentiellement inséparables des champs incontrôlables qu'engendre fatalement l'utilisation des corps d'épreuve chargés.

Je pense que les considérations précédentes font apparaître assez clairement la nature de l'opposition, à première vue si surprenante, entre la théorie de correspondance des électrons et le formalisme de l'électrodynamique quantique. De même que la théorie classique de l'électron est une idéalisation permettant, en dehors de la limitation symbolisée par δ et τ, de traiter les phénomènes atomiques dans la description tant mécanique qu'électromagnétique desquels n'entrent que des quantités d'action grandes par rapport à h, de même l'électrodynamique quantique est une idéalisation qui trouve un domaine d'application légitime dans la description de l'interaction entre des champs électromagnétiques et des corps matériels dont les charges sont grandes par rapport à $\sqrt{hc}$ et dont les dimensions linéaires sont par suite grandes par rapport à $\dfrac{h}{\mathrm{M}c}$, où M est la masse du corps. De son côté, la théorie de correspondance de l'électron est une méthode d'approximation qui n'est rendue possible que par la petitesse effective de la constante ε, et qui s'applique aux nombreux problèmes atomiques tombant en dehors des domaines d'application des deux idéalisations précédentes. L'usage apparemment illogique du concept de champ, caractéristique de cette méthode, est basé exclusivement sur le fait que, d'après la nature même de la méthode, le champ électromagnétique n'est jamais traité en lui-même comme un objet indépendant des particules auxquelles on applique la théorie quantique. C'est ainsi que les champs définissant l'interaction entre les particules des atomes sont traités exclusivement comme

[179]

un attribut de ces particules susceptible d'une définition classique, tandis que les effets de rayonnement des atomes sont traités comme une conséquence de la description quantique de cette interaction. Le fait que cette manière de traiter les effets de rayonnement est compatible tant avec le principe général de superposition de l'optique qu'avec la conservation de l'énergie et de l'impulsion dans les processus individuels de rayonnement, symbolisée par le concept de photon, nous assure en outre que nous ne rencontrerons jamais aucune contradiction avec les conséquences bien définies de l'électrodynamique quantique, pas plus que nous n'en trouverons, d'après la nature même de la méthode de correspondance, avec les applications légitimes de la théorie classique de l'électron.

Je voudrais encore ajouter quelques mots sur la relation entre la théorie de correspondance de l'électron et le problème de la constitution des noyaux. Dans ce domaine un tout nouveau caractère de la théorie atomique nous est offert par l'existence du neutron dont la stabilité est, du point de vue actuel de la théorie atomique, un fait tout aussi élémentaire que l'existence de l'électron. En particulier le rapport $\mu = \dfrac{m}{M}$ entre la masse m de l'électron et la masse M du neutron est une constante naturelle, dont la petitesse vis-à-vis de l'unité est certainement aussi importante pour la constitution des noyaux que la petitesse de la constante ε l'est pour la constitution des configurations électroniques entourant les noyaux. Ce sont, en effet, avant tout les masses relativement grandes des particules nucléaires qui permettent d'expliquer les lois des désintégrations radioactives α et les relations entre les niveaux d'énergie que ces désintégrations permettent d'observer et les spectres de rayons γ, en utilisant les concepts fondamentaux de la théorie quantique de la constitution des atomes, tels que ceux d'états stationnaires et de processus individuels de transition. La seule différence caractéristique entre le problème de la constitution des noyaux et la théorie de la constitution des atomes est que dans le premier cas, par opposition au dernier, nous ne pouvons *a priori* tirer des lois de l'électromagnétisme classique aucun renseignement sûr au sujet des forces agissant entre les particules nucléaires, mais que toutes nos déductions concernant ces forces reposent sur un ensemble d'expériences entièrement nouvelles.

Je désirerais notamment insister sur le fait qu'il n'est nullement possible d'appliquer immédiatement les concepts de la théorie de l'électron dans le domaine propre des phénomènes nucléaires. Que l'on regarde le proton comme formé d'un neutron et d'un positron, ce qui d'après les dernières expériences paraît bien être l'hypothèse la plus naturelle, ou bien qu'on le regarde comme le produit d'une dissociation de neutron accompagnée de la libération d'un électron il s'agit là de processus qui ne peuvent pas être décrits par les moyens actuels, et dont la possibilité doit être cherchée dans le fait que les dimensions empiriquement connues du neutron sont du même ordre de grandeur que le diamètre de l'électron δ, lequel exprime la limite à partir de laquelle les concepts de la théorie classique de l'électron et leur utilisation d'après la méthode de correspondance nous font complètement défaut. A ce sujet on peut aussi remarquer que l'intéressante découverte de Stern, d'après laquelle le moment magnétique du proton s'écarte très sensiblement de la valeur du magnéton multipliée par μ, doit sans doute également trouver son explication dans le fait que le diamètre du neutron, et par suite celui du proton, est sensiblement supérieur à $\mu\lambda$; en effet, comme je l'ai déjà dit, l'application de la théorie de l'électron de Dirac à des effets proprement relativistes présuppose justement que λ soit grand vis-à-vis du diamètre de l'électron δ.

En terminant, je voudrais faire remarquer que si j'ai défendu la nécessité d'envisager sérieusement l'idée que les théorèmes de conservation d'énergie et d'impulsion pourraient bien nous faire défaut dans le cas des spectres continus de rayons β, mon intention était surtout d'insister sur l'insuffisance totale des conceptions classiques pour traiter ce problème, qui pourrait nous réserver encore de grandes surprises. J'apprécie pleinement le poids de l'argument d'après lequel une telle éventualité serait difficilement conciliable avec la théorie de la relativité, et formerait en particulier un contraste peu vraisemblable avec la validité absolue, s'étendant également au domaine des phénomènes nucléaires, du théorème de la conservation de l'électricité, qui

est analogue aux autres d'après la théorie générale des champs.
A ce propos, il faut toutefois remarquer que cette comparaison
même indique combien il serait difficile de démontrer un écart
direct à partir de la théorie de la relativité, même si la masse
et l'énergie totales associées aux particules et aux champs de
rayonnement ne se conservait pas dans les processus nucléaires.
De même que la conservation de l'électricité à l'intérieur d'un
domaine dont la surface n'est pas traversée par des charges est,
tout au moins au point de vue macroscopique, une conséquence
nécessaire de la validité des équations du champ électromagnétique à l'extérieur de cette surface, de même, comme l'a fait
remarquer Landau, il découle nécessairement de la théorie de
la gravitation qu'une variation éventuelle d'énergie à l'intérieur
d'un certain domaine sera accompagnée de variations des forces
de gravitation à l'extérieur de ce domaine qui correspondront
exactement à un transport de masse à travers la surface. Mais
la question qui se pose est de savoir si nous devons nécessairement exiger que toutes ces actions de gravitation soient liées
à des particules atomiques, de la même manière que les quantités
d'électricité sont liées à des électrons. Avant que l'on ait de
nouvelles expériences dans ce domaine, il me paraît par conséquent difficile de prendre position au sujet de l'intéressante suggestion de Pauli, qui propose d'expliquer les paradoxes de l'émission
des rayons β en admettant que les noyaux émettent en même temps
que les électrons des particules neutres, très légères par rapport
aux neutrons. En tout cas l'existence éventuelle de ce « neutrino »
représenterait un élément entièrement nouveau de la théorie
atomique, de l'intervention duquel dans les réactions nucléaires
la méthode de correspondance ne nous offre aucun moyen suffisant
de description.

TRANSLATION

On the Correspondence Method in Electron Theory[1]

The wonderful confirmation which Dirac's theory of the electron has received through the discovery of the positron has shed new light on the paradoxes that for some time seemed to limit in a decisive way the application of correspondence arguments in relativistic quantum mechanics.

The point of departure of this method is the classical electron theory, which represents a direct application of classical mechanics and electrodynamics to systems of charged point particles. It is well known, however, that such an idealization is only applicable to the extent that one may disregard all variations of the forces acting on each particle within space–time regions whose dimensions, in a reference frame in which the electron is momentarily at rest, are given by the so-called "electron diameter"

$$\delta = \frac{e^2}{mc^2},$$

and the corresponding proper time interval

$$\tau = \frac{\delta}{c} = \frac{e^2}{mc^3}.$$

I should like to recall especially that these conditions are necessary in order that the radiation reaction due to the acceleration of the electron be sufficiently small relative to the external forces acting on the electron to permit an unambiguous application of the concept of mass. The fact that δ and τ are very small relative to the region in which the electron is bound even in the heaviest atoms is also essential in allowing an extensive application of classical concepts in the quantum theory of atomic constitution and spectra. The smallness of the coupling between atom and radiation compared to the coupling between the particles of the atom permits in particular a complete neglect of the radiation reaction in the first approximation. All these circumstances are, as is well known, solely due to the smallness of the dimensionless quantity

$$\varepsilon = \frac{e^2}{hc},$$

which is the fundamental condition allowing the reinterpretation of the classical electron theory by the correspondence principle.

---

[1] Elaborated version of some general remarks by the author during the discussion.

The unforeseen paradoxes concerning the appearance of negative energies and of an infinite coupling between an electron and its own field, which the attempts at developing a relativistic quantum mechanics and a general quantum electrodynamics brought to light, led to the suspicion, however, that the applicability of space–time concepts as well as of the field concept in the correspondence method should be subject to stricter limitations than in the classical electron theory. Hence, as is well known, Landau and Peierls, from a study of the possibility of localizing an electron by means of scattering of radiation bundles, have arrived at the conclusion that in the rest system of the electron the length

$$\lambda = \frac{h}{mc},$$

known from the theory of the Compton effect, and the corresponding time

$$\theta = \frac{h}{mc^2}$$

– which was already designated previously by Richardson and Flint as the ultimate proper time interval – should represent absolute limits for any rational use of space–time concepts in the electron theory. Furthermore, the same authors concluded from a study of the measurability of the characteristic field quantities that the classical field concept could never be employed in problems where the quantum of action plays an essential rôle. This would obviously mean that all attempts hitherto made to establish a strictly relativistic quantum extension of electron theory, as well as of field theory, would contain such contradictions that one could not attribute any physical significance to them. As we shall see, these conclusions cannot be maintained, however, and in particular I shall try to show that the situations within the domains of electron theory and field theory are fundamentally different.

As far as the electron theory is concerned, it is true that one cannot carry out space–time measurements within intervals smaller than λ and θ in as simple a way as presupposed in the usual derivation of the Heisenberg uncertainty relations. In this connection, however, we must not forget that as a consequence of the complementary mode of description of quantum phenomena, the question of the interpretation of measurements appears with an essentially different character depending on whether the aim of the measurement is to fix the initial conditions for the problem considered, or whether it is a question of verifying the fundamentally statistical consequences of the theory for a problem with given initial conditions. In the first case we evidently do not encounter any restrictions in principle in relativistic quantum mechanics, which is already a consequence of the invariant form of the uncertainty relations. After a measurement aiming at a space–time localization more accurate than that given by λ and θ, there cannot – because of the complementary uncertainty in the momentum – be any question of fixing a reference system in which the electron has a velocity which is small compared to c. In the second case, where the aim of the measurement is to verify the statistical consequences of the theory, one cannot demand at all that the individual measurement should have an unambiguous interpretation, but only that it should be possible, by repeating the measurement a sufficient number of times under the same initial conditions, to verify the correctness of all well-defined properties of the solution of the problem – like, for example, the space–time dependence of the density function characteristic of the quantum mechanical symbolism. Hence, in problems like those considered by Landau and Peierls, involving a reference system in which the electrons are approximately at rest, one can certainly not interpret the observation of a single Compton effect as a measurement of the position of the electron before the observation with an accuracy greater than λ. However, such an interpretation would also not correspond to the verification of any well-defined prediction of the theory. Still, by performing a sufficiently large number of measurements of the scattering of radiation bundles with a sufficient variation in space and time, one can in principle verify all well-defined properties of the density function for systems like those considered, where all dimensions appearing explicitly in the description of the problem are large compared to λ and θ and where the relativistic modifications only involve comparatively small deviations from the non-relativistic formalism.

As soon as we turn to problems in which space–time quantities of the same order of magnitude as, or smaller than, λ and θ enter explicitly, and thus definitely leave the region where the non-relativistic formalism is even approximately correct, we must for all possible measuring arrangements take into account the creation of pairs of positrons and electrons, which of course implies that any interpretation of such measurements will present great complications, not only regarding the reactions of the measuring instruments on the object, but also concerning the very constitution of the measuring instruments themselves. Since λ and θ represent the smallest dimensions of regions inside which the electron can exist in stationary states in any field of force, as well as the smallest values of the periods of photons which can be emitted by transitions between such states, there is in the first place no question of using as measuring instruments solid bodies which, like ordinary substances, are built from electrons. Furthermore, according to Dirac's theory, any apparatus suited to react with electrons inside regions smaller than λ and θ would incessantly produce pairs of electrons and positrons which not only would disturb the measurements, but also prevent the permanent existence of the instrument itself. In

[185]

this connection we must, however, keep in mind that in all proper relativistic measuring problems, where, as a matter of principle, we are dealing with mean values of the physical quantities over finite space–time regions, we do not at all need strictly stationary measuring instruments, but only instruments whose lifetime is sufficiently large compared to the time intervals involved in the measurement. Even for measuring arrangements of linear dimensions of the same order of magnitude as λ, a lifetime much larger than θ is, at least in principle, ensured just because of the smallness of the quantity ε, as illustrated by the considerable sharpness of even the most penetrating characteristic X-rays. The probability for the creation of pairs of positrons and electrons under the influence of force fields – which, for instance, produces the peculiar polarization of the vacuum that is such an interesting consequence of Dirac's theory – is, as we have seen, simply proportional to ε and therefore, as long as the critical dimensions are large compared to δ and τ, of comparatively secondary importance for the reaction between instrument and object, which is decisive for the precision of the measurements.

When one considers the close verification which the formula for scattering of radiation by electrons, derived by Klein and Nishina on the basis of Dirac's theory, has received for γ-rays of wavelengths considerably smaller than λ, as well as the at least qualitative verification of all the conclusions of the theory concerning the occurrence of positrons, there is indeed no reason whatsoever to doubt that a sufficiently thorough investigation of the measurement problem within relativistic electron theory will reveal an equally complete harmony between the assumptions and the consequences of the theory, as that which, in the idealized case of non-relativistic quantum mechanics, is expressed by the uncertainty principle, and which just as naturally can be incorporated into the general concept of complementarity. It therefore appears to me that from the correspondence point of view, it is permissible to consider Dirac's theory as a consistent application of the fundamental concepts of the classical electron theory, of which the ultimate limits are given by the quantities δ and τ, and to expect that it will yield a good approximation when dealing with problems having characteristic dimensions considerably smaller than λ and θ.

This attitude in no way implies an underestimation of the remarkable difference between the classical electron theory and Dirac's theory, revealed by the spin effect and above all by the occurrence of the positron, but only that we, after Dirac's discovery, recognize all these features as inevitable consequences of the elementary properties of the electron and of the existence of the quantum of action. In this connection it is interesting to remember that there is a great difference between spin effects and the positron phenomenon as regards their unambiguous interpretation by classical concepts. While the concepts of angular momentum and intrinsic magnetic moment of the electron and the positron, like Pauli's exclusion principle, are integral parts of the quantum mechanical formalism, which do not permit any classical definition, the charge and mass of the positron can be measured on a purely classical basis like the corresponding properties of the electron. It is precisely this circumstance which, as especially emphasized by Pauli, permits a completely symmetrical representation of the hole theory in which electrons and positrons from the outset are treated as equally fundamental elements of its classically defined basis.

So far we have completely neglected the paradoxes concerning the coupling between an electron and its own field to which the quantum electrodynamics formalism leads, and which are in complete contradiction not only with the assumption of the postulate of the existence of stationary states, fundamental for the theory of atomic structure and spectra, but even with the experience forming the basis of the classical electron theory. Thus it is clear that properly speaking these paradoxes do not affect the correspondence theory of the electron which, as already mentioned, is an approximation method for treating the radiation problems resting on the assumption that in the first approximation we can neglect the radiation resistance in the calculation of the stationary states of the atoms – in the same way as in all those problems of classical electron theory where the characteristic space–time dimensions are large compared to δ and τ. The theoretical justification for this assumption, whose validity is confirmed by all experiments, depends, as often said, entirely on the smallness of the constant ε. It is altogether a different question how one is to proceed systematically in order to obtain not only the approximate values of the frequencies of the spectral lines, but also the deviations from these values which determine the natural width of the lines, and which are very important for the details of the phenomenon of optical dispersion.

In my opinion this much discussed problem does not present any fundamental difficulty since the determination, carried out already in first approximation by means of correspondence arguments, of the probability of emission of a photon by the transition from one stationary state to another, is quite sufficient to determine also the line width in the approximation to which the problem is defined. Actually, it seems possible and natural to consider the whole question of the line width as an integral part of the simple correspondence treatment of the dispersion problems. As is well known, the line shape deduced from the classical dispersion theory on the basis of the radiation damping of the vibrations corresponds to a distribution law for the energy values of stationary states which has precisely the property that it yields a distribution law for the difference between two arbitrary energy values of two definite states of the same type, with the only difference that the width of the line considered is the sum of the widths corresponding to the two distribution laws. This procedure, as well as the entire dispersion theory that we are dealing with here, is only justified by the extreme smallness of the radiative transition probabilities, which permits us to consider these probabilities as independent additive quantities, in harmony with the general superposition principle of classical radiation theory. It is true that the non-commutative symbolism introduced by Dirac to describe radiation fields has been very useful for the study of the problem of line width, but in consideration of the paradoxes to which the rigorous application of this symbolism leads, one can, for the proper atomic problems, only consider it as an artifice, which allows us to summarize in a convenient manner the fundamentally approximative correspondence arguments.

From this point of view, the only question is, at which point precisely the general quantum electrodynamics, built upon Dirac's symbolism, goes beyond the legitimate domain of applicability of the correspondence method. As already mentioned, Landau and Peierls have attempted to answer this question by asserting that the field concept does not allow any consistent application within the region where the quantum theory plays an essential rôle; in fact, they believed that they could show that no measurement of field quantities within this region is possible. However, in their considerations they used exclusively charged point particles as test bodies, the application of which from the outset appears especially unfavourable. In fact, the condition necessary for the measurements, i.e. that the interaction between the body and the field must be sufficiently large compared to the uncontrollable interaction between this body and the measuring instruments needed in order to establish its space–time coordination, requires that the charge of the test body is large relative to $\sqrt{hc}$, and therefore in the case of a point charge – in contrast to the situation in the correspondence theory of the electron based on the smallness of the constant ε – gives rise to a radiation reaction which in an essential and uncontrollable way disturbs the mechanical behaviour of the test body.

A closer investigation shows, however, that point-like test bodies are not at all suited for field measurements within the domain of quantum electrodynamics. The idealization of field components defined in each space–time point, which is characteristic of classical electrodynamics, is not applicable in the quantum theory, where, as a matter of principle, we always deal with mean values of field quantities over finite space–time regions. For the measurement of such mean values one can of course only employ test bodies of finite extension and with a continuous charge distribution. Although this implies that in all such measurements we have to ignore the atomic constitution of the test bodies, it does not represent any fundamental restriction as regards the verification of the predictions of the quantum theory of fields, since this theory does not contain universal constants other than c and h which by themselves are not sufficient to fix any absolute space–time dimensions. As a matter of fact, in a paper[2] which is going to appear soon, Rosenfeld and I have been able to demonstrate that, to the extent that one can ignore the atomic constitution of the measuring instruments, one obtains really a complete agreement between the possibility for measurement of electromagnetic field quantities and the assumptions and consequences of the quantum theory of radiation. In particular it has become apparent that the peculiar fluctuations of the field quantities in empty space, which are a characteristic consequence of this theory, and in which one has believed to see an argument in favour of the attitude of Landau and Peierls, constitute an element essential for the consistency of the complementary mode of description, since these fluctuations as a matter of principle are inseparable from the uncontrollable fields which the use of the charged test bodies inevitably produces.

I think that the preceding considerations sufficiently clearly elucidate the nature of the contrast, at first sight so surprising, between the correspondence theory of the electron and the quantum electrodynamical formalism. Just as the classical electron theory is an idealization which, apart from the limitation symbolized by δ and τ, permits one to treat atomic phenomena in which, in the mechanical as well as the electromagnetic description, all relevant action quantities are large in comparison with h, so quantum electrodynamics is an idealization which has a legitimate domain of application in the description of the interaction between electromagnetic fields and material bodies whose charges are large in comparison with $\sqrt{hc}$ and whose linear dimensions accordingly must be large in comparison with h/Mc, where M is the mass of the body. As far as the correspondence theory of the electron is concerned, it is an approximation method, made possible solely by the effective smallness of the constant ε, which applies to the numerous atomic problems that fall outside the domains of applicability of each of the two preceding idealizations. The seemingly inconsistent application of the field concept, characteristic of this method, is exclusively based on the fact that the electromagnetic field is never itself treated as an object independent of the particles to which one applies quantum theory. Thus, the fields determining the interaction between particles of the atoms are exclusively treated as an attribute of these particles, susceptible to a classical definition, while the radiative reactions are treated as being a consequence of

---

[2] N. Bohr and L. Rosenfeld, *Zur Frage der Messbarkeit der elektromagnetischen Feldgrössen*, Mat.–Fys. Medd. Dan. Vidensk. Selsk. **12**, no. 8 (1933). [Reproduced on p. [55].]

[189]

the quantum-mechanical description of this interaction. The fact that this treatment of radiative effects is compatible with the general superposition principle of optics, as well as with the conservation of energy and momentum in the individual radiation processes, symbolized by the photon concept, further ensures that we will never get into conflict with the well-defined consequences of quantum electrodynamics nor can there in the nature of the correspondence method ever be any question of a conflict with legitimate applications of the classical electron theory.

I would like to add a few words about the relation between the correspondence theory of the electron and the problem of nuclear constitution. We here meet with an entirely new characteristic of atomic theory in the existence of the neutron, the stability of which, from the current point of view of atomic theory, is as elementary a fact as the existence of the electron. In particular, the ratio $\mu = m/M$ between the electron mass m and the neutron mass M is a constant of nature, the smallness of which compared to unity is certainly as important for the constitution of nuclei as the smallness of the constant ε is for the constitution of the configuration of electrons surrounding the nuclei. Indeed it is, above all, the relatively large masses of the nuclear particles which make it possible to explain the laws of radioactive α-disintegrations and the relations between the energy levels observed in these disintegrations, and the γ-ray spectra by means of the fundamental concepts of the quantum theory of atomic constitution, such as stationary states and individual transition processes. The only characteristic difference between the problem of nuclear constitution and the theory of atomic constitution is that in the former case, as opposed to the latter, we cannot draw a priori any conclusions about the forces between the nuclear particles from the laws of classical electromagnetism, but all deductions about these forces rest on an entirely new empirical basis.

In particular I would like to stress the fact that there is no possibility at all of applying the concepts of electron theory directly within the proper domain of nuclear phenomena. Whether one regards the proton as a combination of a neutron and a positron, which, according to the latest evidence, might well be the most natural hypothesis, or whether one regards it as the product of a dissociation of a neutron with the associated emission of an electron, we are here concerned with processes which cannot be described on the present basis, and the possibility of which has to be sought in the fact that the empirically known dimensions of the neutron are of the same order of magnitude as the electron diameter δ, which represents the limit beyond which the concepts of classical electron theory, and their application according to the correspondence principle, fail completely. In this connection one may also note that the interesting discovery by Stern, according to which the magnetic moment of the proton differs appreciably from the value of the magneton times μ, must no doubt also find its explanation in the fact that the diameter of the neutron, and therefore also that of the proton, is appreciably larger than $\mu\lambda$; indeed, as already mentioned, the application of Dirac's electron theory to proper relativistic effects depends just on λ being large compared to the electron diameter δ.

In conclusion, I would like to remark that if I have advocated that one seriously consider the idea of a possible failure of the theorems of conservation of energy and momentum in connection with the continuous β-ray spectra, my intention was above all to emphasize the total inadequacy of the classical conceptual edifice for treating this problem, which could still hold great surprises for us. I fully appreciate the weight of the argument that such a possibility would be difficult to reconcile with the theory of relativity, and would in particular stand in a rather unlikely contrast to the absolute validity of the theorem, analogous according to general field theory, of conservation of electric charge, which extends also to the region of nuclear phenomena. In this connection one may, however, remark that this comparison itself indicates how difficult it would be to prove a direct deviation from the theory of relativity, even if the total mass and energy associated with the particles and the radiation fields were not conserved in nuclear processes. Just as the conservation of charge inside a region whose boundary is not crossed by charges is, at least macroscopically, a necessary consequence of the validity of the electromagnetic field equations outside this boundary, so, as Landau has pointed out, it is a necessary consequence of the theory of gravitation that any variation of the energy inside a certain region must be accompanied by variations in the gravitational forces outside this region, which would correspond exactly to a mass transport across its boundary. However, the question is whether we must necessarily require that all such gravitational effects are associated with atomic particles in the same way as the electric charges are associated with electrons. Therefore, until we have further experience within this area, it seems to me difficult to assess Pauli's interesting suggestion to resolve the paradoxes of the β-ray emission by assuming that the nuclei emit, together with the electrons, neutral particles, much lighter than the neutrons. In any case, the possible existence of this "neutrino" would represent an entirely new element in atomic theory, and the correspondence method would not offer sufficient help in describing its rôle in nuclear reactions.

[191]

M. Bohr. — Je me demande si, en somme, une vérification expérimentale de ces conséquences de la théorie des lacunes doit être regardée comme possible ou non.

Comme je l'exposerai en détail au cours des remarques générales que j'aurai à faire, la théorie de Dirac tout entière a un caractère essentiellement approximatif, et, dans le domaine dont il est question ici, il faut s'attendre à ce que des effets petits vis-à-vis de $\frac{e^2}{hc}$ n'aient aucune signification bien définie.

Ici se pose encore la question de la validité de la loi de Klein-Nishina. Contrairement à l'opinion que j'ai émise au Congrès de Rome en 1931, je crois maintenant qu'elle cesse d'être valable dans la région où la longueur d'onde du photon devient du même ordre de grandeur que le rayon de l'électron ($h\nu$ à peu près égal à $137 \, mc^2$); en effet, au cours d'une discussion avec M. Landau, je me suis aperçu d'une erreur dans l'argument basé sur la considération de l'effet Compton dans un système de référence où le centre de gravité de l'électron et du photon est au repos, et où la longueur d'onde du photon incident est sensiblement plus grande que dans le système de référence habituel; car dans tout système de référence, c'est dans la région indiquée que les forces de radiation deviennent du même ordre que les forces d'inertie.

Or M. Dirac a dit qu'il prévoit, pour les petites longueurs d'onde, un écart à la formule de Klein-Nishina, dû à ce que la polarisation du vide ne suit plus les oscillations rapides des forces électromagnétiques. N'est-il pas possible que cet écart échappe à une vérification quantitative, en raison de la restriction générale dont je viens de parler ?

TRANSLATION

BOHR: On the whole, I ask myself whether or not an experimental verification of these consequences of the hole theory should be regarded as possible.

As I shall explain in detail during the general remarks* which I shall make, Dirac's entire theory has fundamentally an approximate character and, in the region with which we are dealing here, one must expect that effects which are small compared to e^2/hc do not have any well-defined meaning.

Here we are furthermore faced with the question of the validity of the Klein–Nishina formula. Contrary to the opinion I expressed at the Rome Congress in 1931, I now believe that it ceases to be valid in the region where the wavelength of the photon becomes of the same order of magnitude as the radius of the electron (hv almost equal to $137mc^2$). As a matter of fact, during a discussion with Landau I discovered an error in the argument based on the consideration of the Compton effect in a reference system in which the centre of mass of the electron and of the photon is at rest, and where the wavelength of the incident photon is appreciably larger than in the usual reference system; because, in any reference system, it is in the mentioned region that the radiation forces are of the same order of magnitude as the inertial forces. Now Dirac has said that for small wavelengths he anticipates a deviation from the Klein–Nishina formula due to the fact that the vacuum polarization no longer follows the rapid oscillations of the electromagnetic forces. Is it not possible that this deviation defies a quantitative verification because of the general restriction that I have talked about?

* [See preceding article.]

IV. FIELD AND CHARGE MEASUREMENTS IN QUANTUM THEORY

UNPUBLISHED MANUSCRIPT FROM FOLDER LABELLED *FIELD AND CHARGE MEASUREMENTS IN QUANTUM THEORY A. 1937*

See Introduction to Part I, sect. 3.

[195]

This manuscript is part of the material in the folder with the general title "Field and Charge Measurements" which is preserved in the Niels Bohr Archive (Bohr MSS, microfilm no. 19).

The manuscript consists of 17 typewritten pages and the original page numbers are given in the following by R.1, R.2 etc. in the margin. There are handwritten corrections and remarks, mostly by Bohr, in the manuscript, and, when legible, these have been inserted.

FIELD AND CHARGE MEASUREMENTS IN QUANTUM THEORY. R.1

Introduction

In connexion with the still unsolved problems of quantum electrodynamics and electron theory the question has repeatedly been raised whether the use of classical physical concepts made in present quantum theoretical methods is justifiable. Especially doubt has been expressed whether an unambiguous interpretation of measurements of field quantities and charge densities can be given to the full extent implied by these methods.

It is true that we meet here with a new aspect of the problem of measurements which lies beyond the situation with which we are familiar in ordinary quantum mechanics, and according to which the measurement of a given physical quantity will be accompanied by an uncontrollable change in the value of any other such variable which does not commute with it. In fact, due to the emission of radiation or the creation of electron pairs, which is unavoidable in every measuring process aiming at a sufficiently accurate fixation of a field quantity or charge density, any such process will also involve a modification of the very quantity to be measured, which cannot be wholly compensated for or controlled. Still, as has been proved in a previous paper*, this circumstance does in no way hinder the testing of any well-defined consequence of the quantum theory of fields, at any rate so far as the atomic constitution of the measuring instruments may be neglected – an assumption which is also immediately implied in the way in which the classical concepts are used in present quantum theoretical methods. In the present paper, the investigation will be extended to density measurements, and it will be shown that a similar consistency exists between the possibilities of such measurements and well-defined exigencies of the pair theory of electrons. For this reason an analysis of the measuring possibilities on the one hand cannot give any direct clue to the solution of the yet   R.2
unsettled difficulties of electron theory; but on the other hand it strongly suggests that the solution of these difficulties will claim a still greater limitation in the use of classical concepts than the present method of attack, corresponding to the not yet clarified features of atomicity revealed by the existence of the atomic constituents of matter, and which naturally limits the structure and function of all conceivable measuring instruments.

---

* [N. Bohr and L. Rosenfeld, *Zur Frage der Messbarkeit der elektromagnetischen Feldgrössen*, Mat.–Fys. Medd. Dan. Vidensk. Selsk. **12**, no. 8 (1933), reproduced p. [55].]

*1. *Field Theory.*

The quantities defining the electromagnetic field at a given point of space–time are idealizations derived from the consideration of the forces acting on electrified or magnetized test-bodies filling up finite space regions around the given point during finite time intervals, by letting the space–time regions thus constituted shrink indefinitely towards the given point. While in classical theory this limiting process does not involve any difficulty of principle, it may give rise in the quantum theory of fields to certain paradoxes, connected with the appearance in the fundamental commutation rules for field components of the singular Delta-function. To avoid all such paradoxes it is necessary in quantum theory explicitly to consider, instead of the mentioned idealizations, mean values of field quantities over finite space–time regions; on account of the fact that the universal constants of field theory, the quantum of action h and the velocity of light c, are not sufficient to define units of length or time, no limitation is imposed by the theory on the dimensions of the space–time regions considered.

An important consequence of this fact is that in the construction and use of test-bodies for the purpose of testing the predictions of the theory, any limitation arising from the atomic constitution of matter may entirely be disregarded. The idealized test-bodies will thus be assumed to approximate to any desired accuracy continuous distributions, over the given space regions, of electric or magnetic charge (the last case being obtained by considering each element of the test body as one end of a long flexible magnetized rod). If this distribution is uniform, the test body will be suited to measure the ordinary mean value of the form*

$$\overline{E}_x^{(G)} = \frac{1}{VT} \int\limits_{(T)} dt \int\limits_{(V)} dv \, E_x. \tag{1}$$

We shall first confine ourselves to this case, and later on discuss generalizations.

The mean value under consideration is obtained from the total momentum transferred to the test body by dividing this momentum by the product of time interval and total charge of the test body. The measurements required are thus determinations of the momentum of the test body in the direction considered at the beginning and end of the time interval. These momentum measurements involve uncontrollable displacements of the test body; the accuracy of the momentum determination Δp_x being connected with the displacement Δx

R.3

---

* [The numbering of the formulae is added by the editor for convenience of the reader.]

[198]

by Heisenberg's relation

$$\Delta p_x \cdot \Delta x \geq h. \tag{2}$$

In order that the dimensions of the space region over which the mean value is taken be sufficiently well-defined, it is necessary to arrange the measuring procedure so as sufficiently to reduce the displacement Δx; this can, however, be effected without impairing the accuracy of the field measurement, since the latter depends on the product of Δx [and] the charge of the test body, which may be chosen arbitrarily large. Similarly the duration Δt of the momentum measurements may be assumed as small as desired in comparison with the time interval T.

Some complication, but no essential difficulty, arises from the consideration of the requirements of relativity. Thus, due to the retardation of all forces, it is not permissible to treat the extended test body as a solid body, but we must imagine it subdivided in arbitrarily small, though finite parts, the momentum of each of which is measured independently. By making use of the fact, fundamental in the theory of measurements, that the relative course of the measuring process may always be accurately described in a classical way, it is nevertheless possible to manage to give all partial test bodies exactly equal displacements at exactly the same instant. The uncertainty Δx is due to the impossibility, when aiming at a momentum determination of accuracy Δp_x, of fixing the absolute position of the test bodies with respect to the fixed frame of reference.

It might be thought, however, and it has in fact been suggested (by Landau and Peierls), that a quite essential difficulty should be involved in the measurement of the momentum of a charged body, because the acceleration of this charge during the measurement gives rise to an electromagnetic field, the reaction of which on the test body affects the measured value of the momentum, by an amount depending on the uncontrollable displacement of the test body during the measurement. An estimation of the resulting uncertainty in the result of the momentum determination, carried out by treating the test body as a point charge, has led (Landau and Peierls) to the conclusion that it is so large as to deprive the definition of any field quantity of any unambiguous meaning. When due account is taken, however, of the finite extension of the test bodies, it is easily seen that the perturbation arising from the radiation of the test body during the measurement can always be arbitrarily reduced. By the use of the idealized measuring procedure outlined above, it is thus possible to define with unlimited accuracy the mean values of all field quantities over any finite space–time region. It must also be pointed out that the results of such measurements are strictly reproducible; the repetition of the measurement consists in carrying out a second determination by means of a test body (or system of

R.4

R.5

test bodies) superposed on the first test body in the space region considered, and extending over a time interval infinitesimally displaced with respect to the time interval of the first measurement.

As regards the interpretation of the field quantities thus measured, it must in the first place be noticed that they consist of the sum of the field present before the measurement and of the field created within the region occupied by the test body by the body itself during the whole time interval. Now the latter field, as far as it can be calculated on classical theory, is proportional to the displacement of the test body and can thus be exactly compensated by means of suitable (elastic) contrivances. But this classical part of the field of the test body is only its average value; around this value there are statistical fluctuations due to the fact that the field in question partly consists of a certain number of light quanta, of which only the average value is unambiguously predictable. (*Or*: due to the fact that the field in question partly arises from individual emission processes, of which only the average effect is unambiguously predictable). The results of the measurements will thus unavoidably be affected by these fluctuations.

On the other hand, it is a characteristic consequence of the quantum theory of field[s] that all mean values of field quantities (except in the trivial case where the state of the system is defined just by the specification of these values) are always subject to statistical fluctuations, which do not vanish even in complete absence of charges and electromagnetic energy. These fluctuations become the larger, the smaller the dimensions of the space–time extensions are; in the limiting cases $cT \gg L$ and $cT \ll L$ respectively, they are of the order of magnitude

$$\frac{\sqrt{hc}}{L \cdot cT} \quad \text{and} \quad \frac{\sqrt{hc}}{L^2}. \tag{3}$$

R.6

Now it is readily estimated that, whereas the compensable part of the field of the test body gets larger and larger when the accuracy of the measurement is increased, the fluctuating part of this field tends to a finite limit which has exactly the same expression as the theoretical fluctuations in empty space just mentioned. This may be seen as follows:

The wavelength of the radiation mainly emitted during the displacement of the test body is of the order of magnitude of the linear dimensions L of the body. If F is the value of the field created during the displacement of the test body, the total energy emitted will be F^2V, corresponding to a number*

---

* [Misprint L^2 in original.]

[200]

$$\frac{F^2 V}{hc/L} = \frac{F^2 L^4}{hc} \tag{4}$$

of light quanta. In the case $cT \ll L$, the mean value of the field being F, its mean fluctuation is of the order of magnitude $F/\sqrt{N}$; in the other limiting case, the mean value of the field is reduced in the ratio L/cT, and its fluctuation thus affected in the same manner. In both cases the above-mentioned theoretical expressions are obtained.

This coincidence finds its explanation in the theory in a most elementary way. In fact, it is a general and immediate consequence of the theory that in the case where the sources of the field are described classically (as it must be done for measuring instruments such as test bodies), the solution of the field equations and fundamental commutation rules consists of the superposition of the classical field as calculated from the source distribution, and a purely transversal quantum field, corresponding to a complete absence of light quanta. While this "zero field" of course contains no energy (if the expression for the energy density is defined so as to remove the zero-point energy), its electric and magnetic forces exhibit fluctuations around their average value zero, which are just the above-mentioned fluctuations in empty space. In this connexion it may also be recalled that the so-called spontaneous emission processes can formally be described as "induced" by the zero field just considered.

Somewhat more generally, if besides classically described sources there is R.7 a given distribution of light quanta, the field fluctuations will include, besides terms depending on this distribution, a term exactly equal to the fluctuation of the zero field. In the light of these remarks we are now able completely to solve the question of the interpretation of field measurements. We see, in fact, that the fluctuations of the field created by the measuring process itself cannot be separated from the zero field fluctuations already present before the measurement. The occurrence of such fluctuations does therefore not mean any limitation in the testing of predictions of the quantum theory of fields, but on the contrary is essential for the consistency of the theory, by preventing exactly to the extent wanted the establishing of a connexion on classical lines between the sources of the field and the results of measurements of field quantities.

Apart from the peculiar feature just discussed, where we have to do with an influence of the measuring process on the very quantity to be measured, we meet also in the quantum theory of field[s] with the ordinary features of complementary limitation when we consider measurements of two field quantities, i.e. mean values of two field components over given space–time regions. Also here a detailed investigation shows that the interaction of the test bodies during the measuring process, after having been compensated to the largest possible extent, gives rise to a reciprocal limitation of the accuracies of the results of the measurements, exactly equal to the limitation derived from the commutation relations. This completes the proof of the full harmony existing between the consequences of the quantum theory of fields and the possibilities of measurements, to the extent where the atomic properties of the measuring instruments may be neglected.

R.8

The previous considerations can be immediately extended from the case of ordinary mean values to more complicated functions of the field components. Thus, weighted mean values such as

$$\frac{\int_{(T)} dt \int E_x \cdot S \, dv}{T \cdot \int S \, dv} \tag{5}$$

are defined by means of test bodies with non-uniform charge distribution of density S. Likewise the mean value of the scalar product of the vector of electric force by an arbitrary vector function (S_x, S_y, S_z) of unit length:

$$\frac{1}{TV} \int_{(T)} dt \int_{(V)} (E_x S_x + E_y S_y + E_z S_z) \, dv \tag{6}$$

can be determined by a single measuring process, provided all partial test bodies are connected with an auxiliary body by means of suitable crooked levers transforming the component of momentum of the test body in the direction of the vector S into a momentum of the auxiliary test body in a fixed direction. Quite generally, it is always possible so to adjust the construction of the system of test bodies as to determine the value of any given function of field quantities, which is assumed in the formalism to have a well-defined meaning.

An important special case of the mean values of scalar products just considered is obtained by taking the volume V to represent a closed shell of thickness b, and the vector S to tend, in the limit of infinitely small thickness, towards the direction of the exterior normal to the boundary of the shell. The integral

$$\lim_{b \to 0} \frac{1}{b} \frac{1}{T} \int_{(T)} dt \int_{(V)} (E_x S_x + E_y S_y + E_z S_z) \, dv \tag{7}$$

R.9

then represents the mean flux during the time interval T of the electric force through the boundary, and in quantum theory as well as in classical theory [it] is equal to the total electric charge inside the boundary (more exactly to the mean value of this charge during the time interval T). The consistency of Gauss's theorem in the quantum theory of fields (which is an immediate consequence of the divergence condition imposed on the electric field and the charge density) with the commutation rules for field quantities may be verified by observing that two mean values of scalar products, corresponding to different vectors S and different space regions, but to the same time interval, are always commutable. This property is a direct consequence of the commutability of mean values of two components of the electric (or of the magnetic) force when taken over arbitrary space regions and the same time interval.

Consequently, the measurement of the charge within a given space region can simply be performed, just as in classical theory, by determining the flux of electric force traversing the boundary. Moreover, the value obtained for this flux does not exhibit any fluctuations of the kind discussed above, since the latter are due to a purely transversal field, the flux of which over any closed surface vanishes identically. As long, therefore, as the consequences of the phenomenon of pair creation for measurements are disregarded, the determination of electric charge offers no difficulty whatever. The differences from classical theory due to the atomicity of electricity and to the exclusion principle can adequately be dealt with by help of the quantum mechanical expressions for density and current.

*2. Charge measurements in hole theory.   R.10

The phenomenon of creation and annihilation of pairs of elementary electric charges has necessitated an extension of quantum electrodynamics which, although unable as yet to present an entirely consistent account of the new type of interaction between field and matter involved in this phenomenon, is at least sufficient to disclose unambiguously some general features of this interaction. It was thus pointed out by Heisenberg, that the total electric charge within an enclosure was always (except in the trivial case where just the fixation of a definite value for the charge defines the state of the system at the instant considered) subject, in addition to eventual statistical fluctuations of the usual type, to fluctuations of a peculiar kind, independent of the charge distribution. Strictly speaking, these characteristic fluctuations are even infinite when the boundary of the enclosure is sharply defined; but if instead of the total charge within such a sharply limited enclosure we consider the mean value of this charge taken over a continuous set of enclosures with boundaries distributed about some given surface, the fluctuations of the mean total charge so defined are found to be finite, and in fact to depend, in contrast to the ordinary statistical fluctuations, exclusively on the area of the mean boundary and the breadth of the boundary distribution – at least in the case where this breadth b is small in comparison with the linear dimensions L of the enclosure, a condition re-

[203]

quired to give a well-defined meaning to the expression "mean boundary". In this case, a simple expression is obtained for the order of magnitude of $\overline{\Delta e^2}$ under the further assumption that the time interval cT over which the mean value of the charge is also taken is smaller than the smaller of the lengths b and $\lambda_0 = \dfrac{h}{mc}$, namely

$$\overline{\Delta e^2} = \frac{\varepsilon^2 L^2}{b \cdot cT},\qquad(8)$$

where ε denotes the elementary charge*. If the time interval cT concerned were longer, different expressions would be obtained according to the value of the ratio of λ_0 to the linear dimensions L and breadth b; we will confine ourselves, however, to the detailed discussion of the former case, which might be called the case of an instantaneous mean value of charge, since the complete generality of the argument will at once become apparent.

In view of the occurrence of these peculiar fluctuations, the question immediately arises as to the bearing of this property on the possibility of carrying out accurate charge measurements and consequently making use of the charge concept in theoretical predictions. A preliminary issue was raised by Oppenheimer**, who, after having shown that the charge fluctuations considered could be interpreted as fluctuations of the corresponding mean flux of electric force through the boundary, stated that they therefore were inseparable from the other zero point fluctuations, always present, of the radiation field, which were discussed in *1 above. (This statement is wrong, however, since,) as we pointed out in *1, the flux of the radiation field through a closed boundary does not exhibit any fluctuation. The two kinds of fluctuations are thus entirely uncorrelated, and we are left only with the charge fluctuations just mentioned, characteristic of the theory of pairs; we will presently proceed to show that the situation here met with is entirely parallel to the case of the radiation field fluctuations and their connexion with field measurements, treated in *1. In fact, just in the same way as the latter fluctuations are inseparably connected with the uncontrollable part of the radiation field created by the test bodies during the measuring process, the charge fluctuations will be shown to be related to the creation of pairs during the measurement of the flux of electric force by the test bodies used for this measurement. The existence of such a relation was already concluded in a general way by Heisenberg from the fact that the

---

* [W. Heisenberg, *Über die mit der Entstehung von Materie aus Strahlung verknüpften Ladungsschwankungen*, Ber. d. Sächs. Akad. math.–phys. Kl. **86** (1934) 317–322. Reproduced on p. [239].]
** [J. R. Oppenheimer, *Note on Charge and Field Fluctuations*, Phys. Rev. **47** (1935) 144–145.]

fluctuations in question depended on the boundary area and breadth, and not
on the volume of the enclosure.

In order to discuss the matter more accurately, let us consider the process of measurement of the mean value of the flux of electric force through the boundary of the enclosure, taken on the one hand over a time interval T, on the other hand over a distribution of boundaries of breadth b in the manner defined above. The test body suited for such a measurement will be given a charge density proportional at every point to this distribution function; if, for example, this function were constant within a shell of breadth b and zero outside, the test body would simply fill this shell with constant density; moreover, as we have seen, it is permissible for practical convenience to divide the test body into arbitrary independent elements, since the mean values of the normal field components over any two such elements are always commutable. Now, the displacement and acceleration of the test bodies during the measuring process will give rise, according to the [pair] theory, to a "polarization of the vacuum" in a region extending over the space occupied by the test bodies and a strip of breadth λ_0 around it. The mean value of this polarization, and the resulting mean correction to the result of the charge measurement, can be calculated and even automatically compensated for, but there will always remain a fluctuating residue arising from the fact that the polarization in question is connected with an integral number of elementary processes of pair production. The field acting on the test bodies can thus also here be considered as the superposition of a classical part, due to the total charge inside the enclosure, and modified in a controllable way by the polarization of the vacuum, and of an uncontrollable part, which cannot be unambiguously attributed either to the action of the test bodies during the measurement or to the theoretical fluctuations of charge always present even in a vacuum, i.e. in a state defined by the complete absence of elementary charged particles and of electromagnetic field.

Within its domain of applicability, which will be discussed later, this argument is quite general and complete in itself from a logical point of view; it will be instructive, however, to examine more closely the way in which it can be checked quantitatively by means of the theoretical expressions for the different quantities involved. We shall consider the case previously defined, of an instantaneous charge measurement. Let us first suppose that $cT \ll \lambda_0$. It will be convenient to divide the test body in elements of linear dimensions cT. To estimate the degree of accuracy of our measurement, we shall compare the general expression for the uncertainty of the value of the electric field

$$\Delta E \sim \frac{h}{\Delta x \cdot \rho VT} \qquad (V \sim L^2 b) \qquad (9)$$

with the critical field corresponding to the mean charge fluctuation

$$Q = \frac{\sqrt{\overline{\Delta e^2}}}{L^2} = \varepsilon \frac{1}{\sqrt{L^2 bcT}} \tag{10}$$

and put accordingly*

$$\lambda = \frac{\Delta E}{Q} = \frac{hc}{e^2} \frac{Q}{\rho \Delta x} \tag{11}$$

Let us now calculate the number of pairs created by a measurement of accuracy λ. The displacement Δx of the test body produces a field $F = \rho \Delta x$, corresponding to a potential energy of an elementary particle created by this field

$$\varphi = F \cdot \varepsilon \cdot cT. \tag{12}$$

The probability of pair creation per unit time in such a potential in the volume v of an elementary test body is given by

$$\frac{1}{h} \varphi^2 w, \tag{13}$$

R.14

where w is the number of stationary states per energy range, and is of the order of magnitude

$$w \sim \frac{W^2}{(hc)^3} v \tag{14}$$

if $W \sim h/T$ represents the average energy of one pair. Thus

$$w \sim \frac{T}{h}. \tag{15}$$

During the time T the number of pairs produced is thus

$$\left(\frac{T}{h}\right)^2 \varepsilon^2 F^2 (cT)^2 \tag{16}$$

and since there are V/v test bodies, the total number produced is

* [In the original manuscript, λ has sometimes been changed to χ, perhaps to avoid confusion with λ_0.]

$$N = \left(\frac{T}{h}\right)^2 \frac{\varepsilon^2 F^2 V}{cT} = \frac{\varepsilon^2}{hc} \frac{F^2 VT}{h} = \left(\frac{\varepsilon^2}{hc}\right)^2 \frac{F^2}{\varepsilon^2/V \cdot cT} = \lambda^{-2}. \qquad (17)$$

Just as in the case of field measurements, the number of photons produced by the measurement was the greater, the greater the accuracy, we obtain here an analogous result for the number of pairs produced. The energy of these pairs is

$$N\frac{h}{T} = \frac{\varepsilon^2}{hc} F^2 V, \qquad (18)$$

which may be put into the form FVP, with a "polarization"

$$P = \frac{\varepsilon^2}{hc} \quad F = \frac{Q}{\lambda}. \qquad (19)$$

This polarization transfers to the test body a momentum during the time T, given by

$$P\rho VT = \frac{\varepsilon^2}{hc} \rho \Delta x \cdot \rho VT, \qquad (20)$$

as might also be calculated directly from the theory. At this stage, it may be noted that the polarization P does not affect the accuracy of the momentum measurements, since the momentum transferred to the test body during the time Δt of this measurement is proportional to the product $\Delta x \cdot \Delta t$ and may thus be neglected. While now the mean reaction, just calculated, on the test bodies can be compensated for, there are fluctuations about this mean value, due to the fact that pairs are created according to a probability law. The amount of momentum due to this fluctuation is of the order of magnitude

R.15

$$\rho VT \cdot \frac{P}{\sqrt{N}} = \rho VT \frac{Q}{\lambda} \lambda = \rho VT \cdot Q, \qquad (21)$$

thus exactly corresponding to the theoretical fluctuation Q. This result is of course not due to chance, but is contained, under certain conditions, to be discussed below, in the formalism itself. The preceding considerations bring out in all details a complete analogy with the case of the radiation field fluctuations.

There remains briefly to indicate the modifications affecting the argument in the case $cT \gg \lambda_0$. In this case, it will be convenient to divide the test body into elements of linear dimensions λ_0. The critical field is now given by

$$Q' = Q\sqrt{\frac{\lambda_0}{cT}} \tag{22}$$

and therefore the parameter characterizing the accuracy of the field measurement will be[*]

$$\lambda' = \frac{hc}{\varepsilon^2}\frac{Q'}{F}\frac{cT}{\lambda_0}. \tag{23}$$

Now, pairs are produced chiefly during the time $\frac{\lambda_0}{c}$ and the total number produced in all the elementary test bodies will be obtained from the formula derived above by putting simply $T = \frac{\lambda_0}{c}$; we get therefore

$$N' = \frac{\varepsilon^2}{hc}\frac{F^2 V}{mc^2} = \lambda'^{-2}, \tag{24}$$

the energy of these pairs is

$$N'mc^2 = \frac{\varepsilon^2}{hc}F^2 V, \tag{25}$$

R.16

from which we deduce the expression of the polarization

$$P' = \frac{\varepsilon^2}{hc}F = \frac{Q'}{\lambda'}\frac{cT}{\lambda_0}. \tag{26}$$

The total momentum transferred by this polarization is, in agreement with the direct theoretical calculation

$$P'\rho V\frac{\lambda_0}{c} = \rho VT\frac{\varepsilon^2}{hc}F\frac{\lambda_0}{cT} \tag{27}$$

and its fluctuation

$$\rho V\frac{\lambda_0}{c}\frac{P'}{\sqrt{N'}} = \rho VTQ' \tag{28}$$

confirming in this case also our preceding conclusions.

We must now turn to the examination of the range of validity of our argument. In the first place, we have assumed that the fields with which we were concerned were additive; in order that this condition may be fulfilled with sufficient accuracy, the fields in question must be small in comparison with the critical field strength of the theory

---

[*] [Here again, sometimes λ' has been changed to χ'.]

$$\frac{\varepsilon^2}{hc} \cdot \frac{\varepsilon}{r_0^2}, \quad r_0 = \frac{\varepsilon^2}{mc^2}. \tag{29}$$

This is satisfied provided the dimensions of the space within which the charge is to be measured are large compared with r_0. When this condition is fulfilled, it is also permissible to neglect, as we have done, the modification of the charge fluctuations from their value for empty space due to the field of the test bodies; in other words, we have to this approximation the theorem that the polarization produced by a classical current distribution is subject to the same fluctuations as [in] the empty space. As regards the complications which would arise in the case of enclosures of smaller dimensions, they cannot – insofar as the formalism of the theory of pairs is itself consistent and yields convergent results – involve any difficulty of principle, since the logical argument, which assumes only the consistency of the formalism, would then still be valid.   R.17
In all cases, it is permissible to disregard the atomic structure of the measuring instruments, since we may imagine that these instruments are built up of sufficiently heavy masses. In fact, at the present stage, this assumption lies at the bottom of the whole theory, including the measuring processes, and conditions essentially the possibility of unrestricted use of the space and time concepts in the sense here discussed.

In conclusion, we see that the investigation of the problem of measurement cannot within a consistent formalism lead to any argument tending to disclose logical defects of the theory and the necessity of altering its principles. The only interest of the problem of measurement is to clarify the logical aspect of the definition and use of concepts compatible with the formalism, and thus call our attention to some aspects of this formalism which otherwise would escape investigation, and which are quite fundamental features of its logical consistency. At the same time, the conclusions reached by the analysis of measurements permit a rational solution of the quantum paradoxes. This is already strikingly shown in the case of the uncertainty relations of quantum mechanics, and again with respect to quantum electrodynamics, where a quite new general feature of complementarity is found. As regards the unsolved difficulties of the theory of electrons, the preceding analysis stresses only the consistency of the formalism as a correspondence theory, but leaves entirely open such questions as the possible necessity in a future theory to take into account the atomic constitution of the measuring instruments, which would entirely modify the outlook of the problem of measurements in atomic theory.

V. FIELD AND CHARGE MEASUREMENTS IN QUANTUM ELECTRODYNAMICS

(WITH L. ROSENFELD)

Phys. Rev. **78** (1950) 794–798

See Introduction to Part I, sect. 3.

[211]

Field and Charge Measurements in Quantum Electrodynamics

N. Bohr

Institute for Theoretical Physics, University, Copenhagen, Denmark

AND

L. Rosenfeld

Department of Theoretical Physics, University, Manchester, England

(Received October 19, 1949)

A survey is given of the problem of measurability in quantum electrodynamics and it is shown that it is possible in principle, by the use of idealized measuring arrangements, to achieve full conformity with the interpretation of the formalism as regards the determination of field and charge quantities.

INTRODUCTION

RECENT important contributions[1] to quantum electrodynamics by Tomonaga, Schwinger and others have shown that the problem of the interaction between charged particles and electromagnetic fields can be treated in a manner satisfying at every step the requirements of relativistic covariance. In this formulation, essential use is made of a representation of the electromagnetic field components on the one hand, and of the quantities specifying the electrified particles on the other, corresponding to a vanishing interaction between field and particles. The account of such interaction is subsequently introduced by an approximation procedure based on an expansion in powers of the non-dimensional constant $e^2/\hbar c$. As regards the interpretation of the formalism, this method has the advantage of a clear emphasis on the dualistic aspect of electrodynamics. In fact, an unambiguous definition of the electromagnetic field quantities rests solely on the consideration of the momentum imparted to appropriate test bodies carrying charges or currents, while the charge-current distributions referring to the presence of particles are ultimately defined by the fields to which these distributions give rise.

Just from this point of view the problem of the measurability of field quantities has been discussed by the authors in a previous paper.[2] A similar investigation of the measurability of electric charge density was then also undertaken, but, owing to various circumstances, its publication has been delayed.[3] When recently the work was resumed, it appeared that by making use of the new development as regards the formulation of quantum electrodynamics a more general and ex-

haustive treatment could be obtained.[4] As these considerations may be helpful in the current discussions of the situation in atomic physics, we shall here give a brief account of the implications of present electron theory for measurements of charge-current densities. For this purpose, it will be convenient to start with a summary of our earlier treatment of the measurability of field quantities.[5]

1. MEASUREMENTS OF ELECTROMAGNETIC FIELDS

Classical electrodynamics operates with the idealization of field components $f_{\mu\nu}(x)$ defined at every point (x) of space-time. Although in the quantum theory of fields these concepts are formally upheld, it is essential to realize that only averages of such field components over finite space-time regions R, like

$$F_{\mu\nu}(R) = \frac{1}{R}\int_R f_{\mu\nu}(x)d^4x \tag{1}$$

have a well-defined meaning (I, §2). In the initial step of approximation, in which all effects involving $e^2/\hbar c$ are disregarded, these averages obey commutation relations of the general form

$$[F_{\mu\nu}(R), F_{\kappa\lambda}(R')] = i\hbar c[A_{\mu\nu,\kappa\lambda}(R,R') - A_{\kappa\lambda,\mu\nu}(R',R)], \tag{2}$$

where the expressions of the type $A_{\mu\nu,\kappa\lambda}(R,R')$, defined as integrals over the space-time regions R and R' of certain singular functions, have finite values depending on the shapes and relative situation of the regions R and R'.

The measurement of a field average $F_{\mu\nu}(R)$ demands the control of the total momentum transferred within the space-time region R to a system of movable test bodies with an appropriate distribution of charge or current, of density ρ_ν, covering the whole part of space which at any time belongs to the region R. In the case

[1] S. Tomonaga, Prog. Theor. Phys. **1**, 27 (1946); Phys. Rev. **74**, 224 (1948). J. Schwinger, Phys. Rev. **74**, 1439 (1948); **75**, 651 (1949); **75**, 1912 (1949); **76**, 790 (1949). F. Dyson, Phys. Rev. **75**, 486 (1949); **75**, 1736 (1949). R. Feynman, Phys. Rev. **76**, 749 (1949); **76**, 769 (1949).

[2] N. Bohr and L. Rosenfeld, Kgl. Danske Vid. Sels., Math.-fys. Medd. **12**, No. 8 (1933). This paper will be referred to in the following as I.

[3] An account of the preliminary results of the investigation, which were discussed at several physical conferences in 1938, has recently been included in the monograph by A. Pais, *Developments in the Theory of the Electron* (Princeton University Press, Princeton, New Jersey, 1948).

[4] The bearing of this development on the elucidation of the problem of measurability was brought to the attention of the writers in a stimulating correspondence with Professor Pauli.

[5] A more detailed account of the subject with fuller references to the literature will appear later in the Communications of the Copenhagen Academy.

of an electric field component F_{4l}, we shall take a distribution of charge, with constant density ρ_4, and in the case of a magnetic field component F_{mn}, a uniform distribution of current in a perpendicular direction, with density components ρ_m and ρ_n. The field action of such charge-current distribution, so far as it does not originate from the displacements of the test bodies accompanying the momentum control, can in principle be eliminated by the use of fixed auxiliary bodies carrying a charge-current distribution of opposite sign, and constructed in a way which does not hinder the free motion of the test bodies. In the case of a current distribution, such auxiliary bodies are even indispensable in providing closed circuits for the currents by means of some flexible conducting connection with the test bodies. As a result of this compensation, the field sources of the whole measuring arrangement will thus merely be described by a polarization $P_{\mu\nu}$ arising from the uncontrollable displacements of the test bodies in the course of the field measurements.

If the test bodies are chosen sufficiently heavy, we can throughout disregard any latitude in their velocities, but the control of their momentum will of course imply an essential latitude in their position, to the extent demanded by the indeterminacy relation. Still, it is possible, without violating any requirement of quantum mechanics, not only to keep every test body fixed in its original position except during the time interval within the region R corresponding to this position, but also to secure that, during such time intervals, the displacements of all test bodies in the direction of the momentum transfer to be measured, although uncontrollable, are exactly the same. This common displacement D_μ is described, in the case of the measurement of an electric field, by the component D_l parallel to the field component F_{4l}, and when a magnetic field is measured, by the components D_m and D_n perpendicular to F_{mn}. Without imposing any limit on the accuracy of the field measurement, it is, moreover, possible to keep the displacement D_μ arbitrarily small, if only the charge-current density ρ_ν of the test bodies is chosen sufficiently large. By a further refinement of the composite measuring arrangement described in our earlier paper (I, §3), it is even possible to reduce the measurement of any field average to the momentum control of a single supplementary body, and thus to obtain a still more compendious expression for the ultimate consequences of the general indeterminacy relation.

An essential point in field measurements is, however, the necessity of eliminating so far as possible the uncontrollable contribution to the average field present in R, arising from the displacement of the test bodies in the course of the measurement. In fact, the expectation value of this contribution will vary in inverse proportion to the latitude allowed in the field measurement, since it is proportional to the polarization $P_{\mu\nu} = D_\mu\rho_\nu - D_\nu\rho_\mu$ within the region R. Just this circumstance, however, makes it possible, by a suitable mechanical device, by

which a force proportional to their displacement is exerted on the test bodies, to compensate the momentum transferred to these bodies by the uncontrollable field, insofar as the relation of this field to its sources is expressed by classical field theory. With the compensation procedure described, the resulting measurement of $F_{\mu\nu}(R)$ actually fulfils all requirements of the quantum theory of fields as regards the definition of field averages (I, §5). In fact, the incompensable part of the field action of the test bodies due to the essentially statistical character of the elementary processes involving photon emission and absorption, corresponds exactly to the characteristic field fluctuations which in quantum electrodynamics are superposed on all expectation values determined by the field sources.

When the measurement of two field averages $F_{\mu\nu}(R)$ and $F_{\kappa\lambda}(R')$ is considered, it appears (I, §4) that the expectation value of the average field component $\Phi_{\mu\nu,\kappa\lambda}(R, R')$ which the displacement of the test bodies operated in the region R produces in the region R' is equal to the product of $\frac{1}{2}RP_{\mu\nu}$ with the quantity $A_{\mu\nu,\kappa\lambda}(R, R')$ occurring in the commutation relation (2). Likewise, the expectation value of the average component $\Phi_{\kappa\lambda,\mu\nu}(R', R)$ of the field in R due to the test bodies in R' is equal to $\frac{1}{2}R'P_{\kappa\lambda}'A_{\kappa\lambda,\mu\nu}(R', R)$. When optimum compensation of the momenta transferred to the test bodies by these fields is established by suitable devices, making use of a correlation by light signals transmitted between points of the two regions R and R', it can be deduced from the reciprocal indeterminacy of position and momentum control that the only limitations of the measurability of the two field averages considered correspond exactly to the consequences of the commutation rule (2) for such averages (I, §6, 7). In this connection, it must be stressed that the field fluctuations which are inseparable from the incompensable parts of the fields created by the operation of the test bodies, do not imply any restriction in the measurability of a field component in two asymptotically coinciding space-time regions. In fact, we have here to do with a complete analog to the reproducibility of the fixation of observables in quantum mechanics by immediately repeated measurements.

2. CHARGE-CURRENT MEASUREMENTS IN INITIAL APPROXIMATION

In the formalism of quantum electrodynamics, charge-current densities, like field quantities, are introduced by components $j_\nu(x)$ at every space-time point, but, even in the initial approximation in which such symbols are formally commutable, well-defined expressions are only given by integrals of the type

$$J_\nu(R) = \frac{1}{R} \int_R j_\nu(x) d^4x, \qquad (3)$$

representing the average charge-current density within the finite space-time region R. From the fundamental equations of electrodynamics it follows quite generally that

$$RJ_\nu(R) = \int_R \frac{\partial f_{\nu\mu}}{\partial x_\mu} d^4x = \int_S f_{\nu\mu} d\sigma_\mu, \qquad (4)$$

which expresses the definition of the average charge-current density over the region R in terms of the flux of the electromagnetic field through the boundary S of this region. In this four-dimensional representation, such generalized fluxes comprise, of course, besides the ordinary electric field flux defining the average charge density, other expressions pertaining to the average current densities and representing magnetic field circulations and displacement currents.

In the simple special case in which the region R is defined by a fixed spatial extension V and a constant time interval T, the average charge density, in accordance with (4), will be given, in the ordinary vectorial representation, by

$$J_4(V, T) = \frac{1}{VT} \int_T dt \int_S \mathbf{E} \mathbf{n} d\sigma, \qquad (5)$$

where S is the surface limiting the extension V, and $\mathbf{n}$ the unit vector in the outward normal direction on this surface. In such representation, the average current density will be given by

$$\mathbf{J}(V, T) = \frac{1}{VT} \int_T dt \int_S \mathbf{n} \wedge \mathbf{H} d\sigma - \frac{1}{VT} \int_V \mathbf{E} dv \Big|_{t_1}^{t_2}, \qquad (6)$$

where the first term on the right-hand side represents the time integral of the tangential component of the magnetic field integrated over the surface S, while the last term expresses the difference of the volume integrals of the electric field at the beginning and at the end of the time interval T.

The determination of an average charge-current density $J_\nu(R)$ thus demands the measurement of a field flux through the boundary S of the space-time region R. The approach to the problem of such measurement must rationally start from the consideration of the average flux over a thin four-dimensional shell situated at the boundary S, and which for simplicity we shall assume to have a constant thickness in space-time. As in the situation met with in the measurement of an average field component $F_{\mu\nu}(R)$, we shall require for this purpose a system of movable test bodies, filling the space which belongs to the shell at any time with an appropriate uniform charge-current distribution, and whose field actions are ordinarily neutralized by a distribution of opposite sign on fixed, penetrable, auxiliary bodies. For the measurement of an average charge density J_4, it suffices to take a set of test bodies with a uniform charge distribution of density ρ_4, while in the measurement of a current component, J_l, we shall have to use, besides such test bodies, another independent set of freely movable test bodies with a uniform current distribution ρ_l parallel to the current component to be measured.

In the measurement of an average charge density, the estimation of the flux over the shell demands the determination of the algebraic sum of the momenta transferred to the test bodies in the direction of the normal to the instantaneous spatial boundary. The evaluation of this sum, however, does not require independent measurements of the momenta transferred to the individual test bodies within the time intervals during which their positions belong to the space-time shell, but can be obtained by a composite measuring process in which the positions of all test bodies are correlated by suitable devices to secure during these intervals a displacement of every test body in the normal direction by the same amount. By choosing the product of the thickness of the shell and the charge density of the test bodies sufficiently large, it is possible to keep the uncontrollable common displacement D of all the test bodies in the normal direction arbitrarily small, and still to obtain unlimited accuracy for the average flux over the shell. Like in the measurement of a simple field average, it is further possible to achieve an automatic compensation of the uncontrollable contribution to this average flux, due to the fields created by the displacement of the test bodies, and proportional to $D\rho_4$. This compensation will even be complete, in the initial approximation considered, because the field fluctuations, owing to their source-free character, do not give any contribution to the flux. Since these considerations hold for any given thickness of the shell, it is in principle possible, in the asymptotic limit of a sharp boundary, to measure accurately the average charge density within a well-defined space-time region.

In measurements of an average current component J_l, we have to take into account the magnetic circulation as well as the electric field in the space-time shell. Thus, in the special case in which R is defined by a spatial extension V and a time interval T, we have to do, according to (6), not only with a contribution from the time average over T of the magnetic circulation around the direction l within a thin spatial shell on the boundary of V, but also with a contribution representing the difference between the volume integrals over V of the electric field component in the direction l, averaged over two short time intervals at the beginning and at the end of the interval T. The evaluation of these contributions requires measuring procedures of a similar kind as those described above in the case of measurements of simple field averages. While the measurement of the latter contribution demands the control of the momentum in the direction l transferred to a set of test bodies with uniform charge density ρ, the evaluation of the former contribution demands the control of the momentum normal to the spatial boundary transferred to another set of test bodies with uniform current density ρ_l.

Just as in the field or charge measurements discussed

above, all these operations can be correlated in such a way that the determination of the algebraic sum of the momenta transferred to each test body within the time interval and in the direction required can be reduced to the momentum control of some supplementary body. In such a correlation, all the test bodies of charge density ρ will be subjected during the appropriate time intervals to the same displacement D_l and all the test bodies of current density ρ_l to the same normal displacement D. The interpretation of the current measurement requires further the establishment of a correlation between these two displacements, satisfying the condition $\rho D_l = \rho_l D$. Under such circumstances, it is possible, by choosing ρ and ρ_l sufficiently large, to achieve that the displacements D_l and D be arbitrarily small without imposing any limitation upon the accuracy of the measurement. Moreover, it is possible, by suitable mechanical devices of the kind already mentioned, to obtain a complete automatic elimination of the uncontrollable contributions from the operation of the test bodies to the average current to be measured.

It need hardly be added that the procedure can be extended to quite general space-time regions R, by using an arrangement in which each test body is displaced just in the time interval during which its position belongs to the space-time shell surrounding the region R. In this connection, it may be noted that a compendious four-dimensional description of all the measuring processes pertaining to charge-current components involves a uniform four-vector current distribution in the shell, parallel to the charge-current component to be measured.

Like in charge measurements, all the considerations concerning current measurements are independent of the thickness of the shell, and in principle it is therefore possible, in the initial approximation considered, to determine with unlimited accuracy any average charge-current component $J_\nu(R)$ within a sharply bounded region R. As regards charge-current measurements over two space-time regions, it can easily be seen that, in the limiting case of sharp boundaries, all field actions accompanying the flux measurements will vanish at any point of space-time which does not belong to the boundaries. In conformity with the formalism, there will therefore, to the approximation concerned, be no mutual influence of measurements of average charge-current densities in different space-time regions.

The situation so far described is of course merely an illustration of the compatibility of a consistent mathematical scheme with a strict application of the definition of the physical concepts to which it refers, and is in particular quite independent of the question of the possibility of actually constructing and manipulating test bodies with the required properties. The disregard of all limitations in this respect, which may originate in the atomic constitution of matter, is, however, entirely justified when dealing with quantum electrodynamics in the initial stage of approximation. In fact, at this stage,

the formalism is essentially independent of space-time scale, since it contains only the universal constants c and $\hbar$ which alone do not suffice to define any quantity of the dimensions of a length or time interval.

3. CHARGE-CURRENT MEASUREMENTS IN PAIR THEORY

New aspects of the problem of measurements arise in quantum electrodynamics in the next approximation, in which effects proportional to $e^2/\hbar c$ are taken into consideration, and where we meet with additional features connected with electron pair production induced by the electromagnetic fields. For the commutation rules of the field components, this means in general only a smaller modification expressed by additional terms containing $e^2/\hbar c$. The charge-current quantities, however, will no longer be commutable but will obey commutation relations of the form

$$[J_\nu(R), J_\mu(R')] = i\hbar c[B_{\nu\mu}(R, R') - B_{\mu\nu}(R', R)], \quad (7)$$

where the expressions $B_{\nu\mu}(R, R')$ are integrals of singular functions over the regions R and R'. In contrast to the quantities $A_{\mu\nu, \kappa\lambda}(R, R')$ occurring in (2), which depend only on simple spatio-temporal characteristics of the problem, the B's will, however, besides such characteristics, also essentially involve the length $\hbar/mc$ and the period $\hbar/mc^2$, related to the electron mass m.

To approach the problem of the measurability of a charge-current quantity $J_\nu(R)$ in this approximation, we must again consider systems of electrified test bodies operated in a space-time shell on the boundary of the region R, but we shall now have to examine the effect of the charge-current density appearing as a consequence of actual or virtual electron pair production by the field action of the displacement of the test bodies during the measuring process. As we shall see, these effects, which are inseparably connected with the measurements, do not in any way limit the possibilities of testing the theory.[6]

In the first place, the average effect of the polarization of the vacuum by virtual and actual pair production in the measuring process can be eliminated by a compensation arrangement like that previously described. It is true that a direct estimate of these polarization effects in quantum electrodynamics involves divergent expressions which can only be given finite values by some renormalization or regularization procedure.[7] By such a procedure the average polarization effects will give rise to a contribution to the charge current density which is proportional to the common displacement of the test bodies. Thus in the limit of

---

[6] In a paper by Halpern and Johnson, Phys. Rev. **59**, 896 (1941), arguments are brought forward pointing to a far more restrictive limitation of the field and charge measurements. In these arguments, however, no sufficient separation is made between such actions of the charged test bodies as are directly connected with their use in the measuring procedure and those actions which can be eliminated by appropriate neutralization by auxiliary bodies of opposite charge.

[7] Cf. W. Pauli and F. Villars, Rev. Mod. Phys. **21**, 434 (1949).

a sharp boundary of the region R we get, denoting the surface polarization on the boundary by P_ν, the expression $RP_\nu B_{\mu\nu}(R, R)$, where the last factor represents the value of $B_{\mu\nu}(R, R')$ in (7) for coinciding space-time extensions.

Moreover, the statistical effects caused by actual production of electron pairs in the measurement process are inseparably connected with the interpretation of the fluctuations of average charge-current densities in quantum electrodynamics. While the mean square deviation of the field component $F_{\mu\nu}(R)$ over a sharply bounded space-time region R has a finite value, finite mean-square fluctuations of charge-current quantities can only be obtained, however, by further averaging over an ensemble of regions R whose boundaries are allowed a certain latitude around some given surface.[8]

This feature finds its exact counterpart in the estimate of the statistical effects of the real pairs which are produced in measurements of charge-current quantities by the indicated procedure. In fact, the mean square fluctuations of an average flux will increase indefinitely with decreasing thickness of the shell in which the test bodies are operated, in just the same way as, according to the formalism, the mean-square fluctuation of the corresponding charge-current density will vary with the latitude of the ensemble of space-time extensions over which the averaging is performed. The appearance of an infinite mean-square fluctuation in a sharply limited space-time region is in no way connected with the divergencies which appear in vacuum polarization effects but is a direct consequence of the fundamental assumptions of the theory, according to which the electrons are regarded as point charges.

In the case of measurements of charge-current averages over two space-time regions, it can be shown that the polarization effects of the manipulation of the test bodies used for the measurement of $J_\nu(R)$ will give rise, in the limit of sharp boundaries, to a con-

tribution to the average charge-current density component of index μ in the region R', equal to the product of the quantity $B_{\mu\nu}(R', R)$ occurring in formula (7) with RP_ν, where P_ν is the surface polarization created on the boundary of R during the measuring process. Conversely, the measurement of $J_\mu(R')$ will give a contribution $R'P_{\mu}'B_{\nu\mu}(R, R')$ to the average charge-current density of index ν in R. By similar compensation devices as required for two field measurements, it is therefore possible, as readily seen, to obtain an accuracy of measurements of average charge-current densities in two space-time regions subject only to the reciprocal limitation expressed by the commutation relation (7).

4. CONCLUDING REMARKS

The conformity of the formalism of quantum electrodynamics with the interpretation of idealized field and charge measurements has of course no immediate relation to the question of the scope of the theory and of the actual possibility of measuring the physical quantities with which it deals.

In the present state of atomic physics, the problem of an actual limitation of measurements interpreted by means of the concepts of classical electrodynamics can hardly be fully explored. Still, in view of the great success of quantum electrodynamics in accounting for numerous phenomena, the formal interpretation of which involves space-time coordination of electrons within regions of dimensions far smaller than $\hbar/mc$ and $\hbar/mc^2$, it may be reasonable to assume that measurements within such regions are in principle possible. Indeed, the comparatively heavy and highly charged test bodies of such small dimensions and operated over such short time intervals, which would be required for these measurements, might be conceived to be built up of nuclear particles.

Yet, an ultimate limitation of the consistent application of the formalism is indicated by the necessity of introducing forces of short range in nuclear theory, with no analog in classical electrodynamics, and by the circumstance that the ratio between the electron mass and the rest mass of the quanta of the nuclear field has the same order of magnitude as the fundamental parameter $e^2/\hbar c$ of quantum electrodynamics.[9] The further exploration of such problems may, however, demand a radical revision of the foundation for the application of the basic dual concepts of fields and particles.

---

[8] Cf. W. Heisenberg, Leipziger Ber. **86**, 317 (1934). We are indebted to Drs. Jost and Luttinger for information about their more precise evaluation of charge-current fluctuations, showing that the unlimited increase of the charge-current fluctuations in a space-time region with decreasing latitude in the fixation of its boundary involves only the logarithm of the ratio between the linear dimensions of the region and the width of this latitude. Even a latitude very small compared with $\hbar/mc$ will therefore imply no excessive effect of the charge fluctuations. A situation entirely similar in all such respects to that in electron theory is met with in a quantum electrodynamics dealing with electrical particles of spin zero which obey Bose statistics. We are indebted to Dr. Corinaldesi for the communication of his results regarding the charge-current fluctuations and pair production effects in such a theory.

[9] Cf., e.g., N. Bohr, Report of the Solvay Council (1948).

VI. PROBLEMS OF ELEMENTARY-PARTICLE PHYSICS

Report of an International Conference
on Fundamental Particles and Low Temperatures
held at the Cavendish Laboratory, Cambridge, on 22–27 July 1946,
Volume 1, Fundamental Particles, The Physical Society, London 1947,
pp. 1–4

See Introduction to Part I, sect. 4.

FIRST SESSION

GENERAL INTRODUCTION AND SURVEY

PROBLEMS OF ELEMENTARY-PARTICLE PHYSICS

OPENING ADDRESS

By NIELS BOHR,

Copenhagen

FOREWORD. After expressing, on behalf of the visitors from abroad, gratitude to the Physical Society and the Cavendish Laboratory for the invitation to the Conference, and after referring to the great English traditions in the field of atomic research, the speaker opened the discussions on the present situation in the theory of elementary particles by some general remarks, the essence of which is contained in the following abstract, although certain points have been reconsidered during its preparation.

A S AN INTRODUCTION, a brief reference is made to classical electron theory and, in particular, it is noted that the simple treatment of the electron as a charged mass point is, in principle, limited to the description of phenomena in the anlaysis of which no linear dimensions occur comparable with or smaller than the so-called classical electron radius

$$r_0 = e/mc^2, \qquad \ldots\ldots(1)$$

where e and m are the charge and the mass of the electron, and c is the velocity of light. In this connection, it is stressed that any attempt, however suggestive, at overcoming such restrictions by modifications of the classical theories would seem to involve an element of arbitrariness since, in the region with which we are concerned, phenomena related to the quantum of action are known to play a decisive part.

As regards the quantum-mechanical description of atomic phenomena, emphasis is laid on its consistency within a wide scope and, especially, on the elucidation of the well-known paradoxes as regards the problem of "physical reality" through the recognition that, in the proper quantum effects, we have to do with phenomena where no sharp separation is possible between an independent behaviour of the objects and their interaction with the measuring agencies necessary for the definition of the observable phenomena. Although these cannot be combined in the customary manner of classical physics, they are complementary in the sense that only together they exhaust all knowledge as regards those properties of the objects which are unambiguously definable.

[219]

I

With respect to the properties of elementary particles, the idea of point charges underlying the present quantum mechanical formalism is in the first place justified by the large size of atomic systems compared with r_0, which again depends on the smallness of the non-dimensional constant

$$\alpha = e^2/\hbar c \qquad \qquad \dots\dots(2)$$

expressing the relationship between the quantum of electric charge and the quantum of action. It is especially suggestive as to the character of the situation that the smallness of α depends on the finite value of $\hbar$ in a way which does not allow any asymptotic transition to classical pictures without losing the inherent stability of atomic structures and the specification of essential properties of the elementary particles, like their spin and statistics. The treatment of atomic problems by means of successive approximations involving power series of α entails, however, conspicuous limitations. Although it is possible by such a procedure to separate to a large extent the proper mechanical description of atomic systems and their radiative reaction, we meet, as is well known, with divergencies in the treatment of radiation problems in higher approximations.

A special difficulty of this kind, which has been much discussed, is encountered in the problem of the so-called self-energy of a point charge. Apart from the intricacies, already known in classical electron theory, connected with the coupling between a charged particle and the electromagnetic field it generates, we must in quantum electrodynamics, owing to the fluctuations of the field intensities even in photon-free space, take into account a new contribution to the self-energy. A simple estimate of this contribution gives, in first approximation, an expression of the type

$$W(r) \sim \frac{e^2 \hbar}{mcr^2} = \alpha^{-1} mc^2 \left(\frac{e^2}{mc^2 r}\right)^2, \qquad \dots\dots(3)$$

where $2\pi r$ denotes a lower limit for the wave-lengths of the field components taken into account. Owing to the presence of the factor α^{-1} in this expression, $W(r)$ is very large compared with mc^2 even for $r \sim r_0$, and such calculations, therefore, serve to illustrate the radical difference as regards the self-energy problem in classical and quantum theory.

Further characteristic differences appear when the specific quantum statistics of identical particles are taken into account. In fact, while an expression of the type (3) holds for particles which obey Bose statistics, one finds in first approximation an essentially different formula for an electron in Dirac's hole theory based on the Pauli principle. As first pointed out by Weisskopf, this theory leads, owing to exchange effects between the electron and the negative energy electrons of the "sea", to an expression for the total field self-energy of the type

$$W(r) \sim \alpha mc^2 \log(\hbar/mcr) \qquad \dots\dots(4)$$

and it is most significant that this expression, in contrast to (3), is small compared with mc^2 if $r \sim r_0$ although, of course, it becomes infinite for $r = 0$.

An interesting attempt to avoid this infinity has recently been made by Pais who has shown that, in first approximation, one can obtain in quantum theory a finite expression for the self-energy of a point charge if it is assumed to be coupled, not only to the electromagnetic field, but also to a field of short range giving rise to an attraction between electric charges of same sign as small distances. This idea, which reminds one of the attempts to avoid divergency difficulties in classical electron theory, is especially suggested by the use of short-range fields which has been found necessary in the theory of nuclear constitution. Still, at present, it seems difficult to judge the adequacy of such a hypothesis for the self-energy problem since, in the region concerned, the situation appears in other respects essentially changed by further consequences of the hole theory.

Above all, the recognition that electron pairs can be produced by the inter-action of photons of sufficiently large energy requires a more deep-going analysis of the consequences of the quantum theory of fields in the high-frequency limit. In fact, for wave-lengths comparable with r_0, the calculated cross-sections for photon-photon interaction become so large that the superposition principle, fundamental in present field theories, breaks down. In this region, the whole procedure by successive approximations, therefore, hardly constitutes an adequate approach to the problem of interaction of particles and fields. As was suggested some time ago, especially by Born, the circumstance that the limit for the simple applications of such concepts in quantum theory, in case of electrons and electromagnetic fields, appears comparable with r_0, would even seem to leave open the possibility that a proper treatment of the self-energy problem might account for the electron mass in a manner which, notwithstanding its essentially different basis, resembles the programme originally foreshadowed by classical electron theory.

In case of nucleons, the self-energy problem is primarily connected with their firm coupling with the meson field. Here, the limit for the direct applicability of field concepts should, from similar arguments, lie comparatively higher than for electrons but, still, due to the large nucleon mass, considerably lower than r_0, as it is also assumed in all theories of nuclear constitution, where the nucleons are treated as mass points subjected to the forces derived from simple linear meson fields. In contrast to the view often expressed, that the consistency of a compre-hensive theory of elementary particles will demand a new radical departure involving a universal length of the order of r_0, such an analysis would rather seem to suggest that the restricted scope of the conventional procedure originates in consequences of quantum theory which are indicated already in the present formalism and which may offer clues to the dimensional relations which charac-terize a rational limitation to the simple particle and field concepts.

The problem of field quanta like mesons is in several respects essentially different from those so far discussed. In the first place, the origin of the meson mass cannot primarily be sought in a coupling with the fields acting upon them, but is directly connected with the range of the meson fields themselves. Further, the incompatibility of a localization of field energy with a fixation of the number of field quanta would seem essentially to restrict the particle concept to regions larger than the field range which for mesons is comparable with r_0. Moreover, as regards the problem of the electromagnetic field self-energy of charged mesons obeying Bose statistics which, in first approximation, should be given by a formula of the type (3), an effective breakdown of the electromagnetic field concepts at $r \sim r_0$ leads to a contribution of the order of magnitude of $\alpha \mu c^2$, since the meson mass, μ, is comparable with $\alpha^{-1} m$. The circumstance that this value is small compared with

[221]

the meson rest energy is, in fact, consistent with the assumption underlying present meson theory, that with high approximation the masses of charged and neutral mesons are equal.

However vague such considerations must necessarily be, they may still serve to illustrate how closely the constant α, which plays such a fundamental part in atomic theory, appears correlated with the other non-dimensional constants like the ratios between the electron, meson and nucleon masses, as has been so often suggested. It would, indeed, seem in no way excluded that an analysis on some such lines of the conditions for the consistency of a theory of atomic phenomena would lead to new arguments for the fixation of the actual numerical value of all these constants.

In this connection, attention may also be called to the apparent paradoxes involved in the quantum theory of fields as well as in Dirac's electron theory, which imply the existence in free space of an energy density and electric density, respectively, which, even with the limitations of field and particle concepts suggested, would be far too great to conform to the basis of general relativity theory. While the electric density corresponding to the electron sea may be neutralized by an analogous treatment of the proton problem, a compensation of the negative energy density may require the drawing of the positive zero-point field energy into the picture. At present, it would seem futile to pursue such considerations more closely, but they are here indicated only as a suggestion of how intimately the particle and field concepts may have to be interwoven in a comprehensive theory, and to stress the dualistic character of these concepts, which has its root in the circumstance that the properties of the particles, like their mass and charge, are defined by the fields of force they produce or the effect of fields upon them and, inversely, the fields are themselves only defined through their action on the particles.

Before concluding these tentative remarks, reference must be made to the new line of approach initiated by Heisenberg, the aim of which is to create a frame sufficiently wide for a consistent theory of atomic phenomena by limiting the description more explicitly to directly observable features. No attempt will be made here to enlarge upon these ideas, which have been developed especially by Møller and which will surely be a main topic of the discussions at this meeting, but it may only be mentioned that the question of the limit of " observability " or rather of " definability " is not easily judged before a comprehensive formalism is established, and even for such approach a further analysis of the implications of the features of atomic theory gradually brought to light may, therefore, not be superfluous.

VII. GENERAL DISCUSSION
AT THE EIGHTH SOLVAY CONFERENCE

SOME GENERAL COMMENTS ON
THE PRESENT SITUATION IN ATOMIC PHYSICS

Les particules élémentaires,
Rapports et discussions du huitième Conseil de physique
tenu à Bruxelles du 27 septembre au 2 octobre 1948,
R. Stoops, Bruxelles 1950, pp. 376–380

See Introduction to Part I, sect. 4.

[223]

Some General Comments on the Present Situation in Atomic Physics.

Mr. Bohr. — In connection with the great progress as regards the accumulation of new experimental evidence and the development of theoretical ideas, discussed during this Conference, it may be of interest, as a continuation of the elementary considerations presented at the first session, to make a few comments upon the situation in atomic physics in relation to our conceptional framework.

A question which has often been raised is to what extent the difficulties met with in present theories may have their origin in the application of classical concepts beyond their appropriate scope. In this connection, it should be remembered how useful considerations of idealized experiments, which might serve to measure physical quantities, have been for the clarification of essential aspects of the situation in relativity theory as well as in quantum theory. It must be stressed, however, that such considerations primarily aim at making us familiar with the foundations of the theories and, in general, do not allow us to investigate the correctness of the theoretical expectations which can, of course, only be tested by actual experiment. In fact, the question as to the results to be expected from an imagined experiment can only be judged from a purely theoretical standpoint and, as far as the theory presents a mat6ematically consistent scheme, no conclusions as regards limits of its scope can be derived in this way.

An instructive example is offered by the quantum theory of electromagnetic fields with its apparently paradoxical features as fluctuations of electric and magnetic intensities in empty space. In fact, attempts of tracing the origin of such paradoxes to an inherent limitation in the applicability of field concepts have proved misdirected on closer examination of the logical interpretation of the results which may be obtained by conceivable arrangements for measuring averages of field intensities over definite space-time extensions. In the treatment of this problem we are in the first place justified in disregarding the atomic constitution of the measuring agencies, like test bodies,

376

since the quantum theory of electromagnetic fields contains only two fundamental constants, the velocity of light c and the quantum of action h, which are in themselves not sufficient to specify quantities of the dimension of a length or a time interval. Consequently, the characteristic features of the theory are essentially independent of the space-time scale, and like in proper quantum mechanics all paradoxes find their straightforward explanation in the complementary relationship between phenomena described in terms of space-time coordination and phenomena accounted for by means of dynamical conservation laws.

In the problems encountered in the theory of elementary atomic particles, the situation is of course quite different, since the introduction of the notion of intrinsic charge and rest mass, together with c and h, implies the possibility of specifying space-time quantities, like atomic diameters and periods. On account, however, of the appearance of the non-dimensional constant

$$\alpha = \frac{2\pi e^2}{hc} \ (\sim 1/137) \tag{1}$$

we cannot, without examining special problems, beforehand trace ultimate limitations as regards space-time coordination. Incidentally, it may be noted that all attempts to deduce the value of α by arguments resting upon theories which are presumed to be consistent independently of the value of e and h would seem futile, and that any further elucidation of this problem can only be expected from an examination of the limitation of the theories.

In looking for such limitations, it must be remembered that the present approach to atomic problems on correspondence lines is essentially an approximation procedure in which, as a first step, the constant α is considered to be vanishingly small. At the same time, however, the actual smallness of α is fundamentally connected with the finite value of h, and we must therefore be prepared for a radical departure from such lines of approach. Still, whatever shape the theoretical edifice for comprehending the atomic phenomena may take, the experimental evidence must always be described in classical terms and, since this description must conform with the demands of relativity theory, it will hardly be possible to obtain a consistent scheme, unless the whole formalism, including aspects which defy classical interpretation, exhibits relativistic invariance.

377

[225]

In the treatment of problems involving particles of rest mass m, the length

$$\lambda = \frac{h}{2 \pi mc} \tag{2}$$

presents, as is well known, a measure for the spatial extensions where, in quantum theory, relativistic effects become of decisive importance. Thus, in problems involving lengths and time intervals comparable with or smaller than λ and λ/c, respectively, we are confronted with peculiar features the recognition of which has led to remarkable developments. Above all, in electron theory, Dirac's ideas and their confirmation by the discovery of the phenomena of creation and annihilation of electron pairs have radically changed the situation. Even if certain features like the filling up of phase space by particles of negative energies presents provisional difficulties for our world picture, the eventual removal of which may demand a compensation by means of the zero-point energy of quantum fields, at present also disregarded, our actual possibilities of treating relativistic electron problems have been most decisively augmented.

An especially instructive lesson we have received, as is well known, in connection with the problem of electromagnetic self-energy of charged particles, which already in classical electron theory presents characteristic divergencies. Introducing, for preliminary convergency, a so-called cut-off length a, we get in classical electrodynamics as a measure of the self-energy

$$W \sim \frac{e^2}{a}, \tag{3}$$

provided a is greater than $e^2/mc^2 (= \alpha\lambda)$. By means of (1) and (2), the expression (3) may be written

$$W \sim \alpha mc^2 \, (\lambda/a). \tag{4}$$

In quantum theory, the situation is essentially changed if $a < \lambda$, in which case the self-energy as regards order of magnitude is found to be

$$W \sim \alpha \, mc^2 \log \, (\lambda/a) \tag{5}$$

for particles subjected to the Pauli principle and obeying the Dirac equation, and

$$W \sim \alpha \, mc^2 \, (\lambda/a)^2 \tag{6}$$

378

for particles obeying Bose statistics. For $a \sim \lambda$, the expressions (5) and (6) are small compared with the rest energy mc^2 and comparable with (4), but if a is taken to be small compared with λ the character of the singularities for vanishing a is quite different in the three cases. In particular, it is significant that the degree of the singularities in (λ/a) which, by such a rough cut-off device, appear in the first stage of the approximation procedure may be smaller as well as larger in quantum mechanics than in classical theory according to the statistics of the particles, a notion which defies any interpretation on classical pictures.

As regards electron theory, the situation has been essentially clarified by the ingenious development of the mathematical formalism which, as we have learnt, has also allowed a quantitative explanation of the recent discoveries regarding finer spectral regularities obtained by new powerful experimental methods. As often stressed during the preceding discussions, the state of electron theory is in various respects more complete than has sometimes been assumed and it would seem excluded to specify its limitations on the basis of the theory itself. In fact, decisive progress in electron theory can hardly be obtained without introducing new fundamental features derived from the accumulation of experimental evidence concerning the interaction between the different kinds of elementary particles. In this connection, it is significant that the value of λ for the electron, owing to the comparatively small mass, is considerably larger than the λ-values for other particles and that, therefore, any change which new evidence may entail will not in the first place impede the application of electron theory in the region in which relativistic effects are predominant.

As regards the outlook to further developments, it must be stressed that quite a new stage in atomic theory has been initiated by the recognition that nuclear constitution demands force fields foreign to electromagnetic theory. The conception of short-range forces means indeed the explicit introduction of a microscopic feature in the foundations of the theory, while in the previous description all such features were considered as traceable, on correspondence lines, to the existence of the quantum of action. Notwithstanding the confirmation of Yukawa's ideas by the observation in cosmic radiation of particles with a rest mass intermediate between the masses of the electron and of the nucleons, meson theory is as yet in a most preliminary stage and new viewpoints will obviously be demanded for the theoretical comprehension of the rapidly increasing experimental evidence about the various types of mesons and the conditions for their production.

Even the simple estimate (6) indicates that essential modifications will be demanded for a consistent treatment of particles obeying Bose statistics in problems involving spatial dimensions comparable with the value of λ. In fact, for decreasing values of a this expression for the self-energy increases very quickly and approaches mc^2 for values of a of the same order as the λ corresponding to nucleonic mass. How much stress may be put on such a comparison is difficult to estimate, but further indication that the value of λ for nucleonic masses might constitute a critical limit of spatial coordination is suggested, as emphasized by Heisenberg, by the occurrence of explosive effects in high energy collisions. More over, it must be remembered that it is the position of the heaviest elementary particles which in the first place will define the reference frame in any conceivable measuring arrangement.

In a future, more comprehensive theory of elementary particles, the relation between the elementary unit of electric charge and the universal quantum of action may play a more fundamental rôle than in present theories, as is also indicated by the fact, often commented upon, that the value of α is of the same order as the empirical mass ratios. In conclusion, attention may once more be called to the necessity of removing by compensation effects obvious inconsistencies inherent in present theories. Here we meet new aspects of the duality between the corpuscle and field concepts originating in the very circumstance that, on the one hand, the definition of fields ultimately rests on their action on material corpuscles while, on the other hand, the properties of corpuscles are essentially defined by their field actions. Notwithstanding all additional features of complementarity this duality is in quantum theory as fundamental as in classical physics.

380

APPENDIX

L. LANDAU AND R. PEIERLS

EXTENSION OF THE UNCERTAINTY PRINCIPLE TO RELATIVISTIC QUANTUM THEORY

"Collected Papers of L.D. Landau" (ed. D. ter Haar),
Pergamon Press, Oxford 1965, pp. 40–51

EXTRACT (pp. 43–46)

J. LINDHARD

INDETERMINACY IN MEASUREMENTS BY CHARGED PARTICLES (1991)

UNPUBLISHED MANUSCRIPT

EXTRACT

See Introduction to Part I, sect. 1. For the original paper see L. Landau and R. Peierls, *Erweiterung des Unbestimmtheitsprinzips für die relativistische Quantentheorie*, Z. Phys. **69** (1931) 56–69.

The following extract from the article by Landau and Peierls is included in order to enable the reader to follow in detail the correspondence quoted in the Introduction to Part I and in the Selected Correspondence.

Landau and Peierls consider a collision process between the atomic object and a test body and note the indeterminacy relation referred to as eq. (1)

$$\Delta P(v' - v) > \frac{\hbar}{\Delta t}, \tag{1}$$

where $(v' - v)$ represents the change in velocity of the test body. This was discussed by Bohr already in his Como Lecture* and is quoted again, eq. (29), in the paper on field measurements by Bohr and Rosenfeld (1933)**. Since $(v' - v) \leq c$, Landau and Peierls replace the inequality above by the indeterminacy relation

$$\Delta P \Delta t > \hbar/c, \tag{2}$$

which is formally equivalent to BR eq. (31), except for the fact that Landau and Peierls here and in the following extract apply the symbol Δt in widely different senses.

Footnote nos. 7 and 8 in the facsimile refer to: W. Pauli, "Über das thermische Gleichgewicht zwischen Strahlung und freien Elektronen", Z. Phys. **18** (1923) 272–286, and P. Jordan and V. Fock, "Neue Unbestimmtheitseigenschaften des elektromagnetischen Feldes", Z. Phys. **66** (1930) 206–209, respectively.

---

* Volume 6, p. [109].
** N. Bohr and L. Rosenfeld, *Zur Frage der Messbarkeit der elektromagnetischen Feldgrössen*, Mat.–Fys. Medd. Dan. Vidensk. Selsk. **12**, no. 8 (1933), this volume p. [55]. The abbreviation BR is used for reference to pages and formulae in this paper.

3. Momentum Measurement in the Relativistic Case

We now wish to make use of relativity, i.e. of the finite speed of propagation. There exists as yet no satisfactory relativistic quantum theory, but it is clear that here also we certainly cannot go beyond the limits imposed on the accuracy of measurement by the general principles of wave mechanics.

The scope of the relation just derived for momentum measurement is considerably extended by relativity. In the non-relativistic theory, the definite change of velocity could be made arbitrarily large, and so the momentum could be measured with arbitrary accuracy even in a short time. If, however, we take into account the fact that the velocity cannot exceed c, then $v - v'$ can be at most of the order of c, so that equation (1) gives

$$\Delta P \, \Delta t > \frac{\hbar}{c}. \tag{2}$$

The inequality (2) is particularly easy to derive for the state *after* the measurements. If we assume that the particle had a definite position before the measurement, then after a time Δt, on account of the finite velocity limit, the position is still known with accuracy $c \, \Delta t$. If the momentum after this time were determined more accurately than as given by (2), this would contradict the result $\Delta P \, \Delta q > \hbar$.

On account of (2) the concept of momentum has a precise significance only over long times. Thus, in cases where the momentum changes appreciably within such times, the use of the concept of momentum is purposeless.

In the measurement of momentum of a charged body, in addition to the above-mentioned inaccuracy, a further perturbation of the measurement arises because the body will emit radiation in the necessary change of velocity. We shall consider only the case where the velocity of the body before the measurement is certainly small compared with c. In this case it is favourable to conduct the measurement so that after the measurement the velocity is again considerably less than c. For, if the velocity approaches c, the relation (1) gives very little benefit, while the accuracy is greatly reduced by the emission of radiation. Thus the non-relativistic formula for radiation damping can be used. The energy emitted is then

$$\frac{e^2}{c^3} \int \dot{v}^2 \, \mathrm{d}t,$$

where e is the charge on the body. This energy evidently has its least value for uniform acceleration, i.e. for $\dot{v} = (v' - v)/\Delta t$, so that the energy emitted is

$$\frac{e^2}{c^3} \frac{(v' - v)^2}{\Delta t}.$$

This unknown change of energy has to be taken into account in the energy balance, and there thus arises in the momentum a further inaccuracy:

$$(v' - v)\, \Delta P > \frac{e^2}{c^3} \frac{(v' - v)^2}{\Delta t},$$

or

$$\Delta P \, \Delta t > \frac{e^2}{c^3} (v' - v). \tag{3}$$

For electrons this inequality gives no new information, since even in the most unfavourable case where $v \sim v' + c$ it gives only $\Delta P \, \Delta t > e^2/c^2$, and this is weaker than (2), since $e^2 < \hbar c$. For macroscopic bodies, however, the relation (3) is significant. Multiplication by (1) gives

$$\Delta P \, \Delta t > \frac{\hbar}{c} \sqrt{\frac{e^2}{\hbar c}}, \tag{4}$$

and in this form we shall make use of it later. The inequality (4) is, of course, valid independently of the method of measurement used, and in particular when the measurement is made by means of the charge on the body, as in the case of the Compton effect, where, in addition to the Compton scattered radiation used in the measurement, there is a further radiation corresponding to that discussed above, obtained when higher approximations are taken into account in the perturbation calculation for the interaction between the radiation and the particle[7]. (In the ordinary Compton effect with electrons this effect is of no importance, on account of the smallness of $e^2/\hbar c$.)

[232]

4. FIELD MEASUREMENT

The simplest method of measuring an electric field is to observe the acceleration of a charged test body. In order to avoid interference by magnetic fields, we use a body of very large mass and very small velocity. Let the momentum of the body before the measurement be known, and let the momentum afterwards be measured, again with accuracy ΔP. From this we can deduce the electric field strength with accuracy such that

$$e \, \Delta \mathscr{E} \, \Delta t > \Delta P. \tag{5}$$

In addition, however, the condition (4) must be satisfied in the momentum measurement. Multiplication of equations (4) and (5) gives

$$\Delta \mathscr{E} > \frac{\sqrt{\hbar c}}{(c \, \Delta t)^2}. \tag{6}$$

For the magnetic field strength we easily obtain the same result by considering the motion of a magnetic needle:

$$\Delta \mathscr{H} > \frac{\sqrt{\hbar c}}{(c \, \Delta t)^2}. \tag{6a}$$

If it is desired to measure the electric and magnetic field strengths simultaneously, then, in addition to the effects already discussed, we have to take into account the effect on the needle of the magnetic field due to the charged body and vice versa. This magnetic field is, in order of magnitude,

$$\Delta \mathscr{H} > \frac{e}{(\Delta l)^2} \cdot \frac{v'}{c}, \tag{7}$$

where Δe is the distance between the test body and the needle. If we multiply this inequality by equations (5) and (1), then (with $v = 0$) we have

$$\Delta \mathscr{E} \, \Delta \mathscr{H} > \frac{\hbar c}{(c \, \Delta t)^2} \cdot \frac{1}{(\Delta e)^2}. \tag{6b}$$

This condition differs from the product of (6) and (6a) in that $c \, \Delta t$ in the denominator is partly replaced by Δe.

If follows from (6), (6a) and (6b) that for $\Delta t = \infty$ the measurement can be made arbitrarily accurate for both $\mathscr{E}$ and $\mathscr{H}$. Thus static fields can be completely defined in the classical sense.†

In wave fields (that is, field which are further than $c/v = \lambda$ from the bodies which produce them), it is sufficient to use (6) and (6a), because as a result of the coupling of the space and time variation nothing is discovered about the

† Our thanks are due to Professor Bohr for pointing out this situation and the significance of time in general.

field if the region of measurement has an extent less than $c\,\Delta t$ for a given Δt. Thus here also the measurements of $\mathscr{E}$ and $\mathscr{H}$ do not interfere, and to the extent that the field strengths can be measured in accordance with (6) and (6a) they can be measured simultaneously. Thus the field strengths are in accordance with the classical theory inasfar as they can be defined at all. In the quantum range, on the other hand, the field strengths are not measurable quantities.†

---

† The inaccuracy for the field measurement with an electron found by Jordan and Fock[8] is greater than (6) and therefore proves only that the electron is not a suitable means of measuring the field.

EDITOR'S COMMENT

A crucial point in the analysis of Landau and Peierls is their assumption that the radiated energy is part of an uncertainty in the energy of the charged particle. As we have seen, Pauli soon raised objections to this assumption (see letter to Peierls quoted on p. [10]). An analysis of the fluctuations in the radiated energy is given by Lindhard in the following unpublished manuscript dealing with a situation in which the initial and final velocities have well-defined values.

In an experimental arrangement in which the initial and final velocities are uncertain, the radiation damping involves a corresponding uncertainty, and it may be this situation which Landau and Peierls had in mind and which is also assumed by Bohr and Rosenfeld in their summary of the Landau–Peierls argument, BR p. 24 (see also their comment, BR p. 26, on the arbitrariness of the identification of Δx with L).

The numbering of the equations corresponds to the numbering in the original manuscript.

J. Lindhard
INDETERMINACY IN MEASUREMENTS BY CHARGED PARTICLES

1. Introduction

The present study of uncertainty relations for charged particles is not without connection to a familiar debate which took place during the early stages of the development of quantum field theory. It may therefore be of interest briefly to recall a few aspects of that debate.

In 1931, Landau and Peierls[1] raised doubts about the consistency of the quantum theory of electromagnetic fields, doubts which, if true, were expected to deprive the theory of any physical basis. They maintained that, due to quantal uncertainty relations, it was not possible to measure electromagnetic radiation fields by means of charged particles. Soon after, Bohr and Rosenfeld[2] criticized this derivation and went on to show that electromagnetic fields could indeed be measured if the point-like particles of Landau and Peierls were replaced by spatially extended charge distributions.

In both of these papers, the discussion of limitations in measurements were based on the Heisenberg uncertainty relations, together with the additional indeterminacy due to fluctuations in the radiation damping suffered by a charge during the measuring process. But in the estimate of the latter fluctuation there was introduced a conjecture which does not seem at all obvious. In fact, Landau and Peierls supposed that the indeterminacy due to radiation damping was given by the total magnitude of the classical radiation damping, a quantity not even containing Planck's constant. This part of their argument was not criticized by Bohr and Rosenfeld, who instead found a way to reduce the radiation damping and thereby its fluctuation. Their way out was to employ spatially extended charge distributions.

The problem to be studied in the following is how, for a charged particle, the Heisenberg uncertainty relations are supplemented by the inborn fluctuation due to radiation damping. This problem may also be cast into the form of a criterion for classical orbital pictures in the motion of interacting charged particles, as

---

[1] L. Landau and R. Peierls, *Erweiterung des Unbestimmtheitsprinzips für die relativistische Quantentheorie*, Z. Phys. **69** (1931) 56–69. [Extract reproduced in English translation p. [229].]

[2] N. Bohr and L. Rosenfeld, *Zur Frage der Messbarkeit der elektromagnetischen Feldgrössen*, Mat.–Fys. Medd. Dan. Vidensk. Selsk. **12**, no. 8 (1933). [Reproduced p. [55].]

discussed by Bohr[3] in the elastic case where radiation was neglected.

In the following study of uncertainty relations it should be remembered, first, that we are preferably concerned with situations where the deviations from classical mechanics are comparatively small. Second, a full account is not obtained unless one has a closed phenomenon in the sense described by Bohr, i.e., a definite completed experiment, including of course the initiation of the experiment and the measurement. Still, in several instances only part of the phenomenon need be discussed in detail. In point of fact, the most valuable concepts in the analysis are the Heisenberg uncertainty relations, and they often suffice for an understanding, without reference to the complete phenomenon.

2. Radiation Damping for Point-like Charged Particles

The reaction force on a moving charge is familiar from classical electromagnetic theory. We are not here concerned with the dominating reaction term containing the inertia of the Coulomb energy of the charge, and exhibiting equivalence between energy and mass[4]. Instead, we have to study the radiative damping which, for a point-like charge q and in the momentary rest frame, becomes simply[5]

$$\vec{F}_d = \frac{2}{3} \frac{q^2}{c^3} \ddot{\vec{v}} \qquad (2.1)$$

where $\ddot{\vec{v}}$ is the second derivative of the velocity with respect to time.

The condition that a charged particle is point-like means only that it is small as measured in terms of the characteristics of its motion; the time $\tau = R/c$ that light takes to pass through its charge distribution should be so small that the external force changes only little, or $\tau \ddot{\vec{v}} < \dot{\vec{v}}$, cf. Heitler[5]. Normally, the charged particle is a composite system like a uranium nucleus, or a charged droplet, but then under conditions such that internal excitations would be highly unlikely.

Suppose then that the charged particle is subject to an external force acting within the time interval (t_1, t_2), the force being zero at times t_1 and t_2. The

[3] N. Bohr, *The Penetration of Atomic Particles through Matter*, Mat.–Fys. Medd. Dan. Vidensk. Selsk. **18**, no. 8 (1948). [Reproduced in Vol. 8, p. [423].]

[4] J. Kalckar, J. Lindhard and O. Ulfbeck, *Self-Mass and Equivalence in Special Relativity*, Mat.–Fys. Medd. Dan. Vidensk. Selsk. **40**, no. 11 (1982).

[5] W. Heitler, *The Quantum Theory of Radiation*, Oxford University Press, 1944, 2nd ed., chap. 1, §4.

average momentum transfer from radiation damping is, from (2.1),

$$\langle \vec{P}_d \rangle = -\frac{2}{3} \frac{q^2}{c^3} \int_1^2 dt\, \dddot{\vec{v}}(t) = -\frac{2}{3} \frac{q^2}{c^3} (\ddot{\vec{v}}_2 - \ddot{\vec{v}}_1) = 0. \tag{2.2}$$

This is denoted as an average although the classical result in itself is without fluctuations.

A differential result obtains for the average energy lost by the radiative reaction

$$\langle E_d \rangle = -\frac{2}{3} \frac{q^2}{c^3} \int_1^2 dt\, \dddot{\vec{v}} \cdot \vec{v}(t) = \frac{2}{3} \frac{q^2}{c^3} \int_1^2 dt\, \ddot{\vec{v}}^2 - \frac{2}{3} \frac{q^2}{c^3} (\ddot{\vec{v}}_2 \cdot \vec{v}_2 - \ddot{\vec{v}}_1 \cdot \vec{v}_1)$$

$$= \frac{2}{3} \frac{q^2}{c^3} \int_1^2 dt\, \ddot{\vec{v}}^2(t). \tag{2.3}$$

For given values of $\vec{v}_1$ and $\vec{v}_2$, there is a lower limit to this energy loss

$$\langle E_d \rangle \geq \frac{2}{3} \frac{q^2}{c^3} \frac{(\vec{v}_2 - \vec{v}_1)^2}{t_2 - t_1} \tag{2.4}$$

as also noted by Landau and Peierls.

The preceding results show that the classical radiation damping still allows us to measure an external force on the particle $F_{ext} = F_{ext}(t)$ (or $F_{ext}(r(t),t)$). In fact, the motion of the particle permits a measurement of the total force $F_{tot}(t)$, and according to (2.1) the external force is then determined by the equation

$$\vec{F}_{tot}(t) = \vec{F}_{ext}(t) + \frac{2}{3} \frac{q^2}{c^3} \frac{d}{dt} \vec{F}_{tot}(t), \tag{2.5}$$

since we may use the non-relativistic expression for the time derivative of the acceleration in (2.1).

It is therefore surprising that Landau and Peierls considered the classical expression (2.3) as an unknown quantity, to be treated as an uncertainty. We shall now see how a proper treatment instead leads to a mean square fluctuation proportional to Planck's constant.

The fluctuation in energy loss is arrived at by decomposing $\langle E_d \rangle$ into its classical frequency spectrum. The frequencies are next converted into energies by $E_\omega = \hbar\omega$. The classical energy transfer belonging to a frequency interval $(\omega, \omega + d\omega)$ may be written as $E_\omega p(\omega)\,d\omega$, where $p(\omega)\,d\omega$ is the average number

[237]

of quanta within the interval. We thus have

$$\langle E_d \rangle = \int_0^{\omega_{max}} d\omega \, p(\omega) E_\omega, \tag{2.6}$$

where the cut-off ω_{max} is put in so as to emphasize that $p(\omega)$ is generally varying relatively slowly with ω, but with a comparatively sharp upper limit.

Since the fluctuation in the number of quanta is given by Poisson statistics

$$\langle n_\omega^2 \rangle - \langle n_\omega \rangle^2 = \langle n_\omega \rangle = p(\omega) \, d\omega,$$

the fluctuation in energy has become

$$\sigma_E^2 = \langle E_d^2 \rangle - \langle E_d \rangle^2 = \int_0^{\omega_{max}} d\omega \, p(\omega) E_\omega^2 \lesssim \hbar \omega_{max} \langle E_d \rangle. \tag{2.7}$$

Although eq. (2.7) gives merely a tentative estimate of the fluctuation, it shows that the fluctuation squared is proportional to Planck's constant, as expected. Moreover, the magnitude of the right-hand side can be easily assessed, since $\langle E_d \rangle$ is known classically and also the maximum frequency ω_{max} can be estimated classically.

The fluctuation in momentum transfer is determined by σ_E, since $|\vec{p}_\omega| = \hbar\omega/c$, and because the average momentum transfer is zero, cf. (2.2), the square fluctuation is simply

$$\sigma_p^2 = \langle \vec{p}_d^2 \rangle - \langle \vec{p}_d \rangle^2 = \frac{\sigma_E^2}{c^2}. \tag{2.8}$$

APPENDIX

W. HEISENBERG

ÜBER DIE MIT DER ENTSTEHUNG VON MATERIE
AUS STRAHLUNG VERKNÜPFTEN LADUNGSSCHWANKUNGEN

Ber. Sächs. Akad. math.–phys. Kl. **86** (1934) 317–322

See Introduction to Part I, sect. 3.

Über die mit der Entstehung von Materie aus Strahlung verknüpften Ladungsschwankungen

Von

W. Heisenberg

Die Diracsche Theorie[1]) des Positrons hat gezeigt, daß Materie aus Strahlung entstehen kann, indem z. B. ein Lichtquant sich in ein negatives und ein positives Elektron verwandelt. Dieses auch experimentell bestätigte Ergebnis hat zur Folge, daß überall dort, wo zur Messung eines physikalischen Sachverhalts große elektromagnetische Felder benötigt werden, mit einer bisher nicht beachteten Störung des Beobachtungsobjektes durch das Beobachtungsmittel gerechnet werden muß, nämlich mit der Erzeugung von Materie durch den Meßapparat. So gering diese Störung für die üblichen Experimente auch sein mag, ihre Berücksichtigung ist für das Verständnis der Theorie des Positrons von prinzipieller Bedeutung.

Zu ihrer näheren Untersuchung in einem sehr einfachen Fall betrachten wir ein quantenmechanisches System von freien negativen Elektronen, die sich gegenseitig nicht merklich beeinflussen; ihre Anzahl pro ccm sei $\frac{N}{V}$. In einem Volumen $v (v \ll V)$ wird dann im Mittel die Ladung

$$\bar{e} = - \varepsilon N \frac{v}{V} \tag{1}$$

zu finden sein (ε ist der Absolutbetrag der Elementarladung). Für das mittlere Schwankungsquadrat der Ladung im Volumen v würde man nach der klassischen Statistik den Wert

$$\overline{\Delta e^2} = \varepsilon^2 N \frac{v}{V} \tag{2}$$

erwarten. Es soll nun gezeigt werden, daß sich nach der Diracschen Theorie im allgemeinen ein größeres Schwankungsquadrat ergibt. Der Überschuß gegenüber Gl. (2) ist auf die mögliche Entstehung von Materie bei der Messung der Ladung im Volumen v zurückzuführen.

---

1) P. A. Dirac, The principles of Quantum mechanic . p. 255. Oxford 1930. Proc. Cambr. Phil. Soc. **30**, 150, 1934.

7*

Die Eigenfunktionen der Elektronen hängen vom Ort $\mathfrak{r}$, der Zeit t und der Spinvariable σ ab, sie sollen $u_n\,(\mathfrak{r}, t, \sigma)$ heißen und in einem Volumen V $(V \gg v)$ normiert sein. Die allgemeine Wellenfunktion der Materie wird dann

$$\psi\,(\mathfrak{r}, t, \sigma) = \sum_n a_n u_n\,(\mathfrak{r}, t, \sigma), \tag{3}$$

wobei wegen des Paulischen Ausschließungsprinzips die V.R.

$$a_n^* a_m + a_m a_n^* = \delta_{nm} \tag{4}$$

gelten. Daraus folgt

$$a_n^* a_n = N_n\,; \quad a_n a_n^* = 1 - N_n\,. \tag{5}$$

Für die Zustände negativer Energie $(E_n < 0)$ soll noch eingeführt werden:

$$a_n' = a_n^*\,; \quad a_n'^* = a_n\,; \quad N_n' = 1 - N_n\,. \tag{6}$$

Es bedeutet dann N_n die Anzahl der Elektronen im Zustand n, N_n' die Anzahl der Positronen im Zustand n.

Für die Ladungsdichte ergibt sich nach der Diracschen Theorie der folgende Ausdruck[1]):

$$-\varepsilon \sum_\sigma \left[\sum_{E_n > 0}' N_n u_n^* u_n - \sum_{E_n < 0} N_n' u_n^* u_n + \sum_{n \neq m} a_n^* a_m u_n^* u_m \right]. \tag{7}$$

Für die Ladung e im Volumen v erhält man daher

$$e = -\varepsilon \left[\sum_{E_n > 0} N_n \cdot \frac{v}{V} - \sum_{E_n < 0} N_n' \frac{v}{V} + \sum_{n \neq m} a_n^* a_m \sum_\sigma \int_v d\mathfrak{r}\, u_n^*(\mathfrak{r} t \sigma)\, u_m(\mathfrak{r} t \sigma) \right]. \tag{8}$$

Aus Gründen, die von Bohr und Rosenfeld[2]) ausführlich diskutiert wurden, soll der zeitliche Mittelwert von e über ein endliches Intervall T betrachtet werden, wobei auch die Grenzen des Intervalls eventuell noch unscharf gelassen werden. Wir führen daher eine Funktion $f(t)$ ein, die nur im Bereich $0 \lesssim t \lesssim T$ von Null verschieden ist und für die

$$\int f(t)\, dt = 1\,.$$

Auch die Grenzen des Volumens v sollen eventuell unscharf bleiben; es soll daher $g(\mathfrak{r})$ eine Funktion von $\mathfrak{r}$ bedeuten, die in v bis auf die Umgebung der Grenzen 1 ist, außen verschwindet und für die $\int d\mathfrak{r}\, g\,(\mathfrak{r}) = v$. Der entsprechende zeitliche und räumliche Mittelwert der Ladung e wird dann

---

1) Vgl. z. B. W. H. Furry u. J. R. Oppenheimer, Phys. Rev. 45, 245, 1934.

2) N. Bohr und L. Rosenfeld, Verh. der Kgl. Dän. Gesellsch. d. Wiss. XII, 8, 1933.

$$\bar{e}=-\varepsilon\left[\sum_{E_n>0}N_n\frac{v}{V}-\sum_{E_n<0}N_n'\frac{v}{V}+\sum_{n\neq m}a_n^*a_m\iint dt\,d\mathfrak{r}\sum f(t)\,g(\mathfrak{r})\,u_n^*(\mathfrak{r}t\sigma)\,u_m(\mathfrak{r}t\sigma)\right].\quad(9)$$

Bildet man nun den Erwartungswert $\bar{\bar{e}}$ von $\bar{e}$ für denjenigen Zustand, bei dem die Anzahlen N_n bekannte c-Zahlen sind, so erhält man

$$\bar{\bar{e}}=-\varepsilon\left[\sum_{E_n>0}N_n-\sum_{E_n<0}N_n'\right]\frac{v}{V}.\quad(10)$$

in Übereinstimmung mit Gl. (1). Für den Erwartungswert des Schwankungsquadrats ergibt sich jedoch:

$$\overline{(\Delta\bar{e})^2}=\overline{\bar{e}^2}-\bar{\bar{e}}^2=\varepsilon^2\sum_{n\neq m}\sum_{n'\neq m'}\overline{a_n^*a_m a_{n'}^* a_{m'}'}\sum_{\sigma\sigma'}\int dt\int dt'\int d\mathfrak{r}$$

$$\int d\mathfrak{r}'\,f(t)\,f(t')\,g(\mathfrak{r})\,g(\mathfrak{r}')\,u_n^*(\mathfrak{r}t\sigma)\,u_m(\mathfrak{r}t\sigma)\,u_{n'}^*(\mathfrak{r}'t'\sigma')\,u_{m'}(\mathfrak{r}'t'\sigma')$$

$$=\varepsilon^2\sum_{n\neq m}N_n(1-N_m)\sum_{\sigma\sigma'}\int dt\int dt'\int d\mathfrak{r}\quad(11)$$

$$\int d\mathfrak{r}'\,f(t)\,f(t')\,g(\mathfrak{r})\,g(\mathfrak{r}')\,u_n^*(\mathfrak{r}t\sigma)\,u_n(\mathfrak{r}'t'\sigma')\,u_m^*(\mathfrak{r}'t'\sigma')\,u_m(\mathfrak{r}t\sigma)$$

$$=\varepsilon^2\sum_{n\neq m}N_n(1-N_m)\,J_{nm}.$$

Diesen Ausdruck kann man in drei Teile zerlegen. Der erste wäre bereits vorhanden, wenn die Erwartungswerte der N_n für $E_n>0$ und N_n' für $E_n<0$ alle verschwinden, d. h. im Vakuum (man beachte $J_{nm}=J_{mn}$):

$$\varepsilon^2\sum_{E_n>0;\,E_m<0}J_{nm}.\quad(12)$$

Ein zweiter Teil wird, wenn nur negative Elektronen vorhanden sind ($N'=0$):

$$\varepsilon^2\sum_{E_n>0}N_n\left(\sum_{E_m>0}-\sum_{E_m<0}\right)J_{nm}.\quad(13)$$

Schließlich bleibt noch als dritter Teil

$$-\varepsilon^2\sum_{E_n>0;\,E_m>0}N_n N_m J_{nm}\quad(14)$$

übrig. Dieser dritte Teil enthält jedoch, wie man aus (11) berechnet, den Faktor $\left(\frac{v}{V}\right)^2$; er kann daher gegenüber den beiden ersten vernachlässigt werden, wenn — wie angenommen wurde — $v\ll V$ ist.

Für die Berechnung des zweiten Teils (13) bemerken wir zunächst, daß

$$\left(\sum_{E_m>0}-\sum_{E_m<0}\right)u_m^*(\mathfrak{r}'t'\sigma')\,u_m(\mathfrak{r}t\sigma)\quad(15)$$

$$=\int\frac{d\mathfrak{p}}{h^3}\frac{1}{2}\left\{\left(1-\frac{a_i p_i+\beta mc}{p_0}\right)e^{\frac{i}{\hbar}\left[\mathfrak{p}(\mathfrak{r}-\mathfrak{r}')-p_0 c(t-t')\right]}_{\sigma\sigma'}-\left(1+\frac{a_i p_i+\beta mc}{p_0}\right)e^{\frac{i}{\hbar}\left[\mathfrak{p}(\mathfrak{r}-\mathfrak{r}')+p_0 c(t-t')\right]}_{\sigma\sigma'}\right\}.$$

Es wird also

$$\left(\sum_{E_m>0}-\sum_{E_m<0}\right)J_{nm}=\sum_{\sigma\sigma'}\int dt\int dt'\int d\mathfrak{r}\int d\mathfrak{r}'\,f(t)\,f(t')\,g(\mathfrak{r})\,g(\mathfrak{r}')\cdot\int\frac{d\mathfrak{p}}{h^3}\frac{1}{2}\left\{\ \right\}$$

$$\cdot u_n^*(\mathfrak{r}\,t\,\sigma)\,u_n(\mathfrak{r}'\,t'\,\sigma')=\int\frac{d\mathfrak{p}}{h^3}\sum_{\sigma\sigma'}\frac{1}{2}\left\{\left(1-\frac{a_i\,p_i+\beta\,m\,c}{p_0}\right)|f(p_0-p_0^n)|^2\right.$$

$$\left.-(1+\frac{a_i\,p_i+\beta\,m\,c}{p_0})_{\sigma\sigma'}|f(p_0+p_0^n)|^2\right\}|g(\mathfrak{p}-\mathfrak{p}^n)|^2\,b_n^*(\sigma)\,b_n(\sigma'),$$

wobei

$$f(p_0)=\int dt\,f(t)\,e^{\frac{i}{\hbar}p_0ct}$$

$$g(\mathfrak{p})=\int d\mathfrak{r}\,g(\mathfrak{r})\,e^{\frac{i}{\hbar}\mathfrak{p}\mathfrak{r}}\tag{16}$$

$$u_n(\mathfrak{r}\,t\,\sigma)=b_n(\sigma)\,e^{\frac{i}{\hbar}(\mathfrak{p}^n\mathfrak{r}-p_0^n t)}$$

gesetzt ist. Aus der Wellengleichung folgt dann weiter

$$\left(\sum_{E_m>0}-\sum_{E_m<0}\right)J_{nm}=\frac{1}{V}\int\frac{d\mathfrak{p}}{h^3}\frac{1}{2}\left\{\left(1+\frac{p_i^n\,p_i+(mc)^2}{p_0\,p_0^n}\right)|f(p_0-p_0^n)|^2\right.$$

$$\left.-\left(1-\frac{p_i^n\,p_i+(mc)^2}{p_0\,p_0^n}\right)|f(p_0+p_0^n)|^2\right\}|g(\mathfrak{p}-\mathfrak{p}^n)|^2\,.\tag{17}$$

Wenn die zum Zustand n gehörige Wellenlänge klein ist gegen die räumliche Ausdehnung des Gebiets v und wenn ferner die Zeit, über die gemittelt werden soll, so klein ist, daß die Elektronen in dieser Zeit nur Strecken durchlaufen, die ebenfalls klein sind im Verhältnis zur räumlichen Ausdehnung von v, so ist $g(\mathfrak{p}-\mathfrak{p}^n)$ als sehr schnell veränderlich gegenüber dem Bruch

$$\frac{p_i^n\,p_i+(mc)^2}{p_0\,p_0^n}$$

anzusehen, ferner kann in dieser Annäherung $|f(p_0-p_0^n)|^2\sim 1$ gesetzt werden. Dann wird

$$\left(\sum_{E_m>0}-\sum_{E_m<0}\right)J_{nm}\approx\frac{1}{V}\int\frac{d\mathfrak{p}}{h^3}|g(\mathfrak{p}-\mathfrak{p}^n)|^2=\frac{1}{V}\int d\mathfrak{r}\,g^2(\mathfrak{r})\approx\frac{v}{V}\,,\tag{18}$$

und für den zweiten Teil des Schwankungsquadrats erhält man:

$$\varepsilon^2\left(\sum_{E>0}N_n\right)\cdot\frac{v}{V},\tag{19}$$

in Übereinstimmung mit Gl. (2).

Hierzu kommt nun noch der erste Teil, der auch im Vakuum auftritt:

$$\varepsilon^2\sum_{E_n>0;\,E_m<0}J_{nm}.$$

Ähnlich wie in Gl. (15) findet man:

$$\sum_{\substack{E_n>0;\ E_m<0}} J_{nm} = \int dt \int dt' \int d\mathfrak{r} \int d\mathfrak{r}' \sum_{\sigma\sigma'} f(t)\,f(t')\,g(\mathfrak{r})\,g(\mathfrak{r}') \int \frac{d\mathfrak{p}}{h^3} \int \frac{d\mathfrak{p}'}{h^3} \frac{1}{4} \left(1 - \frac{\alpha_i\,p_i + \beta\,mc}{p_0}\right)$$

$$\left(1 + \frac{\alpha_i\,p_i' + \beta\,mc}{p_0'}\right) e^{\frac{i}{\hbar}[(\mathfrak{p}-\mathfrak{p}')(\mathfrak{r}-\mathfrak{r}') - (p_0+p_0')(t-t')]} \tag{20}$$

$$= \int dt \int dt' \int d\mathfrak{r} \int d\mathfrak{r}'\, f(t)\,f(t')\,g(\mathfrak{r})\,g(\mathfrak{r}') \int \frac{d\mathfrak{p}}{h^3} \int \frac{d\mathfrak{p}'}{h^3} \frac{p_0 p_0' - \mathfrak{p}\mathfrak{p}' - m^2 c^2}{p_0 p_0'} e^{\frac{i}{\hbar}[(\mathfrak{p}-\mathfrak{p}')(\mathfrak{r}-\mathfrak{r}') - (p_0+p_0')(t-t')]}$$

$$= \int \frac{d\mathfrak{p}}{h^3} \int \frac{d\mathfrak{p}'}{h^3} |f(p_0+p_0')|^2 \cdot |g(\mathfrak{p}-\mathfrak{p}')|^2 \frac{p_0 p_0' - \mathfrak{p}\mathfrak{p}' - m^2 c^2}{p_0 p_0'}.$$

Setzt man $g(\mathfrak{r}) = 1$ in einem rechteckigen, scharf umgrenzten Volumen mit den Seiten l_1, l_2, l_3 u. außerhalb Null, so wird

$$g(\mathfrak{p}) = \int d\mathfrak{r}\, g(\mathfrak{r}) e^{\frac{i}{\hbar}\mathfrak{p}\mathfrak{r}} = \int_0^{l_1} dx \int_0^{l_2} dy \int_0^{l_3} dz\, e^{\frac{i}{\hbar}(p_x x + p_y y + p_z z)} = \left(\frac{e^{\frac{i}{\hbar}p_x l_1} - 1}{\frac{i}{\hbar}p_x}\right) \left(\frac{e^{\frac{i}{\hbar}p_y l_2} - 1}{\frac{i}{\hbar}p_y}\right) \left(\frac{e^{\frac{i}{\hbar}p_z l_3} - 1}{\frac{i}{\hbar}p_z}\right) \tag{21}$$

$$\text{u.} \quad |g^2(\mathfrak{p})| = \frac{\hbar^6}{p_x^2 p_y^2 p_z^2}\, 64 \sin^2 \frac{p_x l_1}{2\hbar} \sin^2 \frac{p_y l_2}{2\hbar} \sin^2 \frac{p_z l_3}{2\hbar}.$$

Begrenzt man das Zeitintervall scharf, so wird

$$f(p_0) = \frac{1}{T}\int_0^T dt\, e^{\frac{i}{\hbar}p_0 c t} = \frac{e^{\frac{i}{\hbar}p_0 c T} - 1}{\frac{i}{\hbar}p_0 c T}; \quad |f(p_0)|^2 = 4 \cdot \frac{\sin^2 \frac{p_0 c T}{2\hbar}}{\left(\frac{p_0 c T}{\hbar}\right)^2}. \tag{22}$$

Setzt man die Werte (21) und (22) in Gl. (20) ein, so divergiert das Integral auf der rechten Seite von Gl. (20). Man erkennt dies am einfachsten, indem man als neue Integrationsvariabeln $\mathfrak{p} - \mathfrak{p}' = \mathfrak{k}$ und $\frac{\mathfrak{p} + \mathfrak{p}'}{2} = \mathfrak{P}$ einführt; es wird dann:

$$\sum_{\substack{E_n>0;\ E_m<0}} J_{nm} = \int \frac{d\mathfrak{k}}{h^3} \int \frac{d\mathfrak{P}}{h^3} |f(p_0+p_0')|^2 \cdot |g(\mathfrak{k})|^2 \frac{\sqrt{\left(\mathfrak{P}^2 + \frac{\mathfrak{k}^2}{4} + m^2 c^2\right)^2 - (\mathfrak{P}\mathfrak{k})^2} - \mathfrak{P}^2 + \frac{\mathfrak{k}^2}{4} - m^2 c^2}{\sqrt{\left(\mathfrak{P}^2 + \frac{\mathfrak{k}^2}{4} + m^2 c^2\right)^2 - (\mathfrak{P}\mathfrak{k})^2}}. \tag{23}$$

Bei einem vorgegebenen großen Wert von $\mathfrak{k}\,(k \gg mc)$ ergibt die Integration über $\mathfrak{P}$, die konvergiert, einen Faktor der ungefähren Größe $\frac{k}{\hbar(cT)^2}$ zu $|g(\mathfrak{k})^2|$; die Integration über $\mathfrak{k}$ führt dann im wesentlichen auf das Integral

$$\int \frac{k}{k_x^2 \cdot k_y^2 \cdot k_z^2}\, dk_x\, dk_y\, dk_z\, \sin^2 \frac{k_x l_1}{2\hbar} \sin^2 \frac{k_y l_2}{2\hbar} \sin^2 \frac{k_z l_3}{2\hbar},$$

das divergiert. Das Schwankungsquadrat von $\bar{e}$ wird also unendlich groß, wenn man das Raum-Zeitgebiet, über das man die Mittelung der Ladung ausführt, scharf begrenzt. Dieses Ergebnis entspricht den Resultaten, die sich bei der Untersuchung der Energieschwankungen in einem Strahlungsfeld herausgestellt haben[1]. Auch dort erwies es sich als notwendig, das Raum-

---

[1] Vgl. W. Heisenberg, Verh. d. Sächs. Ak. 83, 3, 1931.

gebiet, in dem die Energie aufgesucht werden soll, unscharf zu begrenzen. Für die Ladungsschwankungen der Gl. (20) genügt es, entweder die Grenzen des Zeitintervalls oder die des Raumintervalls in geeigneter Weise unscharf zu wählen. Nehmen wir z. B. an, daß die Größe $g(\mathfrak{r})$ in einem Gebiet der Breite b in der Umgebung des Randes von v etwa nach der Art einer Gaußschen Fehlerfunktion von 1 auf 0 abnimmt. Dann verschwindet $g(\mathfrak{p})$ für Werte von $p > \frac{\hbar}{b}$ ebenfalls wie die Fehlerfunktion; nimmt man weiter an, daß die Zeit T klein sei gegen $\frac{b}{c}$, so kann man in genügender Approximation den Bruch in (23) entwickeln für $k^2 \ll \mathfrak{P}^2 + m^2 c^2$. Es ergibt sich dann

$$\sum_{\substack{E_n > 0, \\ E_m < 0}} J_{nm} = \int \frac{d\mathfrak{k}}{h^3} \int \frac{d\mathfrak{P}}{h^3} \, |f(2P)|^2 \, |g(\mathfrak{k})|^2 \, \frac{\mathfrak{k}^2 - \frac{(\mathfrak{P}\,\mathfrak{k})^2}{P^2}}{2\,(P^2 + m^2 c^2)} . \tag{24}$$

Bis auf unwesentliche konstante Faktoren wird daraus

$$\frac{1}{\hbar^2 c\, T} \int \frac{d\mathfrak{k}}{h^3} \, |g(\mathfrak{k})|^2 \, \mathfrak{k}^2 \qquad\qquad \text{für } T \ll \frac{\hbar}{m\,c^2} ,$$

$$\frac{1}{\hbar\,(c\,T)^2 \cdot m\,c} \int \frac{d\mathfrak{k}}{h^3} \, |g(\mathfrak{k})|^2 \, \mathfrak{k}^2 \qquad\qquad \text{für } T \gg \frac{\hbar}{m\,c^2} .$$

Für die Größenordnung des ersten Teiles des Schwankungsquadrats erhält man daher $\left(\text{für } l_1 \sim l_2 \sim l_3 = \sqrt[3]{v} \right)$:

$$\overline{(\Delta e)^2} \sim \frac{\varepsilon^2}{\hbar\,c\,T} \frac{\hbar}{b} v^{\frac{2}{3}} = \varepsilon^2 \frac{v^{\frac{2}{3}}}{c\,T \cdot b} \qquad\qquad \text{für } T \ll \frac{\hbar}{m\,c^2}$$

$$\overline{(\Delta e)^2} \sim \qquad\qquad \varepsilon^2 \frac{v^{\frac{2}{3}}\,\hbar}{(c\,T)^2 \cdot m\,c \cdot b} \qquad\qquad \text{für } T \gg \frac{\hbar}{m\,c^2} . \tag{25}$$

Der Faktor $v^{\frac{2}{3}}$ zeigt hier deutlich, daß es sich bei diesen Schwankungen um einen Oberflächeneffekt handelt, der davon herrührt, daß an den Wänden, die das vorgegebene Volumen v abgrenzen, Materie entstehen kann. Er wird um so größer, je kleiner die Zeit gewählt wird, in der die Messung der Ladung vorgenommen werden soll und je schärfer die Begrenzung des Volumens ist. Diese Ergebnisse dürften eng zusammenhängen mit dem Umstand, daß die für die Erzeugung von Materie maßgebende Inhomogenität der Diracschen Gleichung die zweiten Ableitungen der elektromagnetischen Feldstärken enthält, daß also eine Unstetigkeit im ersten Differentialquotienten der Feldstärken bereits zur Entstehung von unendlich viel Materie Anlaß geben könnte.

Zusammenfassend sei festgestellt, daß bei der Messung der Ladung in einem vorgegebenen Raum-Zeitgebiet Schwankungen auftreten, die in der klassischen Theorie kein Analogon besitzen; sie werden verursacht durch die Materie, die an der Oberfläche des betrachteten Raumgebiets bei der Messung entsteht.

PART II

COMPLEMENTARITY: BEDROCK OF THE QUANTAL DESCRIPTION

Niels Bohr, Tisvilde c. 1947 (Photograph by S. Rozental).

[248]

INTRODUCTION

by

JØRGEN KALCKAR

1. FLOURISHES

"The importance of physical science for the development of general philosophical thinking rests not only on its contributions to our steadily increasing knowledge of that nature of which we ourselves are part, but also on the opportunities which time and again it has offered for examination and refinement of our conceptual tools. In our century, the study of the atomic constitution of matter has revealed an unsuspected limitation of the scope of classical physical ideas and has thrown new light on the demands on scientific explanation incorporated in traditional philosophy. The revision of the foundation for the unambiguous application of our elementary concepts, necessary for comprehension of atomic phenomena, therefore has a bearing far beyond the special domain of physical science.

The main point of the lesson given us by the development of atomic physics is, as is well known, the recognition of a feature of wholeness in atomic processes, disclosed by the discovery of the quantum of action. The following articles present the essential aspects of the situation in quantum physics and, at the same time, stress the points of similarity it exhibits to our position in other fields of knowledge beyond the scope of the mechanical conception of nature. We are not dealing here with more or less vague analogies, but with an investigation of the conditions for the proper use of our conceptual means of expression. Such considerations not only aim at making us familiar with the novel situation in physical science, but might on account of the comparatively simple character of atomic problems

[249]

be helpful in clarifying the conditions for objective description in wider fields." [1]

2. LAST COMBAT

Quantum Mechanics and Physical Reality (1935)
Can Quantum-Mechanical Description of Physical Reality be Considered Complete? (1935)

As Bohr relates in "Discussion with Einstein"[2], until the mid-thirties only a narrow circle of physicists had so far taken part in the discussions of the apparent paradoxes through which Einstein time and again attempted to circumvent the indeterminacy relations. In fact, the debates on these questions at the Solvay meetings mainly took place outside the regular sessions. This situation changed radically, however, on the publication of the paper by Einstein, Podolsky and Rosen with the title "Can Quantum-Mechanical Description of Physical Reality be Considered Complete?" (reproduced in the appendix, p. [425]). Not only did it attract the attention of many physicists, but the ensuing discussions aroused interest in the more philosophical aspects of quantum physics far outside the physics community.

In contrast, the small group, Bohr, Heisenberg, Pauli and a few others, who through intense debates during many years had become intimately familiar with all aspects of the quantal description, was mainly astonished that Einstein had found it worthwhile to publish this "paradox" in which they saw nothing but the old problems, resolved long ago, in a new dress. We find this view most dramatically expressed in the following letter to Heisenberg from Pauli – whom Bohr used to characterize as the "conscience of physics". The reader should of course bear in mind that it is a private spontaneous letter to a close friend and colleague. This highly satirical formulation of his harsh judgment was never intended for the eyes of the public[3].

---

[1] Extract of Bohr's introduction to *Atomic Physics and Human Knowledge*, John Wiley & Sons, New York 1958, reissued as *Essays 1933–1957 on Atomic Physics and Human Knowledge, The Philosophical Writings of Niels Bohr, Vol. II*, Ox Bow Press, Woodbridge, Connecticut 1987. The full introduction is reproduced in Vol. 10.

[2] N. Bohr, *Discussion with Einstein on Epistemological Problems in Atomic Physics* in *Albert Einstein: Philosopher–Scientist* (ed. P.A. Schilpp), Library of Living Philosophers, Vol. VII, Evanston, Illinois 1949, pp. 201–241. The article is reproduced on p. [339].

[3] In order that the reader may enjoy Pauli's satirical wit in its genuine form, the German original of this extract is reproduced in Part III, p. [480].

Albert Einstein, Princeton 1940 (Courtesy Time–Life).

"... *Einstein* has once again made a public statement about quantum mechanics, and even in the issue of Physical Review of May 15 (together with Podolsky and Rosen – no good company, by the way). As is well known, that is a disaster whenever it happens. 'Nothing can – so runs his thought – exist, if not exist it ought.' (Morgenstern)[4]

Still, I must grant him that if a student in one of his earlier semesters had raised such objections, I would have considered him quite intelligent and promising. – Since this publication may represent a certain danger of

Pauli to Heisenberg,
15 June 35
PWB II, letter [412]
German
Extract on p. [480]

---

[4] C. Morgenstern, *Alle Galgenlieder*, Berlin 1932. My own attempt at a translation of the German rhyme.

confusing the public opinion – especially in America – it might perhaps be advisable to send an answer to the Physical Review which I would like to persuade *you* to undertake.

Thus it might anyhow be worthwhile if I wasted paper and ink in order to formulate those inescapable facts of quantum mechanics that cause Einstein special mental troubles.

He has now reached the level of understanding, where he realizes that two quantities corresponding to non-commutable operators cannot be measured simultaneously and cannot at the same time be ascribed definite numerical values. But the fact that disturbs him in this connection is the way two systems in quantum mechanics can be coupled to form one single total system.

...

All in all, those elderly gentlemen like *Laue* and *Einstein* are haunted by the notion that quantum mechanics is *correct* but *incomplete*. One could *supplement it by statements it does not contain, without altering the statements it does contain.* (A theory of this kind I should call – in logical respect – *incomplete*. Example: the kinetic theory of gases). Perhaps you could, in connection with an answer to Einstein, make it clear in a definite manner that such a supplement in quantum mechanics is impossible without changing its content."

I feel it important to stress how foreign this tone is to Bohr's own style – although Bohr whole-heartedly enjoyed it when he himself was the target of Pauli's sarcasm.

Heisenberg answered Pauli on 2 July 1935, (PWB II, letter [414]) and told him that he had learned from Copenhagen that Bohr already had a manuscript ready for publication in the Physical Review as a reply to Einstein. He adds that he still feels tempted to elucidate the problem from another angle, and includes an unpublished draft (reproduced in PWB II, Appendix to letter [414]).

As far as Bohr's own reaction to the paper by Einstein, Podolsky and Rosen is concerned, we are fortunate to possess Rosenfeld's vivid description[5] from which it is evident that Bohr fully grasped that the problem raised by Einstein demanded careful analysis:

"No form of approval could be more precious to physicists, not excluding Bohr, than Pauli's benevolent nodding. Bohr's ideas, however, did not

---

[5] L. Rosenfeld, *Niels Bohr in the Thirties* in *Niels Bohr, his life and work as seen by his friends and colleagues* (ed. S. Rozental), North-Holland Publ. Co., Amsterdam 1967, pp. 127–129.

Wolfgang Pauli, Copenhagen 1937.

make such a strong impression on everybody. I had occasion, shortly after the completion of our work on quantum electrodynamics, in 1933, to give a lecture on this subject in Brussels, just at the time when Einstein was staying there, before emigrating to Princeton. He attended the lecture and followed the argument with the closest attention; he made no direct comment on it, but put at once the discussion on the general theme of the meaning of quantum theory. He had no longer any doubt about the logic of Bohr's argumentation; but he still felt the same uneasiness as before ('Unbehagen' was his word) when confronted with the strange consequences of the theory. 'What would you say of the following situation?' he asked me. 'Suppose two particles are set in motion towards each other with the same, very large, momentum, and that they interact with each other for a very short time when they pass at known positions. Consider now an observer who gets hold of one of the particles, far away from the region of interaction, and measures its momentum; then, from the conditions of the experiment, he will obviously be able to deduce the momentum of the other particle. If, however, he chooses to measure the position of the first particle, he will be able to tell where the other particle is. This is a perfectly correct and straightforward deduction from the principles of quantum mechanics; but is it not very paradoxical? How can the final state of the second particle be influenced by a measurement performed on the first, after all physical interaction has ceased between them?'

I had not the impression that Einstein at that time saw in this case, cleverly presented with all the appearances of a paradox, anything else than an illustration of the unfamiliar features of quantum phenomena. Two years later, however, he gave it a much more prominent role in a paper written jointly with Podolsky and Rosen; combined with a 'criterion of reality', it was now used with the intention to expose an essential imperfection of quantum theory. Any attribute of a physical system that can be accurately determined without disturbing the system, thus went the argument, is an 'element of physical reality', and a description of the system can only be regarded as complete if it embodies all the elements of reality which can be attached to it. Now, the example of the two particles shows that the position and the momentum of a given particle can be obtained by appropriate measurements performed on another particle without disturbing the first, and are therefore elements of reality in the sense indicated. Since quantum theory does not allow both to enter into the description of the state of the particle, such a description is incomplete. The paradox which this incomplete description presents, by suggesting an unaccountable influence of the measurement on the state of the particle, would of course not appear in a complete theory.

This onslaught came down upon us as a bolt from the blue. Its effect on Bohr was remarkable. We were then in the midst of groping attempts at exploring the implications of the fluctuations of charge and current distributions, which presented us with riddles of a kind we had not met in electrodynamics. A new worry could not come at a less propitious time. Yet, as soon as Bohr had heard my report of Einstein's argument, everything else was abandoned: we had to clear up such a misunderstanding at once. We should reply by taking up the same example and showing the right way to speak about it. In great excitement, Bohr immediately started dictating to me the outline of such a reply. Very soon, however, he became hesitant: 'No, this won't do, we must try all over again ... we must make it quite clear ...' So it went on for a while, with growing wonder at the unexpected subtlety of the argument. Now and then, he would turn to me: 'What *can* they mean? Do *you* understand it?' There would follow some inconclusive exegesis. Clearly, we were farther from the mark than we first thought. Eventually, he broke off with the familiar remark that he 'must sleep on it'. The next morning he at once took up the dictation again, and I was struck by a change in the tone of the sentences: there was no trace in them of the previous day's sharp expressions of dissent. As I pointed out to him that he seemed to take a milder view of the case, he smiled: 'That's a sign', he said, 'that we are beginning to understand the problem.' And indeed, the real work now began in earnest: day after day, week after week, the whole argument was patiently scrutinized with the help of simpler and more transparent examples. Einstein's problem was reshaped and its solution reformulated with such precision and clarity that the weakness in the critics' reasoning became evident ..."

* * *

The essence of Bohr's reply to Einstein[6] is his demonstration that this new thought experiment does not exhibit *any* features not already inherent in the analysis of the old double-slit experiment debated at the Solvay Conference in 1927[7]. Let us therefore recall the following: If, as suggested by Einstein, a device is brought into action that permits a determination of the recoil momentum of the screen in the y-direction with an accuracy sufficient to decide through which slit the corpuscle passed, the corresponding indeterminacy in the position of the diaphragm excludes the correlation between the position of the photographic plate and the diaphragm required for establishing the interference pattern composed of individual impacts of corpuscles on the photographic plate. After the electron has passed the screen along the line of the x-direction and left its irreversible signature in the photographic emulsion, obviously no later decision can alter this mark. What a later choice *can* and *will* decide is the correlations embodied in the pattern of dots. Indeed, if the position of the diaphragm is measured in each successive experiment, the resulting multitude of dots – each of which is spatially correlated to the diaphragm – will be found to form an interference pattern. In contrast, in a set of experiments in which the momentum of the diaphragm is determined after the passage of each individual corpuscle, the outcome will be a collection of dots which can be divided into two groups, each with a uniform distribution and associated with one of the two values of the momentum transfer that would direct the corpuscle towards one or the other of the two slits. In Bohr's words[8] (this volume p. [370]):

"The problem again emphasizes the necessity of considering the *whole*

---

[6] The reactions to Bohr's answer to Einstein, by Klein, Kramers and Schrödinger, are illuminated through their letters reproduced in the Selected Correspondence, on pp. [459], [461] and [503].

[7] Cf. Vol. 6, Part I.

[8] Since later Bohr himself did not feel quite happy with some of the formulations in his original reply to Einstein (1935), I found that for the guidance of the reader it was preferable to quote his comments from *Discussion with Einstein* (ref. 2). The break in historical chronology must of course be taken into consideration, but I hope it is in Bohr's spirit to attach greater importance to the clarity of exposition.

experimental arrangement, the specification of which is imperative for any well-defined application of the quantum-mechanical formalism. Incidentally, it may be added that paradoxes of the kind contemplated by Einstein are encountered also in such simple arrangements as sketched in Figure 5 [this volume p. [360]]. In fact, after a preliminary measurement of the momentum of the diaphragm, we are in principle offered the choice, when an electron or photon has passed through the slit, either to repeat the momentum measurement or to control the position of the diaphragm and, thus, to make predictions pertaining to alternative subsequent observations. It may also be added that it obviously can make no difference as regards observable effects obtainable by a definite experimental arrangement, whether our plans of constructing or handling the instruments are fixed beforehand or whether we prefer to postpone the completion of our planning until a later moment when the particle is already on its way from one instrument to another."

The analysis of such an experiment with a delayed choice between a determination of the momentum or of the position of the diaphragm is especially simple if its mass is so large that it can be considered at rest during the entire closed process.

In the initial stage of the experiment the diaphragm has a momentum defined with suitable accuracy, but during the passage of the corpuscle there is a momentum exchange between corpuscle and diaphragm. Still, as a consequence of the assumption of an arbitrarily large mass of the latter, its velocity remains vanishingly small, however big a momentum it might have acquired. Thus, the diaphragm remains at rest for any length of time at a position whose absolute value is indeterminate in inverse proportion to the accuracy of the initial definition of its momentum. It might be worth recalling that it is under the same premise of a sufficiently large mass of the electrified test body that Bohr and Rosenfeld, BR, p. 28, (this volume p. [84]) in connection with the measurement of the space–time average of a field component, are able to consider it to be at rest during the entire time interval over which the averaging extends.

The decisive conclusion is formulated by Bohr as follows[9]:

"As regards the special problem treated by Einstein, Podolsky and Rosen, it was next shown that the consequences of the formalism as regards the representation of the state of a system consisting of two interacting atomic objects correspond to the simple arguments mentioned in the preceding

[9] This volume p. [373].

[256]

in connection with the discussion of the experimental arrangements suited for the study of complementary phenomena. In fact, although any pair q and p, of conjugate space and momentum variables obeys the rule of non-commutative multiplication expressed by[10]

$$qp - pq = \sqrt{-1}\frac{h}{2\pi}, \tag{2}$$

and can thus only be fixed with reciprocal latitudes given by

$$\Delta q \cdot \Delta p \approx h, \tag{3}$$

the difference $q_1 - q_2$ between two space-coordinates referring to the constituents of the system will commute with the sum $p_1 + p_2$ of the corresponding momentum components, as follows directly from the commutability of q_1 with p_2 and q_2 with p_1. Both $q_1 - q_2$ and $p_1 + p_2$ can, therefore, be accurately fixed in a state of the complex system[11] and, consequently, we can predict the values of either q_1 or p_1 if either q_2 or p_2, respectively, are determined by direct measurements. If, for the two parts of the system, we take a particle and a diaphragm, like that sketched in Fig. 5 [on p. [360]], we see that the possibilities of specifying the state of the particle by measurements on the diaphragm just correspond to the situation described on p. 220 [on p. [360]] and further discussed on p. 230 [on p. [370]], where it was mentioned that, after the particle has passed through the diaphragm, we have in principle the choice of measuring either the position of the diaphragm or its momentum and, in each case, to make predictions as to subsequent observations pertaining to the particle. As repeatedly stressed, the principal point is here that such measurements demand mutually exclusive experimental arrangements[12]."

Nevertheless Einstein continued to argue[13] (Bohr's quotation):

---

[10] Explicit formulae are inserted here by the editor for the convenience of the reader.

[11] In his letter to Weisskopf dated 5 December 1933 (reproduced on p. [513]), Bohr states explicitly this circumstance in connection with a field measurement.

[12] A measurement of one part of the total system by means of which it is possible to deduce a value for a variable of the other, need not involve a disturbance of this other part of the system in the form of a physical interaction ("mechanical disturbance"). However, it does exclude observations that could yield values for complementarity variables of the other part of the system.

[13] A. Einstein, *Physics and Reality*, Journ. Frankl. Inst. **221** (1936) 349–382 as quoted by Bohr in *Discussion with Einstein*, on p. [375]. This entire exposition by Einstein deserves careful study: As a background for his attitude to the quantal description, Einstein discusses carefully his views on the application of the notion of "physical reality" in Newtonian mechanics and in the special and general theory of relativity.

"... that the quantum-mechanical description is to be considered merely as a means of accounting for the average behaviour of a large number of atomic systems and his attitude to the belief that it should offer an exhaustive description of the individual phenomena is expressed in the following words: 'To believe this is logically possible without contradiction; but it is so very contrary to my scientific instinct that I cannot forego the search for a more complete conception.'"

Bohr formulated his own attitude to Einstein's "scientific instinct" in the following words (this volume p. [375]):

"Even if such an attitude might seem well-balanced in itself, it nevertheless implies a rejection of the whole argumentation exposed in the preceding, aiming to show that, in quantum mechanics, we are not dealing with an arbitrary renunciation of a more detailed analysis of atomic phenomena, but with a recognition that such an analysis is *in principle* excluded. The peculiar individuality of the quantum effects presents us, as regards the comprehension of well-defined evidence, with a novel situation unforeseen in classical physics and irreconcilable with conventional ideas suited for our orientation and adjustment to ordinary experience. It is in this respect that quantum theory has called for a renewed revision of the foundation for the unambiguous use of elementary concepts, as a further step in the development which, since the advent of relativity theory, has been so characteristic of modern science."

It seems to me that in spite of its courteous mildness, this last rejoinder by Bohr reveals a feeling of deep and everlasting surprise and of disappointment that, among all physicists, just Einstein himself did not appreciate how closely Bohr and the younger generation were (cf. Bohr's letter to Einstein, p. [436]) "continuing along the road you have shown us and of realizing the logical presuppositions for the description of the realities".

3. FINAL ANALYSIS

The Causality Problem in Atomic Physics (1939)
On the Notions of Causality and Complementarity (1948)

The two papers dealt with in this section represent steps towards Bohr's final formulation of the complementarity argument, involving the introduction of the phrase "physical phenomenon" to be used in a very precise manner as referring to an entire closed quantal process. By being able to follow this constant sharpening of terminology and refinement of argumentation, we are presented with a valuable opportunity to become aware of the manifold of facets of the critical issues: hitherto unnoticed presuppositions for physical analysis are disclosed, and thereby the conditions for our description of nature and our rôle as observers are fathomed ever more deeply.

In order to illustrate this interesting development in Bohr's analysis, it may be instructive, as a background, to consider the following quotations from the early days. They are from the "Introductory Survey" (1929)[14] (my italics):

"As our knowledge becomes wider, we must always be prepared, therefore, to expect alterations in the points of view best suited for the ordering of our experience. In this connection we must remember, above all, that, as a matter of course, all new experience makes its appearance within the frame of our customary points of view and *forms of perception*. The relative prominence accorded to the various aspects of scientific inquiry depends upon the nature of the matter under investigation. In physics, where our problem consists in the co-ordination of our experience of the external world, the question of the nature of our forms of perception will generally be less acute than it is in psychology where it is our own mental activity which is the object under investigation. Yet occasionally just this 'objectivity' of physical observations becomes particularly suited to emphasize the subjective character of all experience."

Especially surprising is the reference to our "forms of perception" and to the "subjective character of all experience". It is not difficult to understand – in retrospect, at least – that such a manner of speaking could give rise to fundamental misunderstandings of Bohr's position: namely that he should hold as meaningless the very notion of ascribing an objective physical reality to an unobserved atomic system! Later on, when the phrase "forms of perception" was replaced by "experimental arrangement", "the objectivity of physical observations" could be stressed without the somewhat bewildering addition that it could be "particularly suited to emphasize the subjective character of all experience".

Further on in the same paper we are told that (my italics):

---

[14] N. Bohr, *Introductory Survey* in *Atomic Theory and the Description of Nature*, Cambridge University Press, 1934 (1961). Reprinted in *Atomic Theory and the Description of Nature, The Philosophical Writings of Niels Bohr, Vol. I*, Ox Bow Press, Woodbridge, Connecticut 1987. Reproduced in Vol. 6, p. [279].

"The quantum of action has become increasingly indispensable in the ordering of our experimental knowledge of the properties of atoms. At the same time, however, we have been forced step by step to forego a causal description of the behaviour of individual atoms in space and time, and *to reckon with a free choice on the part of nature between various possibilities to which only probability considerations can be applied.*"

It is in the Warsaw Lecture, "The Causality Problem in Atomic Physics", that we first encounter a highly significant shift in terminological emphasis. The Warsaw Conference of the Institute of Intellectual Co-operation, from 30 May to 3 June 1938, was organized in collaboration with the International Union of Physics and the Polish Intellectual Co-operation Committee. Among the speakers, besides Bohr, were de Broglie, Klein and Kramers, and the audience included Brillouin, Darwin, Eddington, Fowler, Gamow, von Neumann and Wigner[15]. In his "Introductory Discourse", the chairman, Professor C. Bialobrzeski from the Joseph Pilsudski University, emphasized "our profound regret at the absence of many German, Italian and Russian colleagues, whom circumstances have prevented from attending in spite of their feeling of brotherhood with scholars all over the world".

* * *

The Warsaw Lecture may be characterized as a watershed in Bohr's exposition of the fundamental features of the viewpoint of Complementarity as a basis for the entire quantal description. Echoes of earlier, more careless, formulations survive here and there, e.g. on p. [305], in the reference to electron transitions between stationary states in an atom (my italics):

"In particular, as regards its possible transitions from a given stationary state to [an]other stationary state, accompanied by the emission of photons of different energies, *the atom may be said to be confronted with a choice* for which, according to the whole character of the description, there is no determining circumstance."

Such lapses are rare, however. This article is of particular interest not only because Bohr here presents for the first time his so-called "phenomenon terminology", but perhaps even more because of his explicit comments on points

---

[15] For the full list of participants, cf. *New Theories in Physics*, International Institute of Intellectual Co-operation, Paris 1939, p. VII, this volume p. [301].

An excursion to Cracow after the Warsaw Conference, 1938.
Front row: –, G. Gamow, E. Hylleraas, L. Rosenfeld, –, A. Eddington, –.
Middle row: J. Detouches, N. Bohr, M. Bohr, E. Wigner, O. Klein.
Back row: R. Smoluchowski, C. Darwin, M. Establier, –, –, –.

otherwise only briefly alluded to. For example, his phrasing provides stuff for reflection as to whether Bohr viewed the Schrödinger equation as being of subordinate significance compared to Heisenberg's formulation, although he found the Schrödinger equation of "utmost importance, not only for the practical use of formalism, but even for the elucidation of essential aspects of its consequences". He continues (p. [307]):

"In this connection, it must also be remembered that in the Schrödinger equation, which may be written

$$H\Psi = \frac{hi}{2\pi} \frac{\partial \Psi}{\partial t},$$ (4a)

where H is a differential operator derived by replacing the p's by the oper-

[261]

ators

$$p_i = \frac{h}{2\pi i} \frac{\partial}{\partial q_i} \qquad (4b)$$

in the Hamiltonian function, the form of this function contains as direct a reference to the corpuscular description of an atomic system as the upholding of the equation[16]

$$\frac{dp_i}{dt} = -\frac{\partial H}{\partial q_i}, \qquad \frac{dq_i}{dt} = \frac{\partial H}{\partial p_i} \qquad (2)$$

in Heisenberg's formalism. Besides, the purely formal character of the suggestive resemblance between atomic mechanics and classical wave problems is perhaps most strikingly illustrated by the fact that this resemblance can only be brought out with the help of the conventional device of complex numbers, which alone permits to condense the abstractions of quantum mechanics in a form as simple as

$$pq - qp = \frac{h}{2\pi i} \qquad (3)$$

and (4). The true significance of the wave formalism as a most practical means of expressing the statistical laws of atomic mechanics was also soon fully realized, especially through the work of Born, Dirac and Jordan and the completeness and self-consistency of the whole formalism is most clearly exhibited by the elegant axiomatic exposition of von Neumann, which in particular makes it evident that the fundamental superposition principle of quantum mechanics logically excludes the possibility of avoiding the non-causal feature of the formalism by any conceivable introduction of additional variables."

Upon von Neumann's comments after Bohr's lecture, "elegantly" phrased, as indeed they were, in the language of formal logic, Bohr is reported to have replied as follows[17]:

"Professor Bohr expressed his admiration for the skill with which Professor von Neumann had treated the fundamental problems of quantum theory from the mathematical and logical point of view. He pointed out at the same time how the very simple experimental cases which he alluded to in his paper shewed, in more elementary form, the same essential points as those

---

[16] Explicit formulae (2) and (3) are inserted here by the editor for the convenience of the reader.
[17] *New Theories in Physics* (ref. 15), pp. 38–39.

which appeared in the mathematical analysis. We must also notice that the question of the logical forms which are best adapted to quantum theory is in fact a practical problem, concerned with the choice of the most convenient manner in which to express the new situation that arises in this domain. Personally, he compelled himself to keep the logical forms of daily life to which actual experiments were necessarily confined. The aim of the idea of complementarity was to allow of keeping the usual logical forms while procuring the extension necessary for including the new situation relative to the problem of observation in atomic physics."

As a matter of fact, I vividly remember myself how Bohr in his later years would shake his head sceptically vis-à-vis von Neumann's "proof" in his well-known book[18]: "How on earth can people believe it possible to deduce such things? Certainly only by smuggling in, from the very beginning, an axiom that in itself entails what you want to prove!". The following quotation (p. [310], my italics) shows how Bohr here is *beginning* to distance himself from expressions like "disturbing the object through observation". (Cf. also the quotation on p. [265].)

"It will hardly be necessary here to discuss more closely typical examples of measuring processes, such as have been treated in detail in the current literature. *The observation problem in quantum theory involves, however, certain novel epistemological aspects as regards the analysis and synthesis of physical experience, which have only been gradually elucidated in recent years*, and which I shall now proceed to discuss.

In the first place, we must recognize that a measurement can mean nothing else than the unambiguous comparison of some property of the object under investigation with a corresponding property of another system, serving as a measuring instrument, and for which this property is directly determinable according to its definition in everyday language or in the terminology of classical physics. While within the scope of classical physics such a comparison can be obtained without interfering essentially with the behaviour of the object, this is not so in the field of quantum theory, where the interaction between the object and the measuring instruments will have an essential influence on the phenomenon itself. Above all, we must realize that this interaction cannot be sharply separated from *an undisturbed behaviour of the object*, since the necessity of basing the description of the

---

[18] J. von Neumann, *Mathematische Grundlagen der Quantenmechanik*, Springer, Berlin 1932.

properties and manipulation of the measuring instruments on purely classical ideas implies the neglect of all quantum effects in that description, and in particular the renunciation of a control of the reaction of the object on the instruments more accurate than is compatible with the relation (5) $[\Delta p \cdot \Delta q \geq h/4\pi]$."

It is typical of Bohr's attitude to the rôle of description in common language that the "phenomenon terminology" is not at all introduced in an abrupt axiomatic manner. The issue is first sounded, as it were, from a distance (p. [312], my italics):

"The essential lesson of the analysis of measurements in quantum theory is thus the emphasis on the necessity, in the account of the phenomena, of taking the whole experimental arrangement into consideration, in complete conformity with the fact that all unambiguous interpretation of the quantum mechanical formalism involves the fixation of the external conditions, defining the initial state of the atomic system concerned and the character of the possible predictions as regards subsequent observable properties of that system. Any measurement in quantum theory can in fact only refer either to a fixation of the initial state or to the test of such predictions, and *it is first the combination of measurements of both kinds which constitutes a well-defined phenomenon*."

Before proceeding along this line, Bohr emphasizes in a most interesting paragraph that we should remember that the situations of observation dealing with interference effects of an electron, on the one hand, and the Compton effect, on the other, represent extreme cases and he comments on the general observational situation (p. [315], my italics):

"Notwithstanding their great importance in illustrating typical aspects of atomic processes, the two kinds of quantum phenomena just discussed represent of course only limiting cases of special simplicity. *It is in fact possible to test the statistical predictions of quantum mechanics referring to any state of the object* defined by the values of suitable functions of the space–time variables and the momentum and energy quantities. Also in such cases, however, it must be remembered that *any well-defined phenomenon involves the combination of several comparable measurements*. The significance of this point is strikingly exemplified by the case, often discussed, of the possible determination of the position of a particle with known momentum by the spot produced by its impact on a photographic plate. Far from meeting any contradiction with the uncertainty relations, we have clearly here to do with a measuring arrangement which is not suited to define a phenomenon involving a test of predictions as regards the location of the object. In conformity with the uncertainty relations, the knowledge of its momentum prevents in fact any unambiguous connection between this object and the frame of reference with respect to which the position of the photographic plate is defined."

Finally Bohr arrives at the definite statement concerning the unambiguous application of the notion of a "physical phenomenon" (p. [316], my italics):

> "The unaccustomed features of the situation with which we are confronted in quantum theory necessitate the greatest caution as regards all questions of terminology. *Speaking, as is often done, of disturbing a phenomenon by observation, or even of creating physical attributes to objects by measuring processes, is, in fact, liable to be confusing,* since all such sentences imply a departure from basic conventions of language which, even though it sometimes may be practical for the sake of brevity, can never be unambiguous. *It is certainly far more in accordance with the structure and interpretation of the quantum mechanical symbolism, as well as with elementary epistemological principles, to reserve the word 'phenomenon' for the comprehension of the effects observed under given experimental conditions.*"

Ten years later, in his "Discussion with Einstein", Bohr was able, in retrospect, to distill his view on these terminological issues in the following concentrated form (p. [363] and p. [377]; my italics, unless otherwise indicated):

> "These problems were instructively commented upon from different sides at the Solvay meeting, in the same session where Einstein raised his general objections. On that occasion an interesting discussion arose also about how to speak of the appearance of phenomena for which only predictions of statistical character can be made. The question was whether, as to the occurrence of individual effects, we should adopt a *terminology proposed by Dirac, that we were concerned with a choice on the part of 'nature' or, as suggested by Heisenberg, we should say that we have to do with a choice on the part of the 'observer' constructing the measuring instruments and reading their recording.* Any such terminology would, however, appear dubious since, on the one hand, it is hardly reasonable to endow nature with volition in the ordinary sense, while, on the other hand, it is certainly not

possible for the observer to influence the events which may appear under the conditions he has arranged. To my mind, there is no other alternative than to admit that, in this field of experience, we are dealing with individual phenomena and that our possibilities of handling the measuring instruments allow us only to make a choice between the different complementary types of phenomena we want to study."

"Meanwhile, the discussion of the epistemological problems in atomic physics attracted as much attention as ever and, in commenting on Einstein's views as regards the incompleteness of the quantum-mechanical mode of description, I entered more directly on questions of terminology[19]. *In this connection I warned especially against phrases, often found in the physical literature, such as 'disturbing of phenomena by observation' or 'creating physical attributes to atomic objects by measurements'.* Such phrases, which may serve to remind of the apparent paradoxes in quantum theory, are at the same time apt to cause confusion, since words like 'phenomena' and 'observations', just as 'attributes' and 'measurements', are used in a way hardly compatible with common language and practical definition.

As a more appropriate way of expression I advocated the application of the word *phenomenon*[20] exclusively to refer to the observations obtained under specified circumstances, including an account of the whole experimental arrangement. In such terminology, the observational problem is free of any special intricacy since, in actual experiments, all observations are expressed by unambiguous statements referring, for instance, to the registration of the point at which an electron arrives at a photographic plate. Moreover, speaking in such a way is just suited to emphasize that the appropriate physical interpretation of the symbolic quantum-mechanical formalism amounts only to predictions, of determinate or statistical character, pertaining to individual phenomena appearing under conditions defined by classical physical concepts."

Finally, let me add that the Warsaw Lecture is unique – as far as I am aware – as regards Bohr's efforts to stretch the parallelism between Relativity and Complementarity to what at that time may have seemed its utmost limits (p. [317]):

"In spite of all differences in the physical problems concerned, relativity theory and quantum theory possess striking similarities in a purely logical

---

[19] In the Warsaw Lecture.
[20] This word is in Bohr's *own* italics.

respect. In both cases we are confronted with novel aspects of the observational problem, involving a revision of customary ideas of physical reality, and originating in the recognition of general laws of nature which do not directly affect practical experience. The impossibility of an unambiguous separation between space and time without reference to the observer, and the impossibility of a sharp separation between the behaviour of objects and their interaction with the means of observation are, in fact, straightforward consequences of the existence of a maximum velocity of propagation of all actions and of a minimum quantity of any action, respectively. The ultimate reason for the unavoidable renunciation as regards the absolute significance of ordinary attributes of objects, and for the recourse to a relative or complementary mode of description respectively, lies also in both cases in the necessity of confining ourselves, in the account of experience, to comparisons between measurements in the interpretation of which relativity refinements and quantum effects respectively have on principle to be neglected."

Even today the question of the scope of this intriguing parallelism appears open.

* * *

The autumn 1948 issue of the Swiss–French journal "Dialectica" was dedicated to "The Concept of Complementarity" with Pauli acting as editor. Among the contributors besides Bohr were Einstein, de Broglie and Heisenberg. From the footnote on the first page of Bohr's paper, as well as from the correspondence quoted below, it is clear that he regarded it more or less as a preparation for his contribution to the Einstein Birthday Volume[21]. Thus, already in May 1947 he writes to Pauli:

"Nevertheless, I have thought a good deal of the Einstein article and looked up some of the old literature and, not least, read a new book about Einstein's life by Philipp Frank, which neither physically nor philosophically may be too good, but certainly gives a most interesting picture of Einstein's personality and the unbalanced situation in Germany between the wars. I feel more and more that it may be a sensible task to give a proper account of the epistemological discussions through the years and, especially, that it

Bohr to Pauli,
16 May 47
English
Full text in
Vol. 6, p. [451]

[21] Ref. 2.

may be a welcome opportunity for us all once more to learn Einstein's reactions. I have just made a start with the writing down and, if I can get it finished, I shall surely be most eager to have your advice and criticism. In the meantime I look forward very much to see your own article.

Of course, I was much interested to learn about the plans for a special issue of 'Dialectica' on the epistemological problems and, if I can manage, it should certainly be a pleasure to me to contribute an article. To my mind, the situation is far more clear than generally assumed, and such tools as three-valued logics I consider rather as complications, since a consistent representation of all axiomatic and dialectic aspects of the situation can be given in simple daily life language. I should, however, be glad to wait with a definite answer until I have come further with the article on which I am working and which perhaps, if it will not be ready in time for the Einstein volume, could be used to some such other purpose."

Bohr's trust in Pauli's prompt advice never wavered and, of course, also on this occasion it proved justified. Pauli replied immediately:

Pauli to Bohr,
29 May 47
English
Full text in
Vol. 6, p. [455]

"I am also very glad for your friendly attitude toward the plan of an issue of 'Dialectica' on complementarity. As a result of your letter, I definitely agreed to overtake the redaction of this particular issue, which, I hope, will appear in one year. I also agree with your view on three-valued logic, but it seems to me good to have a discussion about it in which the physicists should say, why they think that it is superfluous. Could perhaps your original plan of a longer article, which now turned out to be too long for the occasion of the Einstein volume, be used for this issue of the 'Dialectica'?"

As a matter of course Bohr sent a draft manuscript of his article to Pauli, who received it with favour:

Pauli to Bohr,
17 Aug 48
English

Zurich, August 17, 1948

Dear Bohr,

Meanwhile I have received from Rozental the copy of your article for the "Dialectica" and your letter and I am really *very delighted!*

This article will be of great help for me to clarify the content of the other articles in the issue, particularly your definition of "phenomenon". The remark at the conclusion is fitting in very well and everything will be left as it is in printing the article. I shall send it for print at once and the proofs will be sent to you.

It would be a great help for the general redactional comment to the *whole* issue, which I have still to write, if you could return to me the proofs of

Niels Bohr and Wolfgang Pauli, Solvay 1948 (Courtesy CERN).

Einstein's article and also drop me a line on your opinion about the remarks, which I made myself in my letter to Einstein [PWB III, letter [945]] (of which you have a copy). Every criticism is welcome.

In the middle of next week we intend to go in vacation to Southern France, but your letters to me, if sent to the Institute (above address) will be forwarded to me.

With my very best thanks and kindest regards from both of us to yourself, Margrethe and the whole family

Yours ever
W. Pauli

P.S. We had great pleasure with the visit of the Kleins. – I am looking forward to meet you at the Solvay conference.

In spite of Pauli's "delight", Bohr already a month later had to make an elaborate confession to his friend in connection with handing in quite a new version of the manuscript:

[Copenhagen,] September 15, 1948

Dear Pauli,

I was very happy to learn from your letter that you were reasonably satisfied with my article for "Dialectica", in which I made a real effort within the short time available to represent the situation in atomic physics as it appears to me. I have been grateful myself for this opportunity just as for the possibility in the Einstein volume to give an account of many aspects of the situation and its development of which it has been difficult for those interested to get a proper picture.

To confess at once, I have during the last weeks revised both manuscripts, especially with the purpose of obtaining the best correlation between them, mainly because, due to the printing difficulties in the U.S.A., the publication of the Einstein volume has been considerably delayed. In fact, the Editor has written to me that the article, which has till now not been shown to Einstein, will first be put into print in a month's time and that he shall be glad before to receive a revised manuscript.

It is my hope that it will not perturb you too much that I now send you a new version of the "Dialectica" article to replace the old manuscript you received from Rozental. As you will see, many smaller corrections and additions have been introduced; in particular, the language has been thoroughly revised with the help of Pais and Oppenheimer, of whom we were enjoying a visit. If, however, the article is already set in type, the corrections can of course just as well be introduced in the proof, but I wanted to spare you for too many surprise shocks.

I enclose also the proof of Einstein's article to which, of course, I did not make any reference in my own article. In essence, I am quite in agreement with your position in the correspondence with Einstein, but it makes me feel still more strongly that in order to make impression on Einstein or others who are not already convinced of the soundness of quantum mechanics description, it is necessary that the whole situation is treated more explicitly from a dialectical point of view.

As you will appreciate, it has just been my principal aim within the limits of my knowledge and ability to make an effort in this direction and I have for this purpose found it practical to bring the notion of complementarity in the foreground from the very beginning. In this connection, I find especially that the way in which the indeterminacy principle is often presented is apt to create misunderstanding and confusion as regards the logical aspects of the situation in atomic physics.

I need not say that I shall be very interested in your editorial comments which, I am sure, will be of great value to all readers. To bring you quite up to date as to my views, I shall in a few days also send you the revised manuscript for the Einstein volume. I believe that the text has been improved at many points, both as regards conciseness of expression and as regards the endeavours to show how much benefit we all have had of Einstein's concern and subtility, a delicate point which, as you know, is very heavy on my mind. In particular, I have omitted some of the very general considerations at the end, and this has been the reason for adding a few remarks in the last part of the "Dialectica" article.

With kindest regards to Franca and yourself from us all,

Yours ever,
[Niels Bohr]

P.S. September 18. As you will see from the date of this letter, it has even now been delayed due to continued endeavours to approach absolute convergence in the revisional efforts. I am prepared for a scolding at the Solvay conference where I look forward very much to meeting you.

As already mentioned, the Dialectica paper may be regarded, in the main, as a dress rehearsal for the Einstein Celebration article. Still, it presents some striking features of its own. Thus, the Heisenberg relations, $\Delta q \cdot \Delta p \geq \hbar$, are presented with the following terse comment (this volume p. [333], my italics), which certainly appears worth contemplating:

"These *so-called* indeterminacy relations explicitly bear out the *limitation of causal analysis*, but it is important to recognize that *no unambiguous interpretation of such relations can be given in words suited to describe a situation in which physical attributes are objectified in a classical way*.

Thus, a sentence like 'we cannot know both the momentum and the position of an electron' raises at once questions as to the physical reality of such two attributes, which can be answered only by referring to the mutually exclusive conditions for the unambiguous use of space–time coordination, on the one hand, and dynamical conservation laws, on the other."

In any practical solution of the problem of predicting the behaviour of a quantal object within a given experimental arrangement, external agents like mirrors or diaphragms are introduced without any philosophical ado as external classical potentials belonging to the total Hamiltonian, the complete specification of which is an obvious condition for any objective statement concerning the future development of events[22]. This is in complete harmony with Bohr's following "incidental remark" (p. [333], my italics):

"Incidentally, it may be remarked that the construction and the functioning of all apparatus like diaphragms and shutters, serving to define geometry and timing of the experimental arrangements, or photographic plates used for recording the localization of atomic objects, will depend on properties of materials which are themselves essentially determined by the quantum of action. Still, this circumstance is irrelevant for the study of simple atomic phenomena where, *in the specification of the experimental conditions, we may to a very high degree of approximation disregard the molecular constitution of the measuring instruments. If only the instruments are sufficiently heavy compared with the atomic objects under investigation, we can in particular neglect the requirements of relation (3)* $[\Delta q \cdot \Delta p = h/4\pi]$ *as regards the control of the localization in space and time of the single pieces of apparatus relative to each other."*

The whole intricate, but firm line of argumentation is brought to an end with the following definitive conclusion (p. [335]):

"Recapitulating, the impossibility of subdividing the individual quantum effects and of separating a behaviour of the objects from their interaction with the measuring instruments serving to define the conditions under which the phenomena appear implies an ambiguity in assigning conventional attributes to atomic objects which calls for a reconsideration of our attitude towards the problem of physical explanation. In this novel situation, even the old question of an ultimate determinacy of natural phenomena has lost its conceptional basis, and it is against this background that the viewpoint of complementarity presents itself as a rational generalization of the very ideal of causality."

* * *

In spite of this remarkable development in Bohr's manner of presenting the salient features of the complementarity viewpoint, one must be aware of crucial

---

[22] When I was a student Bohr referred me in this connection to the discussion in N.F. Mott and H.S.W. Massey, *The Theory of Atomic Collisions*, Oxford University Press, 1933 (3rd edition 1965), about the solution of the wave equation for a beam of electrons impinging on a diaphragm, in close analogy with ordinary optics.

elements that remained fixed in Bohr's thoughts through these many years. The following passage is typical[23]:

"In this context, we must recognize above all that, even when the phenomena transcend the scope of classical physical theories, the account of the experimental arrangement and the recording of observations must be given in plain language, suitably supplemented by technical physical terminology. This is a clear logical demand, since the very word 'experiment' refers to a situation where we can tell others what we have done and what we have learned."

This argument forms a recurrent theme in article upon article. The central rôle of "plain language" remained the basic tenet for Bohr from beginning to end. In Volume 6 (pp. XVII--XXVI) I tried to outline how ponderings upon the nature of human language, and the conditions for its use to convey unambiguous information, can be traced back to the early times when Bohr was a young student. A main difficulty in grasping Bohr's view on this crucial issue is due to the fact that he considered it to be an irreducible feature, a prerequisite so fundamental for our common description of nature, that any attempt to deduce it from even more basic principles would not only be futile, but would involve a serious logical misconception.

* * *

Einstein was often asking for a precise *definition* of the "Principle of Complementarity", but in this connection words like "definition" and "principle" are hardly suited to convey what Bohr had in mind. Thus, in the "Dialectica" article, the notion of Complementarity is introduced in the following leisurely manner, in connection with contrasting observations obtained by means of mutually exclusive experimental arrangements (p. [332], my italics):

"Such empirical evidence exhibits a novel type of relationship, which has no analogue in classical physics and *which may conveniently be termed 'complementarity'* in order to stress that in the contrasting phenomena we have to do with equally essential aspects of all well-defined knowledge about the objects."

---

[23] N. Bohr, *Unity of Knowledge* in *Atomic Physics and Human Knowledge* (ref. 1), pp. 67–82. Reproduced in Vol. 10.

I recall myself how Bohr in similar terms explained to me as a young student the *convenience* of possessing in our language terms like "relativity" or "complementarity" or "classical", by means of which we in a brief way could refer to an entire network of subtle physical interrelationships. No wonder that Einstein pondered in vain what on earth the "Principle of Complementarity" might mean[24].

In private conversation Bohr allowed himself much more freedom to use colourful phrases to express his points of view than in his carefully polished writings. Especially vividly I remember a remark in which he emphasized that conceptions like "physical reality" are but words in our common language: they have not descended from the Heavens with a label spelling out their meaning once and for all, but whenever we enter new territories in the course of human exploration of nature, we must patiently "learn" (a favourite verb of Bohr's) how to use them unambiguously and meaningfully.

In order to appreciate the urgency Bohr felt to sharpen the use of a word like "phenomenon", consider together the following two quotations, one from the paper by Einstein, Podolsky and Rosen (p. [430], italics in the original), and the other from Bohr's answer:

> "... one would not arrive at our conclusion if one insisted that two or more physical quantities can be regarded as simultaneous elements of reality *only when they can be simultaneously measured or predicted.* On this point of view, since either one or the other, but not both simultaneously, of the quantities P [momentum] and Q [position] can be predicted, they are not simultaneously real. This makes the reality of P and Q depend upon the process of measurement carried out on the first system, which does not disturb the second system in any way. No reasonable definition of reality could be expected to permit this."

Bohr's reply (this volume p. [295], italics in original):

> "... in the phenomena concerned we are not dealing with an incomplete description characterized by the arbitrary picking out of different elements of physical reality at the cost of sacrifying other such elements, but with a rational discrimination between essentially different experimental arrangements and procedures which are suited either for an unambiguous use of

---

[24] Cf. A. Einstein, *Reply to Criticisms* in *Albert Einstein: Philosopher–Scientist* (ref. 2), p. 674: "... Bohr's principle of complementarity, the sharp formulation of which, moreover, I have been unable to achieve despite much effort which I have expended on it."

the idea of space location, or for a legitimate application of the conservation theorem of momentum. Any remaining appearance of arbitrariness concerns merely our freedom of handling the measuring instruments, characteristic of the very idea of experiment.

...

... we are, in the 'freedom of choice' offered by the ... [EPR] arrangement, just concerned with a *discrimination between different experimental procedures which allow of the unambiguous use of complementary classical concepts.*"

Thus, in a situation in which we are dealing with two atomic systems in a quantal state where the sum of their momenta, $p_1 + p_2$, together with their relative distance, $x_1 - x_2$, possess definite values, a determination of p_1 presupposes the establishment of an entire closed process, a *"phenomenon"* that is *different* from the phenomenon with which we are dealing in a determination of x_1. The decisive point here stressed by Bohr is that when dealing with atomic objects whose behaviour is not amenable to a classical physical account, the issue of "physical reality" of variables (like position or momentum) associated with these objects, does not have any content except in connection with a specification of the total experimental arrangement by means of which these quantities are observed. In Bohr's own words[25] from a few years later:

"... in these fields the logical correlations can only be won by a far-reaching renunciation of the usual demands of visualization. It would in particular not be out of place in this connection to warn against a misunderstanding likely to arise when one tries to express the content of Heisenberg's well known indeterminacy relations ... by such a statement as: 'the position and momentum of a particle cannot simultaneously be measured with arbitrary accuracy.' According to such a formulation it would appear as though we had to do with some arbitrary renunciation of the measurement of either the one or the other of the two well-defined attributes of the object, which would not preclude the possibility of a future theory taking both attributes into account on the lines of the classical physics. From the above considerations it should be clear that the whole situation in atomic physics deprives of all meaning such inherent attributes as the idealizations of classical physics would ascribe to the object. On the contrary, the proper rôle of the indeterminacy relations consists in assuring quantitatively the logical

---

[25] N. Bohr, *Causality and Complementarity*, Phil. Sci. **4** (1937) 289–298. Reproduced in Vol. 10.

compatibility of apparently contradictory laws which appear when we use two different experimental arrangements, of which only one permits an unambiguous use of the concept of position, while only the other permits the application of the concept of momentum defined as it is, solely by the law of conservation."

In its happily succinct phrasing this quotation seems to me to convey the very essence of Bohr's view on the conditions for description and observation in quantum physics.

* * *

At this point it may be appropriate to ask – with respect to this so-called "phenomenon terminology": What is its relevance for our own understanding today, when we ponder fundamental issues of quantum mechanics? Surely the notion that Bohr in some manner "found out" what a "physical phenomenon" "means" or "consists of" makes little sense. From the quotation on page [266], it is clear that Bohr advocated the application of the word "phenomenon" for pedagogical reasons, to remind us of the need, in order to avoid ambiguities, always to specify accurately which closed quantal process we are speaking about. Obviously, it does not belong on the same fundamental level of importance as the notion of "complementarity" (cf. p. [273]), which among physicists has acquired an absolutely precise meaning – just as the term "relativity", originally introduced in a novel context by Einstein, has since come to be applied in physics in this definite sense.

4. SERENITY

Discussion with Einstein on Epistemological Problems in Atomic Physics
(1949)
Quantum Physics and Philosophy – Causality and Complementarity (1958)
On Atoms and Human Knowledge (1955)

With the completion of the Warsaw Lecture and the Dialectica paper of 1948, Bohr had found the final form in which to express the conception of complementarity as well as to delineate the causal relationships allowed within the framework of the quantal formalism. As we have seen, this form was developed through a long maturing process over more than twenty years, through ponderings, through analysis in discussions within the circle of the quantal community and through numerous lectures and articles. Little wonder then – and most fortunate indeed for posterity – that he welcomed the preparations for the Einstein

Birthday Volume[26] as a propitious occasion to look back and to relate – in a style wherein the elements of pure physics and personal reminiscences are harmoniously mixed – the story of that part of his life's work which was perhaps closest to his heart.

He lays before us a friendly account of his discussions with Einstein through so many years, he summarizes the essence of the complementarity aspects in the quantal description, exemplifying the argumentation by analysis of several of the apparently paradoxical measuring devices invented by Einstein during their debates. It might be worthwhile to contemplate how much poorer our understanding of the foundations of the quantal description would have been, had only this single work been lacking in Bohr's entire *œuvre*. It is his final summing up, and the most impressive testimony to its mastery is Pauli's reaction: Using only sparingly his characteristic satirical wit, he greeted the manuscript with much warmth and friendly enthusiasm:

Zurich, February 18, 1948

Pauli to Bohr,
18 Feb 48
English

Dear Bohr,

It was a great day the 13th of February, when the manuscript of your article for the Einstein volume together with your letter eventually arrived. Meanwhile I read it twice and there is not much else left to me, than to congratulate you to this fine piece of work. The whole epistemological situation is now very thoroughly analysed and very clearly explained. I was well aware of many refinements in comparison with your earlier articles, especially the constatation that the example treated by Einstein, Podolsky and Rosen does not actually contain more than simpler arrangements with diaphragms and a movable slit.

It is only to avoid the impression that the work cannot be improved at all (a judgment which you have always considered as fatal) that I propose some trifles to be changed: On page 3, line 17 it should be written "probabilities" instead of "possibilities", an obvious typographical error. On page 20 the sentence "Whether, however, we prefer to make such a choice, as to our handling of arrangement, before or after the particle has passed through the diaphragm, is obviously quite indifferent as regards predictions of observable effects" seems to me phrased in a somewhat difficult way, which may easily give rise to the misunderstanding, that it is irrelevant, whether the measurements are made before or after the particle has passed through the diaphragm. I guess what you want to say is that the *time of decision*

---

[26] Cf. ref. 2.

on what one wants to do after the passage of the particle through the diaphragm is irrelevant. Did I "misunderstand you well"? (I like this phrase very much, which I learned from you during my last visit in Copenhagen).

Only to prove that I actually read your manuscript, I may mention that among the "References" at the end, the Newton Tercentenary Celebration is dated 1943 instead of 1946.

The sentence, at the end, on the possible effects of general education on the wisdom of men I read of course with extreme scepticism. But I don't propose at all any change of this conclusion, which expresses so well your views. My own experience as a teacher on the other hand makes me optimistic on the possibility to convey to the younger generation the epistemological significance of the new scientific progress inside physics in such a way that they don't see any more problems where actually they are not as the "classicists" Einstein and Schrödinger still do. (The utterances of the latter, at least in private letters, become continuously worse in this respect). It is getting more and more difficult for me in connection with them, not to forget the complementarity between love and justice.

So I am concluding the part of my letter concerning your article, expressing my hope, that it will be sent to press at once.

I wonder what you will think on the new development of the theory of the small but important electrodynamic effects. Some calculations in this field are well under way in Zurich, and I hope to know a bit more about it in a few weeks. At present I feel both inspiration and uneasiness about it. But Oppenheimer had also this time a positive and stimulating effect on me, namely as an antidote against my old sickness overpessimism.

Shall I meet you at the Solvay Conference in Brussels in April? In any case I shall write to you again in May regarding your promised (short and non historical) article for the "Dialectica". Until now I had to read as redactor much of "nonsense" and of "trivialities", but nearly nothing of "deep truth".

We shall have an international physics conference in July here in Zurich, to which you will soon obtain an official invitation together with the informations of its exact date. We all here would be very glad if you will accept and this time really come here making good our bad luck of 1939, when you had already promised to come, but the meeting had to be cancelled because of the beginning of the war.

Miss Meitner is just in Zurich and she told me about the sickness of your wife during this winter. Franca and I hope, however, that she is well at present and are sending our warmest greetings to both of you and also to Aage.

Please say my regards to all common friends in Princeton (Oppenheimer, Veblen and the others) and last not least, to the main person: Einstein.

Yours ever,
W. Pauli

The passage alluded to by Pauli is presumably that which now begins with the words "It may also be added ..." (this volume p. [370]). Furthermore, in the printed version there is no reference to the Newton Tercentenary Celebration (and the references are not collected at the end). Finally, there is no sentence, at the conclusion, "on the possible effects of general education on the wisdom of men" – the joke about the two kinds of truth could hardly qualify in this respect.

When responding to the contributors in his Birthday Volume, Einstein wrote: "I first attempted to discuss the essays individually. However, I abandoned this procedure because nothing even approximately homogeneous resulted, so that the reading of it could hardly have been either useful or enjoyable. I finally decided, therefore, to order these remarks, as far as possible, according to topical considerations." Hence, Einstein's "Reply to Criticisms" regrettably does not include a specific response to Bohr's article as such[27].

We do possess, however, from the Birthday Volume, in Einstein's "Autobiographical Notes" (pp. 45, 47) his own appreciation of Bohr's creation of the theory of atoms, in the *early* days (square brackets in original):

"All of this was quite clear to me shortly after the appearance of Planck's fundamental work; so that, without having a substitute for classical mechanics, I could nevertheless see to what kind of consequences this law of temperature-radiation leads for the photo-electric effect and for other related phenomena of the transformation of radiation-energy, as well as for the specific heat of (especially) solid bodies. All my attempts, however, to adapt the theoretical foundation of physics to this [new type of] knowledge failed completely. It was as if the ground had been pulled out from under one, with no firm foundation to be seen anywhere, upon which one could have built. That this insecure and contradictory foundation was sufficient to enable a man of Bohr's unique instinct and tact to discover the major laws of the spectral lines and of the electron-shells of the atoms together with their

---

[27] Cf. *Albert Einstein: Philosopher–Scientist* (ref. 2), p. 665.

Gamow's realization of Einstein's experiment involving the weighing of a photon.
The small lamp inside emits photons when the timed shutter opens.
The box was given as a Christmas present to Bohr.

significance for chemistry appeared to me like a miracle – and appears to me as a miracle even today. This is the highest form of musicality in the sphere of thought."

The previous sections of this Introduction already contain extensive comments referring to this article on Bohr's discussions with Einstein, and I shall here only add a single remark concerning Bohr's exposition of the complementarity between energy and time[28].

He discusses the indeterminacy relation between time and energy in connection with his analysis of the "weighing experiment" proposed by Einstein as an

[28] Cf. the unpublished manuscript by Bohr: *Space–Time Continuity and Atomic Physics* (1931), reproduced in Vol. 6, p. [361].

attempt to circumvent this complementary incompatibility. In his Como lecture (1927), Bohr derived the relation

$$\Delta E \cdot \Delta T \gtrsim \hbar \qquad (1)$$

along correspondence lines from Rayleigh's expression for the temporal extension ΔT of a wave train with a spread Δv around the average frequency v. Since the significance of the relation (1) has been much debated[29] it is important to note that in this context Bohr demonstrates with particular clarity that the problem at issue is the unambiguous correlation between the internal time development of a quantal system and the classically described synchronized clocks fixed with respect to the external frame of reference. It should perhaps here be emphasized – in view of misunderstandings brought forth from various quarters – that the rôle of gravity in the experiment at hand does in no way imply that general relativity is essential for the consistency of non-relativistic quantum mechanics – which would also be surprising. In the present context the decisive point is just this: granted that Einstein in his argumentation invokes the equivalence principle, Bohr must in his refutation take into account any immediate consequence of this principle, in casu the gravitational red shift.

From Einstein's "Reply to Criticisms" at the end of the Birthday Volume it is clear, however, that with respect to the conceptual basis of the quantal description, the possibility of a reconciliation between the attitudes of Bohr and Einstein was at this time as remote as ever. That the warmth of their mutual friendship remained unimpaired is testified to by the following exchange of greetings on the occasion of Einstein's 70th birthday:

Princeton, April 4, 1949

Einstein to Bohr, 4 April 49 German text on p. [436]

Dear Bohr,

I thank you most cordially for all the kind efforts you have exerted for my sake on the occasion of an event that in itself is rather insignificant. I also extend my cordial thanks to the members of the Copenhagen Institute for their kind congratulations.

In any case, this is one of those occasions that does not depend on the anxious question as to whether God is really playing dice and whether or not we should hold on to the notion of a reality amenable to physical de-

---

[29] Cf. L. Landau and R. Peierls, *Erweiterung des Unbestimmtheitsprinzips für die relativistische Quantentheorie*, Z. Phys. **69** (1931) 56–69. Extract on p. [229]. L.D. Landau and E.M. Lifshitz, *Quantum Mechanics*, Pergamon Press, London 1965; 2nd edition (1975), §44. Y. Aharonov and D. Bohm, *Time in the Quantum Theory and the Uncertainty Relation for Time and Energy*, Phys. Rev. **122** (1961) 1649–1658.

[281]

scription. In my answer to the articles published in Schilpp's book I have once more sung my lonely old ditty which reminds me of the burden of that little old book[30]:

> At this speech by Master Ed,
> general shaking of the head.

> With cordial greetings
> Yours
> Albert Einstein

[Copenhagen,] April 11, 1949

Dear Einstein,

Many thanks for your friendly lines. It was a great joy for all of us to express our feelings on the occasion of your birthday. To go on in the same jocular vein with respect to that anxious question, I cannot help saying that in my opinion it is not a question of whether or not we shall hold on to a notion of reality amenable to physical description, but rather of continuing along the road you have shown us and of realizing the logical presuppositions for the description of the realities. In my own impertinent manner I might even say that nobody – not even the dear Lord himself – may know what an expression like playing dice means in this connection.

> With cordial greetings.
> Yours
> [Niels Bohr]

As an exposition of the complementary mode of description in quantum physics, the "Discussion with Einstein" stands quite alone, and in his writings during the last decade of his life, Bohr often concentrates on the wider issue of the rôle that complementarity might play in the ordering and analysis of knowledge in other fields of human experience. Among these essays towers the "Unity of Knowledge" which is reproduced in Volume 10. Still, in each of the following essays in the present volume new aspects of the quantal description come to the fore, as for example the question of the unambiguous application of phrases like "objective existence" and "reality". Already in his Como Lecture (1927)[31] Bohr emphasized that the legitimate use of the word "reality",

---

[30] We owe the translation of the rhyme to Helle Bonaparte.

[31] Vol. 6, p. [109].

Albert Einstein, c. 1949 (Photograph by Karsh, courtesy Time–Life).

when referring to stationary states of an atom, in no way differs from the application of the phrase "objective existence" as referring to the electron itself. In an exchange of letters in 1953 between Bohr and Max Born[32] such issues are once more raised for debate.

$$* * *$$

---

[32] Born to Bohr, 28 Jan 1953 and 10 March 1953, Bohr to Born, 2 March 1953 and 26 March 1953. BSC, microfilm no. 27. My attention was drawn to this correspondence through the thoughtful book by Henry Folse, *The Philosophy of Niels Bohr*, North-Holland Publ. Co., Amsterdam 1985. I should like here to acknowledge the many valuable impulses and useful references I have found in that work.

[283]

As regards the last two articles, according to strict chronology, the paper Bohr contributed to "Philosophy in the Mid-Century, A Survey" with the title: "Quantum Philosophy – Causality and Complementarity" should end this volume – and it would be a fitting end. Bohr was happy with it and, in an innocent way, a little proud: he felt that he had here succeeded in formulating the essential points more clearly and concisely than on earlier occasions. In this judgement he might be both right and wrong, and – for reasons later to be explained – I have taken the liberty of reversing the order, so as to make "On Atoms and Human Knowledge" form the conclusion.

∗ ∗ ∗

"Causality and Complementarity", thus being the penultimate article in this volume, is a distillate, so concentrated that it may remind us of one of Bohr's favourite stories; he told it not without a touch of playful self-irony: It happened that a small Jewish community in Poland learned that a famous rabbi was going to lecture in a little town in the vicinity. The members of the community were all rather poor and it was impossible for them all to go and listen to the teachings of the rabbi. Instead they sent the brightest young member as an observer, in order that he would come back to report on the new insights he had imbibed. In due time the young man came back and recounted his experience: "The rabbi gave three lectures. The first was simple and lucid and I understood every word of it. The second talk was even better: deep and subtle: I did hardly understand a single sentence, but to the rabbi it was all transparent and obvious to grasp. The third, however, was by far the greatest and most unforgettable experience: Neither the audience, nor the rabbi understood a single word of it!"

Still, just this lapidary script epitomizes the essence of Bohr's achievement: Through his analysis of a formal symbolic description of entirely new and foreign empirical evidence, "hitherto unnoticed presuppositions" were disclosed in the classical mode of accounting for physical phenomena which *could* and *had to* be abandoned to make room for a consistent quantal formalism. It is essential to recognize *how* different this scrutiny is from any philosophical attempt to establish axioms concerning a notion like "physical reality" within a formal theoretical description with an *already exhaustively ordered* empirical content. As I have earlier emphasized in the Introduction[33] to the Como Lecture, complementarity is not a philosophical superstructure invented by Bohr to

---

[33] Vol. 6, pp. [7]–[51].

be placed as a decoration on top of the quantal formalism: Through their ardent debates, that long winter of 1926 and spring of 1927, Heisenberg was led to establish the indeterminacy relations and Bohr to his *discovery* of complementarity as the bedrock of the quantal description – in the same way that Einstein discovered relativity in classical electrodynamics, thereby exhibiting "hitherto unnoticed presuppositions" for the application of the concept of temporal simultaneity in Newtonian mechanics.

A point of great importance is that Bohr here (cf. p. [390]) states very explicitly the meaning of the phrase "classical terms" in connection with the description of the experimental measuring arrangement (my italics):

> "The decisive point is to recognize that the description of the experimental arrangement and the recording of observations must be given in plain language, suitably refined by the usual physical terminology. This is a simple logical demand, since by the word 'experiment' we can only mean a procedure regarding which we are able to communicate to others what we have done and what we have learnt.
>
> In actual experimental arrangements, *the fulfilment of such requirements is secured by the use, as measuring instruments, of rigid bodies sufficiently heavy to allow a completely classical account of their relative positions and velocities.*"

Thereby the conditions for an unambiguous application of special relativity are also fulfilled.

A curious argument appears on p. [393]:

> "In this connection, the question has even been raised whether recourse to multivalued logics is needed for a more appropriate representation of the situation. From the preceding argumentation it will appear, however, that all departures from common language and ordinary logic are entirely avoided by reserving the word 'phenomenon' solely for reference to unambiguously communicable information, in the account of which the word 'measurement' is used in its plain meaning of standardized comparison. Such caution in the choice of terminology is especially important in the exploration of a new field of experience, where information cannot be comprehended in the familiar frame which in classical physics found such unrestricted applicability."

It seems difficult to understand the immediate relationship between the "phenomenon terminology" and the sufficiency of relying on divalent logic, and it

On the last day of his life Bohr drew on the blackboard in his home a sketch of two interweaving planes to illustrate the ambiguity of language and underneath it a diagram of the "Einstein Box". Thus he left behind a symbolic reminder of the beginning and the end of his thoughts on complementarity (Courtesy AIP).

may be easier, also for others, to grasp the point in Bohr's remark to me as a young student: "We must hope, at least, that these learned people writing textbooks about multivalued logics would explain their deductions in terms of the ordinary logic, embodied in our common language."

***

Margrethe and Niels Bohr, Tisvilde 1962.

The article "On Atoms and Human Knowledge"[34], placed here as the last article, appears to me as an especially harmonious ending. It has an atmosphere of its own: The language has gained a freedom and the style is more "literary" than Bohr otherwise allowed himself. This may be due to the particular occasion: It was originally delivered in Danish, as an address to the Royal Danish Academy of Sciences and Letters, in connection with the celebrations in 1955 of his seventieth birthday and later, in English, as a lecture before the American Academy of Arts and Sciences in 1957. The language of the Danish original is so rich – it is the classic Danish from a vanished golden age, which Bohr spoke so beautifully, but rarely used in print, except in special publications such as "Tale ved Studenter Jubilæet"[35].

In any language, however, into which "On Atoms and Human Knowledge" may be translated, this work by Niels Bohr most beautifully conveys the message that Complementarity, like Relativity, is part of an irrevocable, truly philosophical lesson, taught us by Nature herself.

---

[34] This volume p. [395].

[35] *Speech Given at the 25th Anniversary Reunion of the Student Graduation Class.* Reproduced in Danish and in English translation in Vol. 10.

I. QUANTUM MECHANICS AND PHYSICAL REALITY

Nature **136** (1935) 65

See Introduction to Part II, sect. 2.

(*Reprinted from* NATURE, *Vol.* 136, *page* 65, *July* 13, 1935.)

Quantum Mechanics and Physical Reality

IN a recent article by A. Einstein, B. Podolsky and N. Rosen, which appeared in the *Physical Review* of May 15, and was reviewed in NATURE of June 22, the question of the completeness of quantum mechanical description has been discussed on the basis of a "criterion of physical reality", which the authors formulate as follows : "If, without in any way disturbing a system, we can predict with certainty the value of a physical quantity, then there exists an element of physical reality corresponding to this physical quantity".

Since, as the authors show, it is always possible in quantum theory, just as in classical theory, to predict the value of any variable involved in the description of a mechanical system from measurements performed on other systems, which have only temporarily been in interaction with the system under investigation ; and since in contrast to classical mechanics it is never possible in quantum mechanics to assign definite values to both of two conjugate variables, the authors conclude from their criterion that quantum mechanical description of physical reality is incomplete.

I should like to point out, however, that the named criterion contains an essential ambiguity when it is applied to problems of quantum mechanics. It is true that in the measurements under consideration any direct mechanical interaction of the system and the measuring agencies is excluded, but a closer examination reveals that the procedure of measurements has an essential influence on the conditions on which the very definition of the physical quantities in question rests. Since these conditions must be considered as an inherent element of any phenomenon to which the term "physical reality" can be unambiguously applied, the conclusion of the above-mentioned authors would not appear to be justified. A fuller development of this argument will be given in an article to be published shortly in the *Physical Review*.

N. BOHR.

Institute of Theoretical Physics,
Copenhagen.
June 29.

Printed in Great Britain by FISHER, KNIGHT & Co., LTD., St. Albans

II. CAN QUANTUM-MECHANICAL DESCRIPTION OF PHYSICAL REALITY BE CONSIDERED COMPLETE?

Phys. Rev. **48** (1935) 696–702

See Introduction to Part II, sect. 2.

OCTOBER 15, 1935   PHYSICAL REVIEW   VOLUME 48
Printed in U. S. A.

Can Quantum-Mechanical Description of Physical Reality be Considered Complete?

N. BOHR, *Institute for Theoretical Physics, University, Copenhagen*
(Received July 13, 1935)

It is shown that a certain "criterion of physical reality" formulated in a recent article with the above title by A. Einstein, B. Podolsky and N. Rosen contains an essential ambiguity when it is applied to quantum phenomena. In this connection a viewpoint termed "complementarity" is explained from which quantum-mechanical description of physical phenomena would seem to fulfill, within its scope, all rational demands of completeness.

IN a recent article[1] under the above title A. Einstein, B. Podolsky and N. Rosen have presented arguments which lead them to answer the question at issue in the negative. The trend of their argumentation, however, does not seem to me adequately to meet the actual situation with which we are faced in atomic physics. I shall therefore be glad to use this opportunity to explain in somewhat greater detail a general viewpoint, conveniently termed "complementarity," which I have indicated on various previous occasions,[2] and from which quantum mechanics within its scope would appear as a completely rational description of physical phenomena, such as we meet in atomic processes.

The extent to which an unambiguous meaning can be attributed to such an expression as "physical reality" cannot of course be deduced from *a priori* philosophical conceptions, but—as the authors of the article cited themselves emphasize—must be founded on a direct appeal to experiments and measurements. For this purpose they propose a "criterion of reality" formulated as follows: "If, without in any way disturbing a system, we can predict with certainty the value of a physical quantity, then there exists an element of physical reality corresponding to this physical quantity." By means of an interesting example, to which we shall return below, they next proceed to show that in quantum mechanics, just as in classical mechanics, it is possible under suitable conditions to predict the value of any given variable pertaining to the description of a mechanical system from measurements performed entirely on other systems which previously have been in interaction with the system under investigation. According to their criterion the authors therefore want to ascribe an element of reality to each of the quantities represented by such variables. Since, moreover, it is a well-known feature of the present formalism of quantum mechanics that it is never possible, in the description of the state of a mechanical system, to attach definite values to both of two canonically conjugate variables, they consequently deem this formalism to be incomplete, and express the belief that a more satisfactory theory can be developed.

Such an argumentation, however, would hardly seem suited to affect the soundness of quantum-mechanical description, which is based on a coherent mathematical formalism covering automatically any procedure of measurement like that indicated.* The apparent contradiction in

---

* The deductions contained in the article cited may in this respect be considered as an immediate consequence of the transformation theorems of quantum mechanics, which perhaps more than any other feature of the formalism contribute to secure its mathematical completeness and its rational correspondence with classical mechanics. In fact, it is always possible in the description of a mechanical system, consisting of two partial systems (1) and (2), interacting or not, to replace any two pairs of canonically conjugate variables (q_1p_1), (q_2p_2) pertaining to systems (1) and (2), respectively, and satisfying the usual commutation rules

$$[q_1p_1]=[q_2p_2]=ih/2\pi,$$
$$[q_1q_2]=[p_1p_2]=[q_1p_2]=[q_2p_1]=0,$$

by two pairs of new conjugate variables (Q_1P_1), (Q_2P_2) related to the first variables by a simple orthogonal transformation, corresponding to a rotation of angle θ in the planes (q_1q_2), (p_1p_2)

$$q_1=Q_1\cos\theta-Q_2\sin\theta \quad\quad p_1=P_1\cos\theta-P_2\sin\theta$$
$$q_2=Q_1\sin\theta+Q_2\cos\theta \quad\quad p_2=P_1\sin\theta+P_2\cos\theta.$$

Since these variables will satisfy analogous commutation rules, in particular

$$[Q_1P_1]=ih/2\pi, \quad\quad [Q_1P_2]=0,$$

it follows that in the description of the state of the combined system definite numerical values may not be assigned to both Q_1 and P_1, but that we may clearly assign

---

[1] A. Einstein, B. Podolsky and N. Rosen, Phys. Rev. **47**, 777 (1935).
[2] Cf. N. Bohr, *Atomic Theory and Description of Nature*, I (Cambridge, 1934).

696

fact discloses only an essential inadequacy of the customary viewpoint of natural philosophy for a rational account of physical phenomena of the type with which we are concerned in quantum mechanics. Indeed the *finite interaction between object and measuring agencies* conditioned by the very existence of the quantum of action entails —because of the impossibility of controlling the reaction of the object on the measuring instruments if these are to serve their purpose—the necessity of a final renunciation of the classical ideal of causality and a radical revision of our attitude towards the problem of physical reality. In fact, as we shall see, a criterion of reality like that proposed by the named authors contains—however cautious its formulation may appear—an essential ambiguity when it is applied to the actual problems with which we are here concerned. In order to make the argument to this end as clear as possible, I shall first consider in some detail a few simple examples of measuring arrangements.

Let us begin with the simple case of a particle passing through a slit in a diaphragm, which may form part of some more or less complicated experimental arrangement. Even if the momentum of this particle is completely known before it impinges on the diaphragm, the diffraction by the slit of the plane wave giving the symbolic representation of its state will imply an uncertainty in the momentum of the particle, after it has passed the diaphragm, which is the greater the narrower the slit. Now the width of the slit, at any rate if it is still large compared with the wave-length, may be taken as the uncertainty Δq of the position of the particle relative to the diaphragm, in a direction perpendicular to the slit. Moreover, it is simply seen from de Broglie's relation between momentum and wave-length that the uncertainty Δp of the momentum of the particle in this direction is correlated to Δq by means of Heisenberg's general principle

$$\Delta p \Delta q \sim h,$$

which in the quantum-mechanical formalism is a direct consequence of the commutation relation for any pair of conjugate variables. Obviously the uncertainty Δp is inseparably connected with the possibility of an exchange of momentum between the particle and the diaphragm; and the question of principal interest for our discussion is now to what extent the momentum thus exchanged can be taken into account in the description of the phenomenon to be studied by the experimental arrangement concerned, of which the passing of the particle through the slit may be considered as the initial stage.

Let us first assume that, corresponding to usual experiments on the remarkable phenomena of electron diffraction, the diaphragm, like the other parts of the apparatus,—say a second diaphragm with several slits parallel to the first and a photographic plate,—is rigidly fixed to a support which defines the space frame of reference. Then the momentum exchanged between the particle and the diaphragm will, together with the reaction of the particle on the other bodies, pass into this common support, and we have thus voluntarily cut ourselves off from any possibility of taking these reactions separately into account in predictions regarding the final result of the experiment,—say the position of the spot produced by the particle on the photographic plate. The impossibility of a closer analysis of the reactions between the particle and the measuring instrument is indeed no peculiarity of the experimental procedure described, but is rather an essential property of any arrangement suited to the study of the phenomena of the type concerned, where we have to do with a feature of *individuality* completely foreign to classical physics. In fact, any possibility of taking into account the momentum exchanged between the particle and the separate parts of the apparatus would at once permit us to draw conclusions regarding the "course" of such phenomena,—say through what particular slit of the second diaphragm the particle passes on its way to the photographic plate—which would be quite incompatible with the fact that the probability of the particle reaching a given element of area on this plate is determined not by the presence of any particular slit, but by the positions of all the slits of the second diaphragm within reach

such values to both Q_1 and P_2. In that case it further results from the expressions of these variables in terms of $(q_1 p_1)$ and $(q_2 p_2)$, namely

$$Q_1 = q_1 \cos \theta + q_2 \sin \theta, \qquad P_2 = -p_1 \sin \theta + p_2 \cos \theta,$$

that a subsequent measurement of either q_2 or p_2 will allow us to predict the value of q_1 or p_1 respectively.

of the associated wave diffracted from the slit of the first diaphragm.

By another experimental arrangement, where the first diaphragm is not rigidly connected with the other parts of the apparatus, it would at least in principle* be possible to measure its momentum with any desired accuracy before and after the passage of the particle, and thus to predict the momentum of the latter after it has passed through the slit. In fact, such measurements of momentum require only an unambiguous application of the classical law of conservation of momentum, applied for instance to a collision process between the diaphragm and some test body, the momentum of which is suitably controlled before and after the collision. It is true that such a control will essentially depend on an examination of the space-time course of some process to which the ideas of classical mechanics can be applied; if, however, all spatial dimensions and time intervals are taken sufficiently large, this involves clearly no limitation as regards the accurate control of the momentum of the test bodies, but only a renunciation as regards the accuracy of the control of their space-time coordination. This last circumstance is in fact quite analogous to the renunciation of the control of the momentum of the fixed diaphragm in the experimental arrangement discussed above, and depends in the last resort on the claim of a purely classical account of the measuring apparatus, which implies the necessity of allowing a latitude corresponding to the quantum-mechanical uncertainty relations in our description of their behavior.

The principal difference between the two experimental arrangements under consideration is, however, that in the arrangement suited for the control of the momentum of the first diaphragm, this body can no longer be used as a measuring instrument for the same purpose as in the previous case, but must, as regards its position relative to the rest of the apparatus, be treated, like the particle traversing the slit, as an object of

* The obvious impossibility of actually carrying out, with the experimental technique at our disposal, such measuring procedures as are discussed here and in the following does clearly not affect the theoretical argument, since the procedures in question are essentially equivalent with atomic processes, like the Compton effect, where a corresponding application of the conservation theorem of momentum is well established.

investigation, in the sense that the quantum-mechanical uncertainty relations regarding its position and momentum must be taken explicitly into account. In fact, even if we knew the position of the diaphragm relative to the space frame before the first measurement of its momentum, and even though its position after the last measurement can be accurately fixed, we lose, on account of the uncontrollable displacement of the diaphragm during each collision process with the test bodies, the knowledge of its position when the particle passed through the slit. The whole arrangement is therefore obviously unsuited to study the same kind of phenomena as in the previous case. In particular it may be shown that, if the momentum of the diaphragm is measured with an accuracy sufficient for allowing definite conclusions regarding the passage of the particle through some selected slit of the second diaphragm, then even the minimum uncertainty of the position of the first diaphragm compatible with such a knowledge will imply the total wiping out of any interference effect—regarding the zones of permitted impact of the particle on the photographic plate—to which the presence of more than one slit in the second diaphragm would give rise in case the positions of all apparatus are fixed relative to each other.

In an arrangement suited for measurements of the momentum of the first diaphragm, it is further clear that even if we have measured this momentum before the passage of the particle through the slit, we are after this passage still left with a *free choice* whether we wish to know the momentum of the particle or its initial position relative to the rest of the apparatus. In the first eventuality we need only to make a second determination of the momentum of the diaphragm, leaving unknown forever its exact position when the particle passed. In the second eventuality we need only to determine its position relative to the space frame with the inevitable loss of the knowledge of the momentum exchanged between the diaphragm and the particle. If the diaphragm is sufficiently massive in comparison with the particle, we may even arrange the procedure of measurements in such a way that the diaphragm after the first determination of its momentum will remain at rest in some unknown position relative to the other parts of the apparatus, and the subsequent fixation of this position may therefore simply consist in establishing a rigid connection between the diaphragm and the common support.

My main purpose in repeating these simple, and in substance well-known considerations, is to emphasize that in the phenomena concerned we are not dealing with an incomplete description characterized by the arbitrary picking out of different elements of physical reality at the cost of sacrificing other such elements, but with a rational discrimination between essentially different experimental arrangements and procedures which are suited either for an unambiguous use of the idea of space location, or for a legitimate application of the conservation theorem of momentum. Any remaining appearance of arbitrariness concerns merely our freedom of handling the measuring instruments, characteristic of the very idea of experiment. In fact, the renunciation in each experimental arrangement of the one or the other of two aspects of the description of physical phenomena,—the combination of which characterizes the method of classical physics, and which therefore in this sense may be considered as *complementary* to one another,—depends essentially on the impossibility, in the field of quantum theory, of accurately controlling the reaction of the object on the measuring instruments, i.e., the transfer of momentum in case of position measurements, and the displacement in case of momentum measurements. Just in this last respect any comparison between quantum mechanics and ordinary statistical mechanics,—however useful it may be for the formal presentation of the theory,—is essentially irrelevant. Indeed we have in each experimental arrangement suited for the study of proper quantum phenomena not merely to do with an ignorance of the value of certain physical quantities, but with the impossibility of defining these quantities in an unambiguous way.

The last remarks apply equally well to the special problem treated by Einstein, Podolsky and Rosen, which has been referred to above, and which does not actually involve any greater intricacies than the simple examples discussed above. The particular quantum-mechanical state of two free particles, for which they give an explicit mathematical expression, may be repro-

duced, at least in principle, by a simple experimental arrangement, comprising a rigid diaphragm with two parallel slits, which are very narrow compared with their separation, and through each of which one particle with given initial momentum passes independently of the other. If the momentum of this diaphragm is measured accurately before as well as after the passing of the particles, we shall in fact know the sum of the components perpendicular to the slits of the momenta of the two escaping particles, as well as the difference of their initial positional coordinates in the same direction; while of course the conjugate quantities, i.e., the difference of the components of their momenta, and the sum of their positional coordinates, are entirely unknown.* In this arrangement, it is therefore clear that a subsequent single measurement either of the position or of the momentum of one of the particles will automatically determine the position or momentum, respectively, of the other particle with any desired accuracy; at least if the wave-length corresponding to the free motion of each particle is sufficiently short compared with the width of the slits. As pointed out by the named authors, we are therefore faced at this stage with a completely free choice whether we want to determine the one or the other of the latter quantities by a process which does not directly interfere with the particle concerned.

Like the above simple case of the choice between the experimental procedures suited for the prediction of the position or the momentum of a single particle which has passed through a slit in a diaphragm, we are, in the "freedom of choice" offered by the last arrangement, just concerned with a *discrimination between different experimental procedures which allow of the unambiguous use of complementary classical concepts.* In fact to measure the position of one of the particles can mean nothing else than to establish a correlation between its behavior and some

---

* As will be seen, this description, apart from a trivial normalizing factor, corresponds exactly to the transformation of variables described in the preceding footnote if (q_1p_1), (q_2p_2) represent the positional coordinates and components of momenta of the two particles and if $\theta = -\pi/4$. It may also be remarked that the wave function given by formula (9) of the article cited corresponds to the special choice of $P_2 = 0$ and the limiting case of two infinitely narrow slits.

instrument rigidly fixed to the support which defines the space frame of reference. Under the experimental conditions described such a measurement will therefore also provide us with the knowledge of the location, otherwise completely unknown, of the diaphragm with respect to this space frame when the particles passed through the slits. Indeed, only in this way we obtain a basis for conclusions about the initial position of the other particle relative to the rest of the apparatus. By allowing an essentially uncontrollable momentum to pass from the first particle into the mentioned support, however, we have by this procedure cut ourselves off from any future possibility of applying the law of conservation of momentum to the system consisting of the diaphragm and the two particles and therefore have lost our only basis for an unambiguous application of the idea of momentum in predictions regarding the behavior of the second particle. Conversely, if we choose to measure the momentum of one of the particles, we lose through the uncontrollable displacement inevitable in such a measurement any possibility of deducing from the behavior of this particle the position of the diaphragm relative to the rest of the apparatus, and have thus no basis whatever for predictions regarding the location of the other particle.

From our point of view we now see that the wording of the above-mentioned criterion of physical reality proposed by Einstein, Podolsky and Rosen contains an ambiguity as regards the meaning of the expression "without in any way disturbing a system." Of course there is in a case like that just considered no question of a mechanical disturbance of the system under investigation during the last critical stage of the measuring procedure. But even at this stage there is essentially the question of *an influence on the very conditions which define the possible types of predictions regarding the future behavior of the system.* Since these conditions constitute an inherent element of the description of any phenomenon to which the term "physical reality" can be properly attached, we see that the argumentation of the mentioned authors does not justify their conclusion that quantum-mechanical description is essentially incomplete. On the contrary this description, as appears from the preceding discussion, may be characterized as a rational utilization of all possibilities of unambiguous interpretation of measurements, compatible with the finite and uncontrollable interaction between the objects and the measuring instruments in the field of quantum theory. In fact, it is only the mutual exclusion of any two experimental procedures, permitting the unambiguous definition of complementary physical quantities, which provides room for new physical laws, the coexistence of which might at first sight appear irreconcilable with the basic principles of science. It is just this entirely new situation as regards the description of physical phenomena, that the notion of *complementarity* aims at characterizing.

The experimental arrangements hitherto discussed present a special simplicity on account of the secondary role which the idea of time plays in the description of the phenomena in question. It is true that we have freely made use of such words as "before" and "after" implying time-relationships; but in each case allowance must be made for a certain inaccuracy, which is of no importance, however, so long as the time intervals concerned are sufficiently large compared with the proper periods entering in the closer analysis of the phenomenon under investigation. As soon as we attempt a more accurate time description of quantum phenomena, we meet with well-known new paradoxes, for the elucidation of which further features of the interaction between the objects and the measuring instruments must be taken into account. In fact, in such phenomena we have no longer to do with experimental arrangements consisting of apparatus essentially at rest relative to one another, but with arrangements containing moving parts,—like shutters before the slits of the diaphragms,—controlled by mechanisms serving as clocks. Besides the transfer of momentum, discussed above, between the object and the bodies defining the space frame, we shall therefore, in such arrangements, have to consider an eventual exchange of energy between the object and these clock-like mechanisms.

The decisive point as regards time measurements in quantum theory is now completely analogous to the argument concerning measurements of positions outlined above. Just as the transfer of momentum to the separate parts of the apparatus,—the knowledge of the relative positions of which is required for the description of the phenomenon,—has been seen to be entirely uncontrollable, so the exchange of energy between the object and the various bodies, whose relative motion must be known for the intended use of the apparatus, will defy any closer analysis. Indeed, it is *excluded in principle to control the energy which goes into the clocks without interfering essentially with their use as time indicators.* This use in fact entirely relies on the assumed possibility of accounting for the functioning of each clock as well as for its eventual comparison with other clocks on the basis of the methods of classical physics. In this account we must therefore obviously allow for a latitude in the energy balance, corresponding to the quantum-mechanical uncertainty relation for the conjugate time and energy variables. Just as in the question discussed above of the mutually exclusive character of any unambiguous use in quantum theory of the concepts of position and momentum, it is in the last resort this circumstance which entails the complementary relationship between any detailed time account of atomic phenomena on the one hand and the unclassical features of intrinsic stability of atoms, disclosed by the study of energy transfers in atomic reactions on the other hand.

This necessity of discriminating in each experimental arrangement between those parts of the physical system considered which are to be treated as measuring instruments and those which constitute the objects under investigation may indeed be said to form a *principal distinction between classical and quantum-mechanical description of physical phenomena.* It is true that the place within each measuring procedure where this discrimination is made is in both cases largely a matter of convenience. While, however, in classical physics the distinction between object and measuring agencies does not entail any difference in the character of the description of the phenomena concerned, its fundamental importance in quantum theory, as we have seen, has its root in the indispensable use of classical concepts in the interpretation of all proper measurements, even though the classical theories do not suffice in accounting for the new types of regularities with which we are concerned in atomic physics.

In accordance with this situation there can be no question of any unambiguous interpretation of the symbols of quantum mechanics other than that embodied in the well-known rules which allow to predict the results to be obtained by a given experimental arrangement described in a totally classical way, and which have found their general expression through the transformation theorems, already referred to. By securing its proper correspondence with the classical theory, these theorems exclude in particular any imaginable inconsistency in the quantum-mechanical description, connected with a change of the place where the discrimination is made between object and measuring agencies. In fact it is an obvious consequence of the above argumentation that in each experimental arrangement and measuring procedure we have only a free choice of this place within a region where the quantum-mechanical description of the process concerned is effectively equivalent with the classical description.

Before concluding I should still like to emphasize the bearing of the great lesson derived from general relativity theory upon the question of physical reality in the field of quantum theory. In fact, notwithstanding all characteristic differences, the situations we are concerned with in these generalizations of classical theory present striking analogies which have often been noted. Especially, the singular position of measuring instruments in the account of quantum phenomena, just discussed, appears closely analogous to the well-known necessity in relativity theory of upholding an ordinary description of all measuring processes, including a sharp distinction between space and time coordinates, although the very essence of this theory is the establishment of new physical laws, in the comprehension of which we must renounce the customary separation of space and time ideas.*

---

* Just this circumstance, together with the relativistic invariance of the uncertainty relations of quantum mechanics, ensures the compatibility between the argumentation outlined in the present article and all exigencies of relativity theory. This question will be treated in greater detail in a paper under preparation, where the writer will in particular discuss a very interesting paradox suggested by Einstein concerning the application of gravitation theory to energy measurements, and the solution of which offers an especially instructive illustration of the generality of the argument of complementarity. On the same occasion a more thorough discussion of space-time measurements in quantum theory will be given with all necessary mathematical developments and diagrams of experimental

The dependence on the reference system, in relativity theory, of all readings of scales and clocks may even be compared with the essentially uncontrollable exchange of momentum or energy between the objects of measurements and all instruments defining the space-time system of

reference, which in quantum theory confronts us with the situation characterized by the notion of complementarity. In fact this new feature of natural philosophy means a radical revision of our attitude as regards physical reality, which may be paralleled with the fundamental modification of all ideas regarding the absolute character of physical phenomena, brought about by the general theory of relativity.

arrangements, which had to be left out of this article, where the main stress is laid on the dialectic aspect of the question at issue.

III. THE CAUSALITY PROBLEM IN ATOMIC PHYSICS

New Theories in Physics, Conference organized in collaboration with the
International Union of Physics and the
Polish Intellectual Co-operation Committee,
Warsaw, May 30th – June 3rd 1938,
International Institute of Intellectual Co-operation, Paris 1939, pp. 11–30

See Introduction to Part II, sect. 3.

The proceedings of the conference, *New Theories in Physics*, were also published in French, *Les Nouvelles théories de la Physique*, Institut International de Coopération Intellectuelle, Paris 1939.

Contributions by Bohr to the discussions at the conference are presented on pp. 38–39, 45 (report by N. Bohr), pp. 64–65 (report by L. de Broglie), pp. 97, 98 and 115 (discussion conducted by H.A. Kramers), p. 204 (report by A.S. Eddington), pp. 235–236, 237 (report by P. Langevin) and pp. 242–243 (thanks). Since these are merely summaries it has been decided not to include them here.

The list of participants, pp. VII--VIII, has also been included.

[300]

LIST OF MEMBERS

Chairman :

Professor C. Bialobrzeski,
Joseph Pilsudski University.

Others present :

Professor E. Bauer,
Collège de France.

Professor Niels Bohr,
Copenhagen University.

Professor L. Brillouin,
Collège de France.

Professor L. de Broglie,
University of Paris.

Professor C. Darwin,
University of Cambridge.

Sir Arthur Eddington,
University of Cambridge.

Professor R. H. Fowler,
University of Cambridge.

Professor G. Gamow,
George Washington University.

Professor S. Goudsmit,
Princeton University.

Professor E. Hylleraas,
Oslo University.

Professor O. Klein,
Stockholm University.

Professor H. A. Kramers,
Leyden University.

Professor L. de Kronig,
Groningen University.

Professor P. Langevin,
    Collège de France.

Professor C. Moeller,
    Copenhagen University

Professor J. von Neumann,
    Princeton University.

Professor F. Perrin,
    University of Paris.

Professor L. Rosenfeld,
    Liége University.

Professor W. Rubinowicz,
    Jean Casimir University, Lwow.

Professor S. Szczeniowski,
    Etienne Batory University, Wilno.

Professor J. Weyssenhof,
    Jagellonne University, Cracow.

Professor L. Wertenstein,
    Free University, Warsaw.

Professor E. P. Wigner,
    Princeton University.

Professor F. J. Wisniewski,
    Polytechnic School, Warsaw.

Representatives of the International Organisation for Intellectual Co-operation :

M. Henri Bonnet,
    Director of the International Institute of Intellectual Co-operation.

M. A. Establier,
    Head of the Scientific Relations Service of the International Institute of Intellectual Co-operation.

THE CAUSALITY PROBLEM IN ATOMIC PHYSICS

Report drafted and submitted by N. BOHR.

I. — INTRODUCTION OF STATISTICAL METHODS IN QUANTUM THEORY.

The unrestricted applicability of the causal mode of description to physical phenomena has hardly been seriously questioned until Planck's discovery of the quantum of action, which disclosed a novel feature of atomicity in the laws of nature supplementing in such unsuspected manner the old doctrine of the limited divisibility of matter. Before this discovery statistical methods were of course extensively used in atomic theory but merely as a practical means of dealing with the complicated mechanical problems met with in the attempt at tracing the ordinary properties of matter back to the behaviour of assemblies of immense numbers of atoms. It is true that the very formulation of the laws of thermodynamics involves an essential renunciation of the complete mechanical description of such assemblies and thereby exhibits a certain formal resemblance with typical problems of quantum theory (^{1}). So far there was, however, no question of any limitation in the possibility of carrying out in principle such a complete description ; on the contrary, the ordinary ideas of mechanics and electrodynamics were found to have a large field of application also to proper atomic phenomena, and above all to offer an entirely sufficient basis for the analysis of the experiments leading to the isolation of the electron and the measurement of its charge and mass. Due to the essentially statistical character of the thermodynamical problems which led to the discovery of the quan-

---

(^{1}) See N. Bohr, Faraday Lecture, *Journ. Chem. Soc.*, 1932, p. 349.

tum of action, it was also not to begin with realized, that the insufficiency of the laws of classical mechanics and electrodynamics in dealing with atomic problems, disclosed by this discovery, implied a shortcoming of the causality ideal itself.

The first indication of the seriousness of the situation in this respect resulted from the discussion in the following years of the constitution of radiation in free space. Notwithstanding the adequacy of Einstein's idea of the photon to account for the exchange of energy and momentum in individual radiative processes, it was clear from the outset that the wealth of experience which had led to the acceptance of the wave-picture of light propagation put any revival of a simple corpuscular theory of radiation out of question. This is also at once apparent from the fundamental reactions

$$E = h\nu, \quad \vec{P} = h\,\vec{\sigma} \tag{1}$$

connecting the energy E and the momentum $\vec{P}$ of the photon with the frequency ν and the wave-number vector $\vec{\sigma}$, which refer directly to the idealization of an infinite plane wave, radically contrasting with the concept of a corpuscule. Just this circumstance obviously excludes any possibility of describing the fate of a single photon as a causal event in space and time. Still, this implies in no way a renunciation of a comprehensive account of experience, which is fully exhausted by the statistical laws concerning the occurrence of the individual radiative effects, unambiguously determined by the superposition principe of classical theory and the conservation laws embodied in the photon idea.

That the difficulties of upholding a causal description were not confined to the problem of the structure of radiation, but extended as well to the mechanical problems involved in the constitution of matter, became especially clear after the completion of our ideas of the structure of the atom by Rutherford's discovery of the atomic nucleus. In fact, the extreme simplicity of this structure disclosed at once the total inadequacy of the ordinary laws of mechanics and electrodynamics in accounting for the remarkable stability of atoms, and compelled us to introduce, in addition to the quantum features of the radiative processes, the further assumption that any well-defined change of the electron bindings in an atom consists in a complete transition of the atom from one of its characteristic stationary states to another such state. These so-called quantum postulates are not only totally foreign to classical mechanical ideas, but imply an explicit renunciation of any causal description of such atomic processes. In particular, as regards its possible transitions from a given stationary state to other stationary state, accompanied by the emission of photons of different energies, the atom may be said to be confronted with a choice for which, according to the whole character of the description, there is no determining circumstance.

In a situation like that, all predictions can only concern the probabilities for the various possible courses of the atomic processes open to direct observation ; and just as in the case of proper radiation effects, it is evident that this cannot be attributed to any provisional incompleteness of the pictures in question. Indeed, as adequate as the quantum postulates are in the phenomenological description of the atomic reactions, as indispensable are the basic concepts of mechanics and electrodynamics for the specification of atomic structures and for the definition of fundamental properties of the agencies with which they react. Far from being a temporary compromise in this dilemma, the recourse to essentially statistical considerations is our only conceivable means of arriving at a generalization of the customary way of description sufficiently wide to account for the features of individuality expressed by the quantum postulates and reducing to classical theory in the limiting case where all actions involved in the analysis of the phenomena are large compared with a single quantum. In the search for the formulation of such a generalization, our only guide has just been the so-called correspondence argument, which gives expression for the exigency of upholding the use of classical concepts to the largest possible extent compatible with the quantum postulates.

As is well-known, Einstein succeeded, on the basis of such considerations, to formulate general laws for the emission and absorption

of photons in radiative processes, capable of accounting for Planck's
formula for black body radiation, and in the same way a rational dispersion theory was subsequently developed, especially by Kramers.
The analogy with classical ideas of the interaction between electric
oscillators and radiation fields, from which these results were derived,
proved, however, in spite of a number of promising attempts at
accounting for more specific features of the properties of atoms, insufficient for the establishment of a consistent correspondence treatment. The greater difficulty in this respect of atomic problems compared with the pure radiation effects is due to the absence in classical
mechanics of a superposition principle like that of field theory ; in
fact, although any solution of the mechanical equations of motion can
of course be represented by a superposition of purely harmonic oscillations, these oscillations, due to the non-linearity of the equations,
are not themselves independent solutions of the mechanical problem.
It was indeed the recognition of the necessity to avoid any explicit
use of mechanical motion in connection with the quantum postulates
which led Heisenberg to the establishment of a rational quantum
mechanics, based on suitable formal representation of all kinematic
and dynamical concepts.

In this formalism, the canonical equations of classical mechanics

$$\frac{dp_i}{dt} = -\frac{\partial H}{\partial q_i} \ , \quad \frac{dq_i}{dt} = \frac{\partial H}{\partial p_i} \qquad (2)$$

are maintained unaltered, and the quantum of action is only introduced in the so-called commutation rules

$$p_i \, q_i - q_i \, p_i = \frac{h}{2\pi_i} \qquad (3)$$

for any pair of canonically conjugate variables. While in this way the
whole scheme reduces to classical mechanics in the case $h = o$, all
the exigencies of the correspondence argument are fulfilled also in
the general case through the possibility of combining a purely classical description of the hamiltonian function H, with a proper account
of the quantum characteristics of its behaviour.

In particular is the essentially statistical nature of this account a direct consequence of the fact that the commutation rules prevent us to identify at any instant more than a half of the symbols representing the canonical variables with definite values of the corresponding classical quantities. In the special case of an isolated atomic system, the theory thus allows to predict the possible values of all action variables, which characterize the stationary states of the system, while the conjugate angle variables are left entirely undetermined.

In principle, a suitable use of canonical transformations permits the application of the formalism to any conceivable atomic problem, at any rate as far as relativity refinements are disregarded. On account of the intricate mathematical operations involved, it was, however, of utmost importance, not only for the practical use of the formalism, but even for the elucidation of essential aspects of its consequences, that the treatment of any quantum mechanical problem could be shown to be essentially reducible to the solution of a linear differential equation, allowing to formulate for atomic systems a principle of superposition of states, analogous to that of field theory. Actually, the original establishment of this fundamental equation by Schrödinger was the outcome of an independent attempt at a closer approach to a causal description of atomic phenomena by a radical modification of our ideas on the constitution of matter. As is well-known, this line of argument was initiated by de Broglie, who proposed, in analogy to the photon idea, to associate to the motion of a free particle a wave with frequency and wave number related by the equations (1) to the energy and momentum of the particle. Despite the extreme fruitfulness of these ideas, it is evident however that any concrete wave picture is as unable to account for basic experience regarding the individuality of the electron as a corpuscular picture for the superposition properties of radiation fields. In the two cases we are dealing, in fact, with two complementary aspects of experience.

In this connection, it must also be remembered that in the Schrödinger equation, which may be written

$$H \psi = \frac{h\,i}{2\pi} \frac{\partial \psi}{\partial t},\qquad (4\,a)$$

where H is a differential operator derived by replacing the p's by the operators

$$p_i = \frac{h}{2\pi i} \frac{\partial}{\partial q_i} \qquad (4\ b)$$

in the hamiltonian function, the form of this function contains as direct a reference to the corpuscular description of an atomic system as the upholding of the equation (2) in Heisenberg's formalism. Besides, the purely formal character of the suggestive resemblance between atomic mechanics and classical wave problems is perhaps most strikingly illustrated by the fact that this resemblance can only be brought out with the help of the conventional device of complex numbers, which alone permits to condense the abstractions of quantum mechanics in a form as simple as (3) and (4). The true significance of the wave formalism as a most practical means of expressing the statistical laws of atomic mechanics was also soon fully realized, especially through the work of Born, Dirac and Jordan and the completeness and self-consistency of the whole formalism is most clearly exhibited by the elegant axiomatic exposition of von Neumann, which in particular makes it evident that the fundamental superposition principle of quantum mechanics logically excludes the possibility of avoiding the non-causal feature of the formalism by any conceivable introduction of additional variables [1].

It needs hardly be emphasized that the whole argumentation here outlined for the necessity of renouncing a causal description of atomic phenomena is in no essential way affected by the preliminary disregard of relativity exigencies in quantum mechanics already alluded to.

---

[1] A complete statement of this development will be found in the numerous and excellent treatises of quantum mechanics. See for example : M. Born and P. Jordan, *Elementare Quantenmechanik*, Berlin, 1930. — P. Dirac, *Principles of quantum mechanics*, Oxford, 1935. — J. von Neumann, *Mathematische Grundlagen der Quantenmechanik*, Berlin, 1932. — W. Pauli, *Quantentheorie*, Handbuck der Phys., vol. XXIV/1, Berlin, 1933. — H. Kramers, *Theorien des Aufbaues der Materie*, Hand-und Jahrbuch der Chemischen Physik, vol. I, Leipzig, 1933.

Moreover the absence of any contradiction in principle between the principle of relativity and quantum theory is secured beforehand by the relativistic invariance of action quantities which, at a very early date, was emphasized by Planck himself and which, in the hands of Einstein and de Broglie, proved so powerful an argument for the generality of the relation, expressed by (1), between kinematical and dynamical quantities.

Also the wonderful success of Dirac's relativistic quantum theory of the electron, based on an ingenious generalization of the quantum mechanical formalism with the help of still more complex abstractions, entails the greatest encouragement to proceed on such lines. That for the moment the paradoxes connected with the use of the idealization of point charge for the electron are preventing the development on correspondence lines of a comprehensive relativistic quantum electrodynamics, must indeed rather be imputed to our failure, already acute in classical electron theory, to grasp some deeper feature of the stability of the individual particles themselves than to any lack of soundness of the general lines on which the incorporation of the quantum of action in atomic theory has been achieved. Quite independently of such open problems of atomic theory, the apparent paradoxes unavoidably connected with this incorporation, have also in recent years received a complete elucidation through a renewed revision of the very problem of observation in this field which, in disclosing unsuspected presuppositions for the unambiguous application of our most elementary concepts, reminds us in several respects of the great lesson of relativity theory.

2. — THE OBSERVATION PROBLEM IN QUANTUM THEORY.

At all stages of the development of quantum theory, the implications of the individuality of the quantum effects for the coordination of experimental evidence have of course played a prominent part in the argumentation. Still, a clarification of the situation as regards the observation problem in quantum theory, comparable to that which was the starting point of relativity theory, was first achieved after

[309]

the establishment of a rational quantum mechanical formalism. As is well-known, it was Heisenberg who first called attention to the fact that due to the intervention of the quantum of action in any conceivable measuring process, the information about the state of an atomic system, obtainable by direct measurements, must always imply a reciprocal latitude as regards the values of any pair of canonically conjugate variables, which just corresponds to the limitation imposed on the fixation of the values of such variables in quantum mechanics by the lack of commutability of the corresponding symbols.

This last limitation is expressed, as a direct consequence of the commutation rules (3), by the so-called uncertainty relations

$$\Delta p. \ \Delta q \geqslant \frac{h}{4\,\pi} \tag{5}$$

where $(\Delta p)^2$, $(\Delta q)^2$ represent the mean square deviations of p and q from their average values in the state of the system considered. In the especially simple case of a photon or a free particle, the relations (5) represent indeed the minimum reciprocal latitude in the fixation of position and momentum compatible with the relations (1), as follows immediately from the analysis of a wave packet in its harmonic components. The ultimate reason that in no conceivable measurement conjugate quantities can be fixed with a greater accuracy than that given by (5) is indeed the complementary character of the pictures employed in the description of every such auxiliary agency used in the measuring process. Any such process, aiming at the fixation of one quantity, say q, with a given accuracy Δq will in fact on account of the intervention of the quantum of action in this process, not only involve a finite change of the conjugate quantity p, but this change will even be uncontrollable to an extent Δp related to Δq by the inequality (5). It will hardly be necessary here to discuss more closely typical examples of measuring processes, such as have been treated in detail in the current literature ([1]). The observa-

---

([1]) Cf. W. Heisenberg, *Die physikalischen Grundlagen der Quantentheorie*, Leipzig, 1930; N. Bohr, *Atomtheorie und Naturbeschreibung*, Berlin, 1931.

tion problem in quantum theory involves, however, certain novel epistemological aspects as regards the analysis and synthesis of physical experience, which have only been gradually elucidated in recent years, and which I shall now proceed to discuss.

In the first place, we must recognize that a measurement can mean nothing else than the unambiguous comparison of some property of the object under investigation with a corresponding property of another system, serving as a measuring instrument, and for which this property is directly determinable according to its definition in everyday language or in the terminology of classical physics. While within the scope of classical physics such a comparison can be obtained without interfering essentially with the behaviour of the object, this is not so in the field of quantum theory, where the interaction between the object and the measuring instruments will have an essential influence on the phenomenon itself. Above all, we must realize that this interaction cannot be sharply separated from an undisturbed behaviour of the object, since the necessity of basing the description of the properties and manipulation of the measuring instruments on purely classical ideas implies the neglect of all quantum effects in that description, and in particular the renunciation of a control of the reaction of the object on the instruments more accurate than is compatible with the relation (5).

This last circumstance clearly shows that the statistical character of the uncertainty relations in no way originates from any failure of measurements to discriminate within a certain latitude between classically describable states of the object, but rather expresses an essential limitation of the applicability of classical ideas to the analysis of quantum phenomena. The significance of the uncertainty relations is just to secure the absence, in such an analysis, of any contradiction between different imaginable measurements. The very fact that in quantum phenomena no sharp separation can be made between an independent behaviour of the objects and their interaction with the measuring instruments, lends indeed to any such phenomenon a novel feature of individuality which evades all attempts at analysis on classical lines, because every imaginable experimental arrangement aiming at a subdivision of the phenomenon will be incompatible with its appearance and give rise, within the latitude indicated by the uncertainty relations, to other phenomena of similar individual character.

The essential lesson of the analysis of measurements in quantum theory is thus the emphasis on the necessity, in the account of the phenomena, of taking the whole experimental arrangement into consideration, in complete conformity with the fact that all unambiguous interpretation of the quantum mechanical formalism involves the fixation of the external conditions, defining the initial state of the atomic system concerned and the character of the possible predictions as regards subsequent observable properties of that system. Any measurement in quantum theory can in fact only refer either to a fixation of the initial state or to the test of such predictions, and it is first the combination of measurements of both kinds which constitutes a well-defined phenomenon.

Before entering more closely into the epistemological aspect of this situation as regards the synthesis of experience and its implications for our attitude to the causality problem, it must be emphasized that we have here in no way to do with any arbitrariness in the interpretation of experiments or incompleteness of the quantum mechanical formalism. As is well known, this point has been contested by several physicists who have taken a leading part in the development of quantum theory, and especially Einstein (¹) has argued that, on account of our freedom in the use of the measuring instruments, we should apparently be capable of attributing to the objects properties which, according to quantum mechanics, should be mutually incompatible. In fact, even after a contact between the object and some body serving as measuring instrument has taken place, we are left with a choice between different manipulations of the body concerned, permitting alternative predictions as regards subsequent measurements. Since the object itself is not interfered with in the interval between the first contact with the auxiliary body and these later

---

(¹) A. Einstein, B. Podolsky and N. Rosen, *Phys. Rev.*, **47**, 777, 1935.

measurements, it might indeed at first sight seem that all such predictions should refer to the same state of the object and had therefore
to be included in an exhaustive account of physical reality.

If, for instance, before the contact with the object, the momentum
of the auxiliary body had been controlled with great accuracy, leaving a correspondingly large latitude in the knowledge of its position
relatively to the other parts of the apparatus, we have still after the
contact the choice either to measure its momentum with the same
accuracy again, or to control its position relative to the other apparatus with any given accuracy. In the first case we have obviously to
do with a direct measurement of the momentum exchanged with the
object, allowing us to calculate its momentum after the contact if
it was known before. In the other case, if we only assume the body
to be so heavy that the latitude in the knowledge of its momentum
will not appreciably affect its velocity, we are able to calculate the
position of the body at the moment of the contact with the object,
and therefore to obtain a knowledge of the space-time co-ordination
of the latter with any desired accuracy. Now, since in both cases the
object itself has been treated alike, we should, as mentioned, apparently be able to assign to one and the same state of the object two
well-defined physical attributes in a way incompatible with the
uncertainty relations.

Such an argument, which is certainly very illuminating as regards
the inadequacy of the ordinary ideas of mechanics when applied to
quantum phenomena, is not suited, however, to disclose any defi-
ciency in the quantum mechanical method of description ([1]). In fact,
the paradox finds its complete solution within the frame of the quantum mechanical formalism, according to which no well-defined use
of the concept of « state » can be made as referring to the object separate from the body with which it has been in contact, until the external conditions involved in the definition of this concept are unambiguously fixed by a further suitable control of the auxiliary body.
Instead of disclosing any incompleteness of the formalism, the argu-

---

([1]) Cf. N. Bohr, *Phys. Rev.*, **48**, 696, 1935.

ment outlined entails in fact an unambiguous prescription as to how
this formalism is rationally applied under all conceivable manipulations of the measuring instruments. The complete freedom of the
procedure in experiments common to all investigations of physical
phenomena, is in itself of course contained in our free choice of the
experimental arrangement, which again is only dictated by the particular kind of phenomena we wish to investigate.

The apparent contrast between different types of quantum phenomena, the description of which involves different classical ideas,
like space-time coordination or momentum and energy conservation,
finds in fact its straight-forward explanation in the mutually exclusive
character of the different experimental arrangements demanded for
the appearance of such phenomena. Thus, any phenomenon in
which we are concerned with tracing a displacement of some atomic
object in space and time necessitates the establishment of several
coincidences between the object and the rigidly connected bodies
and movable devices which, in serving as scales and clocks respectively, define the space-time frame of reference to which the phenomenon
in question is referred. Just this situation implies, however, a renunciation of any sharp control of the amount of momentum or energy
exchanged during each coincidence between the object and the separate bodies entering into the experimental arrangement. Inversely,
every phenomenon in which we are essentially concerned with momentum and energy exchanges — and which therefore necessitates an
experimental arrangement allowing at least two successive determinations of momentum and energy quantities — will, in principle, imply a renunciation of the control of any precise space-time co-ordination of the objects in the time intervals between these measurements.

Instructive examples of this situation are offered respectively by
the interference effects of electrons and by the Compton effect, which
are equally paradoxical from the point of view of classical physics.
In the former case, the phenomenon is in fact only defined when the
relative positions of all scattering bodies and photographic plates are
known with an accuracy excluding the possibility, by means of a
control of momentum transfer, of discriminating between various

imaginable paths of the electron, to an extent incompatible with the very idea of interference. In the latter case, a control of the space-time co-ordination of the scattering process irreconcilable with the definition of momentum and energy quantities is excluded in advance by any arrangement allowing a test of the momentum and energy conservation such as is implied in the specification of the phenomenon itself. In neither case is there indeed any question of a simple replacement of the classical particle picture of electrons and wave picture of light with the electron wave idea or the photon concept respectively ; rather we have to do with individual phenomena, which cannot be analyzed on classical lines, and which exhibit the peculiar complementary relationship of superposition principle and conservation laws in quantum theory.

Notwithstanding their great importance in illustrating typical aspects of atomic processes, the two kinds of quantum phenomena just discussed represent of course only limiting cases of special simplicity. It is in fact possible to test the statistical predictions of quantum mechanics referring to any state of the object defined by the values of suitable functions of the space-time variables and the momentum and energy quantities. Also in such cases, however, it must be remembered that any well-defined phenomenon involves the combination of several comparable measurements. The significance of this point is strikingly exemplified by the case, often discussed, of the possible determination of the position of a particle with known momentum by the spot produced by its impact on a photographic plate. Far from meeting any contradiction with the uncertainty relations, we have clearly here to do with a measuring arrangement which is not suited to define a phenomenon involving a test of predictions as regards the location of the object. In conformity with the uncertainty relations, the knowledge of its momentum prevents in fact any unambiguous connection between this object and the frame of reference with respect to which the position of the photographic plate is defined.

In the system to which the quantum mechanical formalism is applied, it is of course possible to include any intermediate auxiliary agency employed in the measuring processes. Since, however, all those properties of such agencies which, according to the aim of the measurement, have to be compared with corresponding properties of the object, must be described on classical lines, their quantum mechanical treatment will for this purpose be essentially equivalent with a classical description. The question of eventually including such agencies within the system under investigation is thus purely a matter of practical convenience, just as in classical physical measurements ; and such displacements óf the section between object and measuring instruments can therefore never involve any arbitrariness in the description of a phenomenon and its quantum mechanical treatment. The only significant point is that in each case some ultimate measuring instruments, like the scales and clocks which determine the frame of space-time coordination — on which, in the last resort, even the definitions of momentum and energy quantities rest — must always be described entirely on classical lines, and consequently kept outside the system subject to quantum mechanical treatment.

The unaccustomed features of the situation with which we are confronted in quantum theory necessitate the greatest caution as regards all questions of terminology. Speaking, as is often done, of disturbing a phenomenon by observation, or even of creating physical attributes to objects by measuring processes, is, in fact, liable to be confusing, since all such sentences imply a departure from basic conventions of language which, even though it sometimes may be practical for the sake of brevity, can never be unambiguous. It is certainly far more in accordance with the structure and interpretation of the quantum mechanical symbolism, as well as with elementary epistemological principles, to reserve the word « phenomenon » for the comprehension of the effects observed under given experimental conditions.

These conditions, which include the account of the properties and manipulation of all measuring instruments essentially concerned, constitute in fact the only basis for the definition of the concepts by which the phenomenon is described. It is just in this sense that phenomena defined by different concepts, corresponding to mutually exclusive experimental arrangements, can unambiguously be regarded as complementary aspects of the whole obtainable evidence concerning the objects under investigation. The view-point of complementarity allows us indeed to avoid any futile discussion about an ultimate determinism or indeterminism of physical events, by offering a straightforward generalization of the very ideal of causality, which can aim only at the synthesis of phenomena describable in terms of a behaviour of objects independent of the means of observation (¹).

3. — RELATIVITY AND COMPLEMENTARITY.

In spite of all differences in the physical problems concerned, relativity theory and quantum theory possess striking similarities in a purely logical respect. In both cases we are confronted with novel aspects of the observational problem, involving a revision of customary ideas of physical reality, and originating in the recognition of general laws of nature which do not directly affect practical experience. The impossibility of an unambiguous separation between space and time without reference to the observer, and the impossibility of a sharp separation between the behaviour of objects and their interaction with the means of observation are, in fact, straightforward consequences of the existence of a maximum velocity of propagation of all actions and of a minimum quantity of any action, respectively. The ultimate reason for the unavoidable renunciation as regards the absolute significance of ordinary attributes of objects, and for the recourse to a relative or complementary mode of description respectively, lies also in both cases in the necessity of confining ourselves, in the account of experience, to comparisons between measurements in the interpretation of which relativity refinements and quantum effects respectively have on principle to be neglected.

Even the formalisms, which in both theories within their scope offer adequate means of comprehending all conceivable experience, exhibit deepgoing analogies. In fact, the astounding simplicity of the gene-

---

(¹ Cf. N. Bohr, *Erkenntnis*, 6, 293, 1936.

ralizations of classical physical theories, which are obtained by the use of multi-dimensional geometry and non-commutative algebra respectively, rests in both cases essentially on the introduction of the conventional symbol $\sqrt{-1}$. The abstract character of the formalisms concerned is indeed on closer examination as typical of relativity theory as it is of quantum mechanics, and it is in this respect purely a matter of tradition if the former theory is considered as a completion of classical physics rather than as a first fundamental step in the thoroughgoing revision of our conceptual means of comparing observations, which the modern development of physics has forced upon us. No more in atomic theory than in classical theories of mechanics and electrodynamics was such a revision contemplated from the beginning, and just as the understanding of the relativity of even our most elementary concepts, the recognition of the complementary aspects of atomic phenomena has only gradually developed from the frustration of all endeavours to cope with new experience on classical lines.

In this situation it is above all of decisive importance that the two theories, despite their entirely different starting points, are constituted in such a way that they harmoniously unite into a logically consistent structure. Just the relativistic invariance of action quantities ensures in fact, as already mentionned in § 1, that all basic quantum relations are quite independent of the frame of reference. Naturally we meet paradoxes still more acute from the classical point of view in theories in which relativity as well as quantum requirements are inseparably taken into account, as is especially apparent in the formal use of negative energy values in such theories. This is, however, only apt to emphasize the inherent limitations of an analysis and synthesis on classical lines of the phenomena in question ; limitations appearing so clearly even in any phenomenon involving only the one or the other kind of departure from classical physics. In fact, all argumentation regarding the adequacy and completeness of the comprehension of experience can be concerned only with an examination of the unambiguous interpretation of measurements and their comparison ; and just in this respect, the relativistic invariance of the uncertainty relations (5) allows us to uphold the whole argumentation outlined in § 2 without essential alteration.

It is true that under the impression of the paradoxes alluded to the opinion has been expressed from various sides that simple relativistic arguments should impose limitations in the use of kinematical and dynamical variables and of electromagnetic field concepts more stringent than those which follow from the commutation rules of the quantum mechanical and electrodynamical formalisms. On close examination, it appears, however, that such arguments are based on the consideration of special measuring agencies, which, just due to the complementary character of the phenomena concerned, are not suited to the purpose contemplated. Actually, it is always possible, in conformity with the above argumentation, to conceive experimental arrangements which, at least in principle, would allow to test the predictions of any consistent relativistic formalism, like Dirac's quantum theory of the electron or the quantum theory of electromagnetic fields in free space (1). In such considerations, we are naturally only concerned with the implications of the so-called special theory of relativity, since in proper atomic problems, the gravitational interactions of the small masses involved are entirely negligible. It is nevertheless interesting to remark that all arguments concerning the observational problem in quantum theory are in complete harmony with the general principle of relativity.

This is most instructively shown by the example of a typical experimental arrangement, suggested by Einstein and discussed at a meeting of the Solvay Council of Physics in 1930. In this arrangement, the exchange of energy during a contact between an atomic object and a measuring body is determined by the difference in weight of this body before and after the contact. At first sight it might appear that the weighing of the body could be performed without hindering the accurate control of the time at which the contact takes place, as registered by a clock rigidly connected with the body concerned.

---

(1) Cf. N. Bohr and L. Rosenfeld, *Copenhagen Acad. Math.-phys. Comm.*, 12, Nr. 8, 1933.

Obviously, however, this would furnish information about energy
and time variables referring to the object with greater accuracy than
i s compatible with quantum mechanics ; and on closer examination,
it is also seen that the use of any weighing device with a given accuracy will exclude such a control of the time of contact. In fact, the
necessary latitude in our knowledge of the position of the body in
the gravitational field will imply an uncertainty in the regulation
of the clock, connected with the assigned accuracy of the energy
measurement just by the quantum mechanical uncertainty relation.
This result was of course to be expected, in view of the way in which
the equivalence arguments of general relativity are based on purely
kinematical and dynamical principles rationally incorporated in
quantum mechanics.The main lesson of the whole discussion is rather
to enforce again the necessity of describing entirely on classical lines
all ultimate measuring instruments which define the external conditions of the phenomenon, and therefore of keeping them outside
the system for the treatment of which the quantum of action is to be
taken essentially into account.

Entirely new aspects of the observation problem in atomic theory
are met with, however, in connection with the still unsolved diff-
culties, already alluded to in § 1, of incorporating harmoniously into
relativistic quantum theory the essentially atomistic features of matter and electricity which lie beyond the scope of classical electrodynamics. It must in fact be realized that the whole present attack on
quantum theory is based on the presupposition that it is possible,
without regard to the atomic constitution of all matter, to have recourse at least in principle to measuring instruments of arbitrary
fineness of construction, which can be manipulated without restriction. The adequacy of this idealization — essentially legitimate for
all experience regarding atomic phenomena which can be co-ordinated
by the methods discussed in §§ 1 and 2 — will, in all cases where
these methods fail, naturally have to be examined more closely ([1]).
In connection with the complications encountered in the gradua

---

([1]) Cf. N. Bohr, Report of the Solvay Council, 1933, p. 216.

development of electron theory, such an examination has, however, so far only disclosed unsuspected potentialities of present methods, leading above all to the final establishment of the theory of electron pairs, which in the last few years has proved so fruitful in accounting for the remarkable phenomena of electron showers in cosmic rays. Still, quite recently, Heisenberg [1] has suggested that other features of these phenomena point to an essential instability of all matter under extreme conditions, which may be brought into connection with the existence of an ultimate limit for the application of space-time concepts within four-dimensional extensions of a certain size, extremely small in comparison with the spacial dimensions and periods involved in the account of ordinary atomic phenomena.

Quite apart from any prospect of mastering hitherto unsolved problems of atomic theory, which such considerations may entail, there can of course be no question in further developments of returning to a description of atomic phenomena in closer conformity with the causality ideal. Rather are we here confronted with the necessity of a still more radical departure from accustomed modes of description of natural phenomena implying a further extension of the viewpoint of complementarity; just as the gradual development of cosmological theories, far from involving a limitation of the applicability of the relativity principles, has led to ever further widening of their scope. In cosmology as well as in atomic theory our starting point can indeed be nothing else than the conventions suited to account for ordinary practical experience, and the different character of the generalizations of the methods of classical physics, with which we are concerned in the two fields, is quite naturally connected with the two opposite lines along which experience is extended, and which place us before essentially different epistemological situations.

In concluding, I should like to remark that, however unfamiliar the aspects of the observation problem met with in atomic theory may appear on the background of classical physics, they are by no means new in other fields of science. I need only recall the impossibi⁻

---

[1] W. Heisenberg, *Ann. d. Phys.*, **32**, 20, 1938.

lity, in the analysis of psychological experience, of separating sharply between objective content and subjective consciousness of perception. The mutually exclusive relation ship between different psychological phenomena, denoted by such words as « thoughts » and « feelings », or « reason » and « instinct », reminds us in fact most strikingly of the complementary relationship between atomic phenomena described by kinematical and dynamical concepts respectively. Although such analogies are of course of a purely formal character, they may nevertheless be profitably studied by scientists in other fields. Not only is it indeed to be hoped that rational attempts at mastering the comparatively simple physical problems concerned may suggest new methods of comprehending the results of the more intricate psychological researches ; but also physicists may derive some help and encouragement from the recognition that the novel situation as regards the causality problem in which they find themselves is not unique.

* * *

IV. ANALYSIS AND SYNTHESIS IN ATOMIC PHYSICS

ANALYSE OG SYNTESE INDENFOR ATOMFYSIKKEN
Overs. Dan. Vidensk. Selsk. Virks. Juni 1941 – Maj 1942, p. 30

Communication to the Royal Danish Academy on 12 December 1941

ABSTRACT

TEXT AND TRANSLATION

NIELS BOHR giver en Meddelelse: *Analyse og Syntese indenfor Atomfysikken.*

Atomfysikkens Udvikling, og især Opdagelsen af Virkningskvantet, har vist, at den klassiske Fysiks Beskrivelsesmaade er en Idealisation, der kun kan anvendes, saalænge man ved Redegørelsen af Fænomenerne kan se fuldstændig bort fra den med Iagttagelsen forbundne Vekselvirkning mellem Objekterne og Maaleinstrumenterne. I Foredraget vil den derved opstaaede Situation blive nærmere undersøgt, og det vil blive eftervist, med hvilken Forsigtighed alle tilvante Begreber maa anvendes for at sikre Harmonien imellem de atomare Fænomeners Analyse og Syntese.

Vil blive trykt i Mat.-fys. Medd.

TRANSLATION

Communication by Niels Bohr: *Analysis and Synthesis within Atomic Physics.*

The development of atomic physics, and in particular the discovery of the quantum of action, has revealed that the mode of description of classical physics is an idealization which is applicable only as it is possible, in the account of the phenomena, to neglect completely the interaction between the objects and the measuring instruments associated with the observation. In the lecture, the situation thus arising will be discussed in greater detail, and it will be shown how carefully we need to apply the customary concepts in order to ensure the harmony between analysis and synthesis of atomic phenomena.

To appear in Mat.–Fys. Medd.*

* [Never published.]

V. ON THE NOTIONS OF CAUSALITY AND COMPLEMENTARITY

Dialectica **2** (1948) 312–319

See Introduction to Part II, sect. 3.

ON THE NOTIONS OF CAUSALITY AND COMPLEMENTARITY (1948)

Versions published in English

A   Dialectica **2** (1948) 312–319
B   8th Solvay Conference, *Les Particules élémentaires*, Bruxelles, 27.09.–2.10.1948, R. Stoops, Bruxelles 1950, pp. 9–17
C   Science **111** (1950) 51–54
D   *Nature des problèmes en philosophie (Entretiens d'été – Lund 1947)*, Actualités scientifiques et industrielles, vol. V.II: *Logique et sciences de la nature*, Hermann & Cie, Paris 1949, pp. 67–76

All these versions agree with each other. *A* is the only version which has a summary in English, French and German. *B* has a short introduction. *D* is the proceedings of a conference in Lund where Bohr gave a talk with the title *Causality and Complementarity* on 11 June 1947.

QUARTERLY REVIEW   REVUE TRIMESTRIELLE   VIERTELJAHRESSCHRIFT

DIALECTICA

7/8

INTERNATIONAL REVIEW OF PHILOSOPHY OF KNOWLEDGE
REVUE INTERNATIONALE DE PHILOSOPHIE DE LA CONNAISSANCE
INTERNATIONALE ZEITSCHRIFT FÜR PHILOSOPHIE DER ERKENNTNIS
Dialectica   Vol. 2 No 3/4, pp. 305-424 Neuchâtel, Suisse, 15. 8.-15. 11. 1948

DIALECTICA

Board of Directors   Comité directeur   Leitendes Komitee

FERDINAND GONSETH
GASTON BACHELARD   PAUL BERNAYS

Consulting board   Comité consultatif   Beratendes Komitee

M. BARZIN, Bruxelles. R. BAYER, Paris. A.-C. BLANC, Roma. G. BOULIGAND, Paris. G. DAVY
Paris. J.-L. DESTOUCHES, Paris. Ph. DEVAUX, Liège et Anvers. E. DUPRÉEL, Bruxelles
K. DÜRR, Zürich. H. FISCHER. Zürich. S. GAGNEBIN, Neuchâtel. H. GOLDMANN, Bern
E. HADORN, Zürich. H. HOPF, Zürich. H. KÖNIG, Bern. P. LARDY, Zürich. F.-E. LEHMANN
Bern. J. LUGEON, Zürich. A. MERCIER, Bern. P. NIGGLI, Zürich. W. PAULI, Zürich
J. PIAGET, Genève. M. PLANCHEREL, Zürich. H.-J. POS, Amsterdam. A. REYMOND
Lausanne. K. REIDEMEISTER, Marburg. P.-H. ROSSIER, Zürich. F. STÜSSI, Zürich

Manager   Administrateur   Geschäftsleiter

PADROT NOLFI

Editorial correspondents   Correspondants de la rédaction
Korrespondierende Redaktionsmitglieder

Belgique :   CH. PERELMAN, 32, rue de la Pêcherie, Uccle-Bruxelles.

England :   K. R. POPPER, London School of Economics
(University of London), London W. C. 2.

France :   Mme P. DESTOUCHES-FÉVRIER, rue Thénard 4,
Paris Ve.

Italia :   S. CECCATO, via G. Colombo 79, Milano.

Neederland :   J. CLAY, Plantage Franschelaer 13A, Amsterdam.

Board of Editors   Comité de rédaction   Redaktions-Komitee

Frl. M. ÆBI, B. ECKMANN, F. FIALA, E. GAGNEBIN, M. GEX, G. HIRSCH,
R.-L. JEANNERET, A. LINDER, M. MONNIER, A. PREISSMANN, J. ROSSEL,
E. SPECKER, E. J. WALTER.

Editorial secretary   Secrétaire de la rédaction   Redaktionssekretär

H.-S. GAGNEBIN

Communications should be addressed to   Adresser la correspondance à
Alle Zuschriften sind zu adressieren an

DIALECTICA, ZÜRICH 33
Switzerland   Suisse   Schweiz

Publisher   Editeur   Verleger

PRESSES UNIVERSITAIRES DE FRANCE - PARIS
ÉDITIONS DU GRIFFON - NEUCHATEL - SUISSE

[328]

DIALECTICA

VOL. 2 ⋆ No. 3/4

15. 8.-15.11. 1948

THE CONCEPT OF COMPLEMENTARITY *DIE IDEE DER KOMPLEMENTARITÄT*
    L'IDÉE DE COMPLÉMENTARITÉ

This issue is edited by   Dieses Heft ist redigiert von
    Fascicule publié sous la direction de

    Wolfgang PAULI

Contents   Sommaire   Inhalt

PRESSES UNIVERSITAIRES DE FRANCE PARIS
ÉDITIONS DU GRIFFON NEUCHATEL SUISSE

[329]

ON THE NOTIONS OF
CAUSALITY AND COMPLEMENTARITY [1]

The causal mode of description has deep roots in the conscious endeavours to utilize experience for the practical adjustment to our environments, and is in this way inherently incorporated in common language. By the guidance which analysis in terms of cause and effect has offered in many fields of human knowledge, the principle of causality has even come to stand as the ideal for scientific explanation.

In physics, causal description, originally adapted to the problems of mechanics, rests on the assumption that the knowledge of the state of a material system at a given time permits the prediction of its state at any subsequent time. However, already here the definition of state requires special consideration and it need hardly be recalled that an adequate analysis of mechanical phenomena was only possible after the recognition that, in the account of a state of a system of bodies, not merely their location at a given moment but also their velocities have to be included.

In classical mechanics, the forces between bodies were assumed to depend simply on the instantaneous positions and velocities; but the discovery of the retardation of electromagnetic effects made it necessary to consider force fields as an essential part of a physical system, and to include in the description of the state of the system at a given time the specification of these fields in every point of space. Yet, as is well known, the establishment of the differential equations connecting the rate of variation of electromagnetic intensities in space and time has made possible a description of electromagnetic phenomena in complete analogy to causal analysis in mechanics.

It is true that, from the point of view of relativistic argumentation, such attributes of physical objects as position and velocity of material bodies, and even electric or magnetic field intensities, can no longer be given an absolute content. Still, relativity theory, which has endued classical physics with unprecedented unity and scope, has just through its elucidation of the conditions for the unambiguous use of elementary physical concepts allowed a concise formulation of the principle of causality along most general lines.

---

[1] The purpose of this article is to give a very brief survey of some epistemological problems raised in atomic physics. A fuller account of the historical development, illustrated by typical examples which have served to clarify the general principles, will appear soon as a contribution by the writer to the Einstein volume in the series « Living Philosophers ».

However, a wholly new situation in physical science was created through the discovery of the universal quantum of action, which revealed an elementary feature of « individuality » of atomic processes far beyond the old doctrine of the limited divisibility of matter originally introduced as a foundation for a causal explanation of the specific properties of material substances. This novel feature is not only entirely foreign to the classical theories of mechanics and electromagnetism, but is even irreconcilable with the very idea of causality.

In fact, the specification of the state of a physical system evidently cannot determine the choice between different individual processes of transition to other states, and an account of quantum effects must thus basically operate with the notion of the probabilities of occurrence of the different possible transition processes. We have here to do with a situation which is essentially different in character from the recourse to statistical methods in the practical dealing with complicated systems that are assumed to obey laws of classical mechanics.

The extent to which ordinary physical pictures fail in accounting for atomic phenomena is strikingly illustrated by the well-known dilemma concerning the corpuscular and wave properties of material particles as well as of electromagnetic radiation. It is further important to realize that any determination of Planck's constant rests upon the comparison between aspects of the phenomena which can be described only by means of pictures not combinable on the basis of classical physical theories. These theories indeed represent merely idealizations of asymptotic validity in the limit where the actions involved in any stage of the analysis of the phenomena are large compared with the elementary quantum.

In this situation, we are faced with the necessity of a radical revision of the foundation for description and explanation of physical phenomena. Here, it must above all be recognized that, however far quantum effects transcend the scope of classical physical analysis, the account of the experimental arrangement and the record of the observations must always be expressed in common language supplemented with the terminology of classical physics. This is a simple logical demand, since the word «experiment » can in essence only be used in referring to a situation where we can tell others what we have done and what we have learned.

The very fact that quantum phenomena cannot be analysed on classical lines thus implies the impossibility of separating a behaviour of atomic objects from the interaction of these objects with the measuring instruments which serve to specify the conditions under which the phenomena appear. In particular, the individuality of the typical quantum effects finds proper expression in the circumstance that any attempt at subdividing the phenomena will demand a change in the experimental arrangement, introducing new sources of uncontrollable interaction between objects and measuring instruments.

In this situation, an inherent element of ambiguity is involved in assigning conventional physical attributes to atomic objects. A clear example of such an ambiguity is offered by the mentioned dilemma as to the properties of electrons or photons, where we are faced with the contrast revealed by the comparison between observations regarding an atomic object, obtained by means of different experimental arrangements. Such empirical evidence exhibits a novel type of relationship, which has no analogue in classical physics and which may conveniently be termed « complementarity » in order to stress that in the contrasting phenomena we have to do with equally essential aspects of all well-defined knowledge about the objects.

An adequate tool for the complementary mode of description is offered by the quantum-mechanical formalism, in which the canonical equations of classical mechanics are retained while the physical variables are replaced by symbolic operators subjected to a non-commutative algebra. In this formalism Planck's constant enters only in the commutation relations

$$qp - pq = \sqrt{-1}\,\frac{h}{2\pi} \tag{1}$$

between the symbols q and p standing for a pair of conjugate variables, or in the equivalent representation by means of the substitutions of the type

$$p = -\sqrt{-1}\,\frac{h}{2\pi}\frac{\partial}{\partial q} \tag{2}$$

by which one of each set of conjugate variables is replaced by a differential operator. According to the two alternative procedures, quantum-mechanical calculations may be performed either by representing the variables by matrices with elements referring to the individual transitions between two states of the system or by making use of the so-called wave equation, the solutions of which refer to these states and allow us to derive probabilities for the transitions between them.

The entire formalism is to be considered as a tool for deriving predictions, of definite or statistical character, as regards information obtainable under experimental conditions described in classical terms and specified by means of parameters entering into the algebraic or differential. equations of which the matrices or the wave-functions, respectively, are solutions. These symbols themselves, as is indicated already by the use of imaginary numbers, are not susceptible to pictorial interpretation; and even derived real functions like densities and currents are only to be regarded as expressing the probabilities for the occurrence of individual events observablé under well-defined experimental conditions.

A characteristic feature of the quantum-mechanical description is that the representation of a state of a system can never imply the accurate determination of both members of a pair of conjugate variables q and p. In fact, due to the non-commutability of such variables, as expressed by (1) and (2), there will always be a reciprocal relation

$$\Delta q \cdot \Delta p = \frac{h}{4\pi} \qquad (3)$$

between the latitudes Δq and Δp with which these variables can be fixed. These so-called indeterminacy relations explicitly bear out the limitation of causal analysis, but it is important to recognize that no unambiguous interpretation of such relations can be given in words suited to describe a situation in which physical attributes are objectified in a classical way.

Thus, a sentence like « we cannot know both the momentum and the position of an electron » raises at once questions as to the physical reality of such two attributes, which can be answered only by referring to the mutually exclusive conditions for the unambiguous use of space-time coordination, on the one hand, and dynamical conservation laws, on the other. In fact, any attempt at locating atomic objects in space and time demands an experimental arrangement involving an exchange of momentum and energy, uncontrollable in principle, between the objects and the scales and clocks defining the reference frame. Conversely, no arrangement suitable for the control of momentum and energy balance will admit precise description of the phenomena as a chain of events in space and time.

Strictly speaking, every reference to dynamical concepts implies a classical mechanical analysis of physical evidence which ultimately rests on the recording of space-time coincidences. Thus, also in the description of atomic phenomena, use of momentum and energy variables for the specification of initial conditions and final observations refers implicitly to such analysis and therefore demands that the experimental arrangements used for the purpose have spatial dimensions and operate with time intervals sufficiently large to permit the neglect of the reciprocal indeterminacy expressed by (3). Under these circumstances it is, of course, to a certain degree a matter of convenience to what extent the classical aspects of the phenomena are included in the proper quantum-mechanical treatment where a distinction in principle is made between measuring instruments, the description of which must always be based on space-time pictures, and objects under investigation, about which observable predictions can in general only be derived by the non-visualizable formalism.

Incidentally, it may be remarked that the construction and the functioning of all apparatus like diaphragms and shutters, serving to define geometry and timing of the experimental arrangements, or photographic plates used for recording the localization of atomic objects, will depend on properties of materials which are themselves essentially determined by the quantum of action. Still, this circumstance is irrelevant for the study of simple atomic phenomena where, in the specification of the experimental conditions, we may to a very high degree of approximation disregard the molecular constitution of the measuring instruments. If only the instruments are sufficiently heavy compared with the atomic objects under investigation, we can in particular neglect the requirements of relation (3) as regards the control of the localization in space and time of the single pieces of apparatus relative to each other.

In representing a generalization of classical mechanics suited to allow for the existence of the quantum of action, quantum mechanics offers a frame sufficiently wide to account for empirical regularities which cannot be comprised in the classical way of description. Besides the characteristic features of atomic stability, which gave the first impetus to the development of quantum mechanics, we may here refer to the peculiar regularities exhibited by systems composed of identical entities, such as photons or electrons, and determining for radiative equilibrium or essential properties of material substances. As is well known, these regularities are adequately described by the symmetry properties of the wave-functions representing the state of the whole systems. Of course, such problems cannot be explored by any experimental arrangement suited for the tracing in space and time of each of the identical entities separately.

It is furthermore instructive to consider the conditions for the determination of positional and dynamical variables in a state of a system with several atomic constituents. In fact, although any pair, q and p, of conjugate space and momentum variables obeys the rule of non-commutative multiplication expressed by (1), and thus can only be fixed with reciprocal latitudes given by (3), the difference $q_1 - q_2$ between the space coordinates referring to two constituents of a system will commute with the sum $p_1 + p_2$ of the corresponding momentum components, as follows directly from the commutability of q_1 with p_2 and of q_2 with p_1. Both $q_1 - q_2$ and $p_1 + p_2$ can, therefore, be accurately fixed in a state of the complex system and we can consequently predict the values of either q_1 or p_1 if either q_2 or p_2, respectively, are determined by direct measurements. Since at the moment of measurement the direct interaction between the objects may have ceased, it might thus appear that both q_1 and p_1 were to be regarded as well-defined physical attributes of the isolated object and that, therefore, as has been argued, the quantum-mechanical representation of a state should not offer an adequate means of a complete description of physical reality. With regard to such an argumentation, however, it must be stressed that any two arrangements which admit accurate measurements of q_2 and p_2 will be mutually exclusive and that therefore predictions as regards q_1 or p_1, respectively, will pertain to phenomena which basically are of complementary character.

As regards the question of the completeness of the quantum-mechanical mode of description, it must be recognized that we are dealing with a mathematically consistent scheme which is adapted within its scope to every process of measurement and the adequacy of which can only be judged from a comparison of the predicted results with actual observations. In this connection, it is essential to note that, in any well-defined application of quantum mechanics, it is necessary to specify the whole experimental arrangement and that, in particular, the possibility of disposing of the parameters defining the quantum-mechanical problem just corresponds to our freedom of constructing and handling the measuring apparatus, which in turn means the freedom to choose between the different complementary types of phenomena we wish to study.

In order to avoid logical inconsistencies in the account of this unfamiliar situation, great care in all questions of terminology and dialectics is obviously imperative. Thus, phrases often found in the physical literature, as «disturbance of phenomena by observation» or «creation of physical attributes of objects by measurements» represent a use of words like «phenomena» and «observation» as well as «attribute» and «measurement» which is hardly compatible with common usage and practical definition and, therefore, is apt to cause confusion. As a more appropriate way of expression, one may strongly advocate limitation of the use of the word *phenomenon* to refer exclusively to observations obtained under specified circumstances, including an account of the whole experiment.

With this terminology, the observational problem in atomic physics is free of any special intricacy, since in actual experiments all evidence pertains to observations obtained under reproducible conditions and is expressed by unambiguous statements referring to the registration of the point at which an atomic particle arrives on a photographic plate or to a corresponding record of some other amplification device. Moreover, the circumstance that all such observations involve processes of essentially irreversible character lends to each phenomenon just that inherent feature of completion wich is demanded for its well-defined interpretation within the framework of quantum mechanics.

Recapitulating, the impossibility of subdividing the individual quantum effects and of separating a behaviour of the objects from their interaction with the measuring instruments serving to define the conditions under which thc phenomena appear implies an ambiguity in assigning conventional attributes to atomic objects which calls for a reconsideration of our attitude towards the problem of physical explanation. In this novel situation, even the old question of an ultimate determinacy of natural phenomena has lost its conceptional basis, and it is against this background that the viewpoint of complementarity presents itself as a rational generalization of the very ideal of causality.

The complementary mode of description does indeed not involve any arbitrary renunciation on customary demands of explanation but, on the contrary, aims at an appropriate dialectic expression for the actual conditions of analysis and synthesis in atomic physics. Incidentally, it would seem that the recourse to three-valued logic, sometimes proposed as means for dealing with the paradoxical features of quantum theory, is not suited to give a clearer account of the situation, since all well-defined experimental evidence, even if it cannot be analysed in terms of classical physics, must be expressed in ordinary language making use of common logic.

The epistemological lesson we have received from the new development in physical science, where the problems enable a comparatively concise formulation of principles, may also suggest lines of approach in other domains of knowledge where the situation is of essentially less accessible character. An example is offered in biology where mechanistic and vitalistic arguments are used in a typically complementary manner. In sociology too such dialectics may often be useful, particularly in problems confronting us in the study and comparison of human cultures, where we have to cope with the element of complacency inherent in every national culture and manifesting itself in prejudices which obviously cannot be appreciated from the standpoint of other nations.

Recognition of complementary relationship is not least required in psychology, where the conditions for analysis and synthesis of experience exhibit striking analogy with the situation in atomic physics. In fact, the use of words like « thoughts » and « sentiments », equally indispensable to illustrate the diversity of psychical experience, pertain to mutually exclusive situations characterized by a different drawing of the line of separation between subject and object. In particular, the place left for the feeling of volition is afforded by the very circumstance that situations where we experience freedom of will are incompatible with psychological situations where causal analysis is reasonably attempted. In other words, when we use the phrase « I will » we renounce explanatory argumentation.

Altogether, the approach towards the problem of explanation that is embodied in the notion of complementarity suggests itself in our position as conscious beings and recalls forcefully the teaching of ancient thinkers that, in the search for a harmonious attitude towards life, it must never be forgotten that we ourselves are both actors and spectators in the drama of existence. To such an utterance applies, of course, as well as to most of the sentences in this article from the beginning to the end, the recognition that our task can only be to aim at communicating experiences and views to others by means of language, in which the practical use of every word stands in a complementary relation to attempts of its strict definition.

Niels BOHR.

Summary

A short exposition is given of the foundation of the causal description in classical physics and the failure of the principle of causality in coping with atomic phenomena. It is emphasized that the individuality of the quantum processes excludes a separation between a behaviour of the atomic objects and their interaction with the measuring instruments defining the conditions under which the phenomena appear. This circumstance forces us to recognize a novel relationship, conveniently termed complementarity, between empirical evidence obtained under different experimental conditions. An appropriate tool for a complementary mode of description is provided by the quantum-mechanical formalism which allows us to account for regularities of definite or statistical character beyond the grasp of classical physical explanation. — N. B.

Résumé

Un court exposé est donné du fondement de la description causale en physique classique et de l'incapacité du principe de causalité à maîtriser les phénomènes atomiques. On souligne que l'individualité des processus quantiques exclut toute séparation entre un comportement des objets atomiques et leur interaction avec les appareils de mesure définissant les conditions sous lesquelles apparaissent les phénomènes. Cette circonstance nous oblige à reconnaître une nouvelle relation qu'on peut appeler complémentarité entre les résultats empiriques obtenus dans des conditions expérimentales différentes. Un instrument approprié à un mode de description complémentaire est fourni par le formalisme de la mécanique quantique qui nous permet de rendre compte de régularités de caractère défini ou statistique au delà de la portée des explications physiques classiques. — N. B.

Zusammenfassung

Die Grundlagen der kausalen Beschreibungsweise in der klassischen Physik und das Versagen des Kausalitätsprinzipes beim Erfassen atomarer Phänomene werden kurz erörtert. Grundsätzlich schliesst die Individualität der Quantenprozesse eine Trennung aus zwischen einem Verhalten atomarer Objekte und der Wechselwirkung mit den Messinstrumenten, welche die Bedingungen festlegen, unter denen die Phänomene erscheinen. Dieser Umstand zwingt uns zur Konzeption einer neuartigen Beziehung zwischen Beobachtungsergebnissen, die unter verschiedenen experimentellen Bedingungen gewonnen werden; diese Beziehung wird als Komplementarität bezeichnet. Ein angemessenes Mittel für eine komplementäre Beschreibungsweise liefert der quantenmechanische Formalismus, der es uns ermöglicht, Gesetzmässigkeiten jenseits des klassisch-physikalischer Erklärung zugänglichen Bereiches zu beschreiben. — N. B.

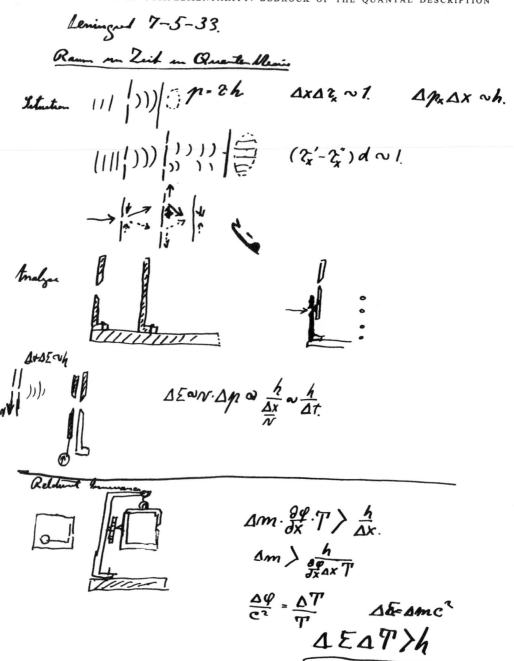

Facsimile, Leningrad, May 1934. (The date on the facsimile seems to be wrong. Bohr gave a lecture in Leningrad on 7 May 1934, whereas in May 1933 he was in Princeton, U.S.A.)

VI. DISCUSSION WITH EINSTEIN ON EPISTEMOLOGICAL PROBLEMS IN ATOMIC PHYSICS

"Albert Einstein: Philosopher–Scientist" (ed. P.A. Schilpp),
The Library of Living Philosophers, Vol. VII, Evanston, Illinois 1949,
pp. 201–241

See Introduction to Part II, sect. 4.

DISCUSSION WITH EINSTEIN ON EPISTEMOLOGICAL PROBLEMS IN ATOMIC PHYSICS (1949)

Versions published in English, German and Danish

English: Discussion with Einstein on epistemological problems in atomic physics
A   "Albert Einstein: Philosopher–Scientist" (ed. P.A. Schilpp), The Library of Living Philosophers, Vol. VII, Evanston, Illinois 1949, pp. 201–241
B   "Atomic Physics and Human Knowledge", John Wiley & Sons, New York 1958, pp. 32–66 (reprinted 1987 in: "Essays 1933–1957 on Atomic Physics and Human Knowledge, The Philosophical Writings of Niels Bohr, Vol. II", Ox Bow Press, Woodbridge, Connecticut 1987, pp. 32–66)

German: Diskussion mit Einstein über erkenntnistheoretische Probleme in der Atomphysik
C   "Albert Einstein als Philosoph und Naturforscher" (ed. P.A. Schilpp), W. Kohlhammer Verlag, Stuttgart 1955, pp. 115–150
D   "Atomphysik und menschliche Erkenntnis", Friedr. Vieweg & Sohn, Braunschweig 1958, pp. 32–67

Danish: Diskussion med Einstein om erkendelsesteoretiske Problemer i Atomfysikken
E   "Atomfysik og menneskelig erkendelse", J.H. Schultz Forlag, Copenhagen 1957, pp. 45–82

All these versions agree with each other.

7

DISCUSSION WITH EINSTEIN ON
EPISTEMOLOGICAL PROBLEMS
IN ATOMIC PHYSICS

WHEN invited by the Editor of the series, "Living Philosophers," to write an article for this volume in which
contemporary scientists are honouring the epoch-making contributions of Albert Einstein to the progress of natural philosophy and are acknowledging the indebtedness of our whole
generation for the guidance his genius has given us, I thought
much of the best way of explaining how much I owe to him for
inspiration. In this connection, the many occasions through the
years on which I had the privilege to discuss with Einstein
epistemological problems raised by the modern development of
atomic physics have come back vividly to my mind and I have
felt that I could hardly attempt anything better than to give
an account of these discussions which, even if no complete concord has so far been obtained, have been of greatest value and
stimulus to me. I hope also that the account may convey to
wider circles an impression of how essential the open-minded
exchange of ideas has been for the progress in a field where new
experience has time after time demanded a reconsideration of
our views.

————

From the very beginning the main point under debate has
been the attitude to take to the departure from customary principles of natural philosophy characteristic of the novel development of physics which was initiated in the first year of this century by Planck's discovery of the universal quantum of action.
This discovery, which revealed a feature of atomicity in the laws
of nature going far beyond the old doctrine of the limited divisibility of matter, has indeed taught us that the classical theories

201

of physics are idealizations which can be unambiguously applied only in the limit where all actions involved are large compared with the quantum. The question at issue has been whether the renunciation of a causal mode of description of atomic processes involved in the endeavours to cope with the situation should be regarded as a temporary departure from ideals to be ultimately revived or whether we are faced with an irrevocable step towards obtaining the proper harmony between analysis and synthesis of physical phenomena. To describe the background of our discussions and to bring out as clearly as possible the arguments for the contrasting viewpoints, I have felt it necessary to go to a certain length in recalling some main features of the development to which Einstein himself has contributed so decisively.

As is well known, it was the intimate relation, elucidated primarily by Boltzmann, between the laws of thermodynamics and the statistical regularities exhibited by mechanical systems with many degrees of freedom, which guided Planck in his ingenious treatment of the problem of thermal radiation, leading him to his fundamental discovery. While, in his work, Planck was principally concerned with considerations of essentially statistical character and with great caution refrained from definite conclusions as to the extent to which the existence of the quantum implied a departure from the foundations of mechanics and electrodynamics, Einstein's great original contribution to quantum theory (1905) was just the recognition of how physical phenomena like the photo-effect may depend directly on individual quantum effects.[1] In these very same years when, in developing his theory of relativity, Einstein laid a new foundation for physical science, he explored with a most daring spirit the novel features of atomicity which pointed beyond the whole framework of classical physics.

With unfailing intuition Einstein thus was led step by step to the conclusion that any radiation process involves the emission or absorption of individual light quanta or "photons" with energy and momentum

$$E = h\nu \quad \text{and} \quad P = h\sigma \tag{1}$$

[1] A. Einstein, *Ann. d. Phys.*, 17, 132, (1905).

respectively, where h is Planck's constant, while ν and σ are the number of vibrations per unit time and the number of waves per unit length, respectively. Notwithstanding its fertility, the idea of the photon implied a quite unforeseen dilemma, since any simple corpuscular picture of radiation would obviously be irreconcilable with interference effects, which present so essential an aspect of radiative phenomena, and which can be described only in terms of a wave picture. The acuteness of the dilemma is stressed by the fact that the interference effects offer our only means of defining the concepts of frequency and wavelength entering into the very expressions for the energy and momentum of the photon.

In this situation, there could be no question of attempting a causal analysis of radiative phenomena, but only, by a combined use of the contrasting pictures, to estimate probabilities for the occurrence of the individual radiation processes. However, it is most important to realize that the recourse to probability laws under such circumstances is essentially different in aim from the familiar application of statistical considerations as practical means of accounting for the properties of mechanical systems of great structural complexity. In fact, in quantum physics we are presented not with intricacies of this kind, but with the inability of the classical frame of concepts to comprise the peculiar feature of indivisibility, or "individuality," characterizing the elementary processes.

The failure of the theories of classical physics in accounting for atomic phenomena was further accentuated by the progress of our knowledge of the structure of atoms. Above all, Rutherford's discovery of the atomic nucleus (1911) revealed at once the inadequacy of classical mechanical and electromagnetic concepts to explain the inherent stability of the atom. Here again the quantum theory offered a clue for the elucidation of the situation and especially it was found possible to account for the atomic stability, as well as for the empirical laws governing the spectra of the elements, by assuming that any reaction of the atom resulting in a change of its energy involved a complete transition between two so-called stationary quantum states and that, in particular, the spectra were emitted by a step-like process in which each transition is accompanied by the emission of a monochromatic light quantum of an energy just equal to that of an Einstein photon.

These ideas, which were soon confirmed by the experiments of Franck and Hertz (1914) on the excitation of spectra by impact of electrons on atoms, involved a further renunciation of the causal mode of description, since evidently the interpretation of the spectral laws implies that an atom in an excited state in general will have the possibility of transitions with photon emission to one or another of its lower energy states. In fact, the very idea of stationary states is incompatible with any directive for the choice between such transitions and leaves room only for the notion of the relative probabilities of the individual transition processes. The only guide in estimating such probabilities was the so-called correspondence principle which originated in the search for the closest possible connection between the statistical account of atomic processes and the consequences to be expected from classical theory, which should be valid in the limit where the actions involved in all stages of the analysis of the phenomena are large compared with the universal quantum.

At that time, no general self-consistent quantum theory was yet in sight, but the prevailing attitude may perhaps be illustrated by the following passage from a lecture by the writer from 1913:[2]

I hope that I have expressed myself sufficiently clearly so that you may appreciate the extent to which these considerations conflict with the admirably consistent scheme of conceptions which has been rightly termed the classical theory of electrodynamics. On the other hand, I have tried to convey to you the impression that—just by emphasizing so strongly this conflict—it may also be possible in course of time to establish a certain coherence in the new ideas.

Important progress in the development of quantum theory was made by Einstein himself in his famous article on radiative equilibrium in 1917,[3] where he showed that Planck's law for thermal radiation could be simply deduced from assumptions

[2] N. Bohr, *Fysisk Tidsskrift*, *12*, 97, (1914). (English version in *The Theory of Spectra and Atomic Constitution*, Cambridge, University Press, 1922).
[3] A. Einstein, *Phys. Zs.*, *18*, 121, (1917).

conforming with the basic ideas of the quantum theory of atomic constitution. To this purpose, Einstein formulated general statistical rules regarding the occurrence of radiative transitions between stationary states, assuming not only that, when the atom is exposed to a radiation field, absorption as well as emission processes will occur with a probability per unit time proportional to the intensity of the irradiation, but that even in the absence of external disturbances spontaneous emission processes will take place with a rate corresponding to a certain *a priori* probability. Regarding the latter point, Einstein emphasized the fundamental character of the statistical description in a most suggestive way by drawing attention to the analogy between the assumptions regarding the occurrence of the spontaneous radiative transitions and the well-known laws governing transformations of radioactive substances.

In connection with a thorough examination of the exigencies of thermodynamics as regards radiation problems, Einstein stressed the dilemma still further by pointing out that the argumentation implied that any radiation process was "unidirected" in the sense that not only is a momentum corresponding to a photon with the direction of propagation transferred to an atom in the absorption process, but that also the emitting atom will receive an equivalent impulse in the opposite direction, although there can on the wave picture be no question of a preference for a single direction in an emission process. Einstein's own attitude to such startling conclusions is expressed in a passage at the end of the article (*loc. cit.*, p. 127 f.), which may be translated as follows:

These features of the elementary processes would seem to make the development of a proper quantum treatment of radiation almost unavoidable. The weakness of the theory lies in the fact that, on the one hand, no closer connection with the wave concepts is obtainable and that, on the other hand, it leaves to chance (*Zufall*) the time and the direction of the elementary processes; nevertheless, I have full confidence in the reliability of the way entered upon.

When I had the great experience of meeting Einstein for the first time during a visit to Berlin in 1920, these fundamental questions formed the theme of our conversations. The discussions, to which I have often reverted in my thoughts, added to all my admiration for Einstein a deep impression of his detached attitude. Certainly, his favoured use of such picturesque phrases as "ghost waves (*Gespensterfelder*) guiding the photons" implied no tendency to mysticism, but illuminated rather a profound humour behind his piercing remarks. Yet, a certain difference in attitude and outlook remained, since, with his mastery for co-ordinating apparently contrasting experience without abandoning continuity and causality, Einstein was perhaps more reluctant to renounce such ideals than someone for whom renunciation in this respect appeared to be the only way open to proceed with the immediate task of co-ordinating the multifarious evidence regarding atomic phenomena, which accumulated from day to day in the exploration of this new field of knowledge.

————

In the following years, during which the atomic problems attracted the attention of rapidly increasing circles of physicists, the apparent contradictions inherent in quantum theory were felt ever more acutely. Illustrative of this situation is the discussion raised by the discovery of the Stern-Gerlach effect in 1922. On the one hand, this effect gave striking support to the idea of stationary states and in particular to the quantum theory of the Zeeman effect developed by Sommerfeld; on the other hand, as exposed so clearly by Einstein and Ehrenfest,[4] it presented with unsurmountable difficulties any attempt at forming a picture of the behaviour of atoms in a magnetic field. Similar paradoxes were raised by the discovery by Compton (1924) of the change in wave-length accompanying the scattering of X-rays by electrons. This phenomenon afforded, as is well known, a most direct proof of the adequacy of Einstein's view regarding the transfer of energy and momentum in radiative processes; at the same time, it was equally clear that no simple picture of a corpuscular collision could offer an exhaustive description of the phenomenon. Under the impact of such difficulties, doubts

[4] A. Einstein and P. Ehrenfest, *Zs. f. Phys.*, *11*, 31, (1922).

were for a time entertained even regarding the conservation of energy and momentum in the individual radiation processes;[5] a view, however, which very soon had to be abandoned in face of more refined experiments bringing out the correlation between the deflection of the photon and the corresponding electron recoil.

The way to the clarification of the situation was, indeed, first to be paved by the development of a more comprehensive quantum theory. A first step towards this goal was the recognition by de Broglie in 1925 that the wave-corpuscle duality was not confined to the properties of radiation, but was equally unavoidable in accounting for the behaviour of material particles. This idea, which was soon convincingly confirmed by experiments on electron interference phenomena, was at once greeted by Einstein, who had already envisaged the deep-going analogy between the properties of thermal radiation and of gases in the so-called degenerate state.[6] The new line was pursued with the greatest success by Schrödinger (1926) who, in particular, showed how the stationary states of atomic systems could be represented by the proper solutions of a wave-equation to the establishment of which he was led by the formal analogy, originally traced by Hamilton, between mechanical and optical problems. Still, the paradoxical aspects of quantum theory were in no way ameliorated, but even emphasized, by the apparent contradiction between the exigencies of the general superposition principle of the wave description and the feature of individuality of the elementary atomic processes.

At the same time, Heisenberg (1925) had laid the foundation of a rational quantum mechanics, which was rapidly developed through important contributions by Born and Jordan as well as by Dirac. In this theory, a formalism is introduced, in which the kinematical and dynamical variables of classical mechanics are replaced by symbols subjected to a non-commutative algebra. Notwithstanding the renunciation of orbital pictures, Hamilton's canonical equations of mechanics are kept unaltered and

[5] N. Bohr, H. A. Kramers and J. C. Slater, *Phil. Mag.*, 47, 785, (1924).
[6] A. Einstein, *Berl. Ber.*, (1924), 261, and (1925), 3 and 18.

Planck's constant enters only in the rules of commutation

$$qp - pq = \sqrt{-1}\,\frac{h}{2\pi} \qquad (2)$$

holding for any set of conjugate variables q and p. Through a representation of the symbols by matrices with elements referring to transitions between stationary states, a quantitative formulation of the correspondence principle became for the first time possible. It may here be recalled that an important preliminary step towards this goal was reached through the establishment, especially by contributions of Kramers, of a quantum theory of dispersion making basic use of Einstein's general rules for the probability of the occurrence of absorption and emission processes.

This formalism of quantum mechanics was soon proved by Schrödinger to give results identical with those obtainable by the mathematically often more convenient methods of wave theory, and in the following years general methods were gradually established for an essentially statistical description of atomic processes combining the features of individuality and the requirements of the superposition principle, equally characteristic of quantum theory. Among the many advances in this period, it may especially be mentioned that the formalism proved capable of incorporating the exclusion principle which governs the states of systems with several electrons, and which already before the advent of quantum mechanics had been derived by Pauli from an analysis of atomic spectra. The quantitative comprehension of a vast amount of empirical evidence could leave no doubt as to the fertility and adequacy of the quantum-mechanical formalism, but its abstract character gave rise to a widespread feeling of uneasiness. An elucidation of the situation should, indeed, demand a thorough examination of the very observational problem in atomic physics.

This phase of the development was, as is well known, initiated in 1927 by Heisenberg,[1] who pointed out that the knowledge obtainable of the state of an atomic system will always involve a peculiar "indeterminacy." Thus, any measurement of the position of an electron by means of some device,

[1] W. Heisenberg, *Zs. f. Phys.*, 43, 172, (1927).

like a microscope, making use of high frequency radiation, will, according to the fundamental relations (1), be connected with a momentum exchange between the electron and the measuring agency, which is the greater the more accurate a position measurement is attempted. In comparing such considerations with the exigencies of the quantum-mechanical formalism, Heisenberg called attention to the fact that the commutation rule (2) imposes a reciprocal limitation on the fixation of two conjugate variables, q and p, expressed by the relation

$$\Delta q \cdot \Delta p \approx h, \tag{3}$$

where Δq and Δp are suitably defined latitudes in the determination of these variables. In pointing to the intimate connection between the statistical description in quantum mechanics and the actual possibilities of measurement, this so-called indeterminacy relation is, as Heisenberg showed, most important for the elucidation of the paradoxes involved in the attempts of analyzing quantum effects with reference to customary physical pictures.

The new progress in atomic physics was commented upon from various sides at the International Physical Congress held in September 1927, at Como in commemoration of Volta. In a lecture on that occasion,[8] I advocated a point of view conveniently termed "complementarity," suited to embrace the characteristic features of individuality of quantum phenomena, and at the same time to clarify the peculiar aspects of the observational problem in this field of experience. For this purpose, it is decisive to recognize that, *however far the phenomena transcend the scope of classical physical explanation, the account of all evidence must be expressed in classical terms.* The argument is simply that by the word "experiment" we refer to a situation where we can tell others what we have done and what we have learned and that, therefore, the account of the experimental arrangement and of the results of the observations must be expressed in unambiguous language with suitable application of the terminology of classical physics.

This crucial point, which was to become a main theme of the

[8] Atti del Congresso Internazionale dei Fisici, Como, Settembre 1927 (reprinted in *Nature*, *121*, 78 and 580, 1928).

[349]

discussions reported in the following, implies the *impossibility of any sharp separation between the behaviour of atomic objects and the interaction with the measuring instruments which serve to define the conditions under which the phenomena appear*. In fact, the individuality of the typical quantum effects finds its proper expression in the circumstance that any attempt of subdividing the phenomena will demand a change in the experimental arrangement introducing new possibilities of interaction between objects and measuring instruments which in principle cannot be controlled. Consequently, evidence obtained under different experimental conditions cannot be comprehended within a single picture, but must be regarded as *complementary* in the sense that only the totality of the phenomena exhausts the possible information about the objects.

Under these circumstances an essential element of ambiguity is involved in ascribing conventional physical attributes to atomic objects, as is at once evident in the dilemma regarding the corpuscular and wave properties of electrons and photons, where we have to do with contrasting pictures, each referring to an essential aspect of empirical evidence. An illustrative example, of how the apparent paradoxes are removed by an examination of the experimental conditions under which the complementary phenomena appear, is also given by the Compton effect, the consistent description of which at first had presented us with such acute difficulties. Thus, any arrangement suited to study the exchange of energy and momentum between the electron and the photon must involve a latitude in the space-time description of the interaction sufficient for the definition of wave-number and frequency which enter into the relation (1). Conversely, any attempt of locating the collision between the photon and the electron more accurately would, on account of the unavoidable interaction with the fixed scales and clocks defining the space-time reference frame, exclude all closer account as regards the balance of momentum and energy.

As stressed in the lecture, an adequate tool for a complementary way of description is offered precisely by the quantum-mechanical formalism which represents a purely symbolic scheme permitting only predictions, on lines of the correspondence principle, as to results obtainable under conditions specified by means of classical concepts. It must here be remembered that even in the indeterminacy relation (3) we are dealing with an implication of the formalism which defies unambiguous expression in words suited to describe classical physical pictures. Thus, a sentence like "we cannot know both the momentum and the position of an atomic object" raises at once questions as to the physical reality of two such attributes of the object, which can be answered only by referring to the conditions for the unambiguous use of space-time concepts, on the one hand, and dynamical conservation laws, on the other hand. While the combination of these concepts into a single picture of a causal chain of events is the essence of classical mechanics, room for regularities beyond the grasp of such a description is just afforded by the circumstance that the study of the complementary phenomena demands mutually exclusive experimental arrangements.

The necessity, in atomic physics, of a renewed examination of the foundation for the unambiguous use of elementary physical ideas recalls in some way the situation that led Einstein to his original revision on the basis of all application of space-time concepts which, by its emphasis on the primordial importance of the observational problem, has lent such unity to our world picture. Notwithstanding all novelty of approach, causal description is upheld in relativity theory within any given frame of reference, but in quantum theory the uncontrollable interaction between the objects and the measuring instruments forces us to a renunciation even in such respect. This recognition, however, in no way points to any limitation of the scope of the quantum-mechanical description, and the trend of the whole argumentation presented in the Como lecture was to show that the viewpoint of complementarity may be regarded as a rational generalization of the very ideal of causality.

––––––––

At the general discussion in Como, we all missed the presence of Einstein, but soon after, in October 1927, I had the opportunity to meet him in Brussels at the Fifth Physical Conference of the Solvay Institute, which was devoted to the theme "Electrons and Photons." At the Solvay meetings, Einstein had from their beginning been a most prominent figure, and several

of us came to the conference with great anticipations to learn
his reaction to the latest stage of the development which, to our
view, went far in clarifying the problems which he had himself
from the outset elicited so ingeniously. During the discussions,
where the whole subject was reviewed by contributions from
many sides and where also the arguments mentioned in the
preceding pages were again presented, Einstein expressed, however, a deep concern over the extent to which causal account in
space and time was abandoned in quantum mechanics.

To illustrate his attitude, Einstein referred at one of the sessions[9] to the simple example, illustrated by Fig. 1, of a particle
(electron or photon) penetrating through a hole or a narrow
slit in a diaphragm placed at some distance before a photographic plate. On account of the diffraction of the wave con-

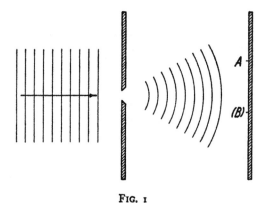

FIG. 1

nected with the motion of the particle and indicated in the figure
by the thin lines, it is under such conditions not possible to
predict with certainty at what point the electron will arrive at
the photographic plate, but only to calculate the probability
that, in an experiment, the electron will be found within any
given region of the plate. The apparent difficulty, in this description, which Einstein felt so acutely, is the fact that, if in the
experiment the electron is recorded at one point A of the plate,

[9] Institut International de Physique Solvay, *Rapport et discussions* du 5ᵉ Conseil, Paris 1928, 253ff.

then it is out of the question of ever observing an effect of this electron at another point (B), although the laws of ordinary wave propagation offer no room for a correlation between two such events.

Einstein's attitude gave rise to ardent discussions within a small circle, in which Ehrenfest, who through the years had been a close friend of us both, took part in a most active and helpful way. Surely, we all recognized that, in the above example, the situation presents no analogue to the application of statistics in dealing with complicated mechanical systems, but rather recalled the background for Einstein's own early conclusions about the unidirection of individual radiation effects which contrasts so strongly with a simple wave picture (cf. p. 205). The discussions, however, centered on the question of whether the quantum-mechanical description exhausted the possibilities of accounting for observable phenomena or, as Einstein maintained, the analysis could be carried further and, especially, of whether a fuller description of the phenomena could be obtained by bringing into consideration the detailed balance of energy and momentum in individual processes.

To explain the trend of Einstein's arguments, it may be illustrative here to consider some simple features of the momentum and energy balance in connection with the location of a particle in space and time. For this purpose, we shall examine the simple case of a particle penetrating through a hole in a diaphragm without or with a shutter to open and close the hole, as indicated in Figs. 2a and 2b, respectively. The equidistant parallel lines to the left in the figures indicate the train of plane waves corresponding to the state of motion of a particle which, before reaching the diaphragm, has a momentum P related to the wave-number σ by the second of equations (1). In accordance with the diffraction of the waves when passing through the hole, the state of motion of the particle to the right of the diaphragm is represented by a spherical wave train with a suitably defined angular aperture ϑ and, in case of Fig. 2b, also with a limited radial extension. Consequently, the description of this state involves a certain latitude Δp in the momentum component of the particle parallel to the diaphragm and, in the case of a diaphragm with a shutter, an additional latitude ΔE of the kinetic energy.

Since a measure for the latitude Δq in location of the particle in the plane of the diaphragm is given by the radius a of the hole, and since $\vartheta \approx (1/\sigma a)$, we get, using (1), just $\Delta p \approx \vartheta P \approx (h/\Delta q)$, in accordance with the indeterminacy relation (3). This result could, of course, also be obtained directly by noticing that, due to the limited extension of the wave-field at the place of the slit, the component of the wave-number parallel to the plane of the diaphragm will involve a latitude $\Delta \sigma \approx (1/a) \approx (1/\Delta q)$. Similarly, the spread of the frequencies

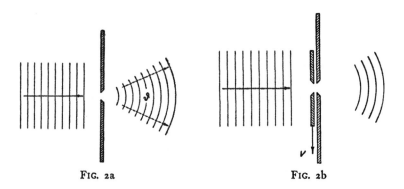

FIG. 2a   FIG. 2b

of the harmonic components in the limited wave-train in Fig. 2b is evidently $\Delta \nu \approx (1/\Delta t)$, where Δt is the time interval during which the shutter leaves the hole open and, thus, represents the latitude in time of the passage of the particle through the diaphragm. From (1), we therefore get

$$\Delta E \cdot \Delta t \approx h, \qquad (4)$$

again in accordance with the relation (3) for the two conjugated variables E and t.

From the point of view of the laws of conservation, the origin of such latitudes entering into the description of the state of the particle after passing through the hole may be traced to the possibilities of momentum and energy exchange with the diaphragm or the shutter. In the reference system considered in Figs. 2a and 2b, the velocity of the diaphragm may be disregarded and only a change of momentum Δp between the particle and the diaphragm needs to be taken into consideration. The shutter, however, which leaves the hole opened during the time Δt, moves with a considerable velocity $v \approx (a/\Delta t)$, and a momentum transfer Δp involves therefore an energy exchange with the particle, amounting to $v\Delta p \approx (1/\Delta t)\ \Delta q\ \Delta p \approx (h/\Delta t)$, being just of the same order of magnitude as the latitude ΔE given by (4) and, thus, allowing for momentum and energy balance.

The problem raised by Einstein was now to what extent a control of the momentum and energy transfer, involved in a location of the particle in space and time, can be used for a further specification of the state of the particle after passing through the hole. Here, it must be taken into consideration that the position and the motion of the diaphragm and the shutter have so far been assumed to be accurately co-ordinated with the space-time reference frame. This assumption implies, in the description of the state of these bodies, an essential latitude as to their momentum and energy which need not, of course, noticeably affect the velocities, if the diaphragm and the shutter are sufficiently heavy. However, as soon as we want to know the momentum and energy of these parts of the measuring arrangement with an accuracy sufficient to control the momentum and energy exchange with the particle under investigation, we shall, in accordance with the general indeterminacy relations, lose the possibility of their accurate location in space and time. We have, therefore, to examine how far this circumstance will affect the intended use of the whole arrangement and, as we shall see, this crucial point clearly brings out the complementary character of the phenomena.

Returning for a moment to the case of the simple arrangement indicated in Fig. 1, it has so far not been specified to what use it is intended. In fact, it is only on the assumption that the diaphragm and the plate have well-defined positions in space that it is impossible, within the frame of the quantum-mechanical formalism, to make more detailed predictions as to the point

of the photographic plate where the particle will be recorded.
If, however, we admit a sufficiently large latitude in the knowledge of the position of the diaphragm it should, in principle, be
possible to control the momentum transfer to the diaphragm
and, thus, to make more detailed predictions as to the direction
of the electron path from the hole to the recording point. As
regards the quantum-mechanical description, we have to deal
here with a two-body system consisting of the diaphragm as
well as of the particle, and it is just with an explicit application
of conservation laws to such a system that we are concerned in
the Compton effect where, for instance, the observation of the
recoil of the electron by means of a cloud chamber allows us to
predict in what direction the scattered photon will eventually
be observed.

The importance of considerations of this kind was, in the
course of the discussions, most interestingly illuminated by the
examination of an arrangement where between the diaphragm
with the slit and the photographic plate is inserted another

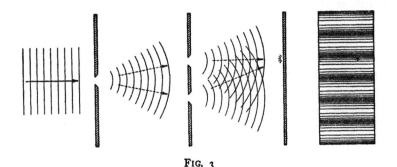

FIG. 3

diaphragm with two parallel slits, as is shown in Fig. 3. If a
parallel beam of electrons (or photons) falls from the left on
the first diaphragm, we shall, under usual conditions, observe on
the plate an interference pattern indicated by the shading of
the photographic plate shown in front view to the right of the
figure. With intense beams, this pattern is built up by the accumulation of a large number of individual processes, each
giving rise to a small spot on the photographic plate, and the
distribution of these spots follows a simple law derivable from

the wave analysis. The same distribution should also be found in the statistical account of many experiments performed with beams so faint that in a single exposure only one electron (or photon) will arrive at the photographic plate at some spot shown in the figure as a small star. Since, now, as indicated by the broken arrows, the momentum transferred to the first diaphragm ought to be different if the electron was assumed to pass through the upper or the lower slit in the second diaphragm, Einstein suggested that a control of the momentum transfer would permit a closer analysis of the phenomenon and, in particular, to decide through which of the two slits the electron had passed before arriving at the plate.

A closer examination showed, however, that the suggested control of the momentum transfer would involve a latitude in the knowledge of the position of the diaphragm which would exclude the appearance of the interference phenomena in question. In fact, if ω is the small angle between the conjectured paths of a particle passing through the upper or the lower slit, the difference of momentum transfer in these two cases will, according to (1), be equal to $h\sigma\omega$ and any control of the momentum of the diaphragm with an accuracy sufficient to measure this difference will, due to the indeterminacy relation, involve a minimum latitude of the position of the diaphragm, comparable with $1/\sigma\omega$. If, as in the figure, the diaphragm with the two slits is placed in the middle between the first diaphragm and the photographic plate, it will be seen that the number of fringes per unit length will be just equal to $\sigma\omega$ and, since an uncertainty in the position of the first diaphragm of the amount of $1/\sigma\omega$ will cause an equal uncertainty in the positions of the fringes, it follows that no interference effect can appear. The same result is easily shown to hold for any other placing of the second diaphragm between the first diaphragm and the plate, and would also be obtained if, instead of the first diaphragm, another of these three bodies were used for the control, for the purpose suggested, of the momentum transfer.

This point is of great logical consequence, since it is only the circumstance that we are presented with a choice of either tracing the path of a particle or observing interference effects, which

[357]

allows us to escape from the paradoxical necessity of concluding that the behaviour of an electron or a photon should depend on the presence of a slit in the diaphragm through which it could be proved not to pass. We have here to do with a typical example of how the complementary phenomena appear under mutually exclusive experimental arrangements (cf. p. 210) and are just faced with the impossibility, in the analysis of quantum effects, of drawing any sharp separation between an independent behaviour of atomic objects and their interaction with the measuring instruments which serve to define the conditions under which the phenomena occur.

Our talks about the attitude to be taken in face of a novel situation as regards analysis and synthesis of experience touched naturally on many aspects of philosophical thinking, but, in spite of all divergencies of approach and opinion, a most humorous spirit animated the discussions. On his side, Einstein mockingly asked us whether we could really believe that the providential authorities took recourse to dice-playing (". . . *ob der liebe Gott würfelt*"), to which I replied by pointing at the great caution, already called for by ancient thinkers, in ascribing attributes to Providence in every-day language. I remember also how at the peak of the discussion Ehrenfest, in his affectionate manner of teasing his friends, jokingly hinted at the apparent similarity between Einstein's attitude and that of the opponents of relativity theory; but instantly Ehrenfest added that he would not be able to find relief in his own mind before concord with Einstein was reached.

———

Einstein's concern and criticism provided a most valuable incentive for us all to reexamine the various aspects of the situation as regards the description of atomic phenomena. To me it was a welcome stimulus to clarify still further the rôle played by the measuring instruments and, in order to bring into strong relief the mutually exclusive character of the experimental conditions under which the complementary phenomena appear, I tried in those days to sketch various apparatus in a pseudorealistic style of which the following figures are examples. Thus, for the study of an interference phenomenon of the type indicated in Fig. 3, it suggests itself to use an experimental arrangement like that shown in Fig. 4, where the solid parts of the apparatus, serving as diaphragms and plate-holder, are

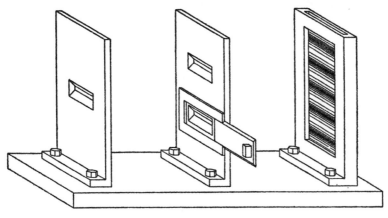

FIG. 4

firmly bolted to a common support. In such an arrangement, where the knowledge of the relative positions of the diaphragms and the photographic plate is secured by a rigid connection, it is obviously impossible to control the momentum exchanged between the particle and the separate parts of the apparatus. The only way in which, in such an arrangement, we could insure that the particle passed through one of the slits in the second diaphragm is to cover the other slit by a lid, as indicated in the figure; but if the slit is covered, there is of course no question of any interference phenomenon, and on the plate we shall simply observe a continuous distribution as in the case of the single fixed diaphragm in Fig. 1.

In the study of phenomena in the account of which we are dealing with detailed momentum balance, certain parts of the whole device must naturally be given the freedom to move independently of others. Such an apparatus is sketched in Fig. 5, where a diaphragm with a slit is suspended by weak springs from a solid yoke bolted to the support on which also other immobile parts of the arrangement are to be fastened. The scale on the diaphragm together with the pointer on the bearings of the yoke refer to such study of the motion of the diaphragm, as may be required for an estimate of the momentum transferred to it, permitting one to draw conclusions as to the deflection suffered by the particle in passing through the slit. Since, however, any reading of the scale, in whatever way performed, will

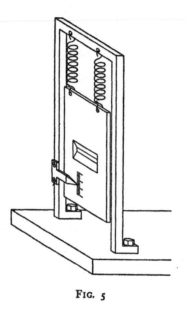

FIG. 5

involve an uncontrollable change in the momentum of the diaphragm, there will always be, in conformity with the indeterminacy principle, a reciprocal relationship between our knowledge of the position of the slit and the accuracy of the momentum control.

In the same semi-serious style, Fig. 6 represents a part of an arrangement suited for the study of phenomena which, in contrast to those just discussed, involve time co-ordination explicitly. It consists of a shutter rigidly connected with a robust clock resting on the support which carries a diaphragm and on which further parts of similar character, regulated by the same clock-work or by other clocks standardized relatively to it, are also to be fixed. The special aim of the figure is to underline that a clock is a piece of machinery, the working of which can completely be accounted for by ordinary mechanics and will be affected neither by reading of the position of its hands nor by the interaction between its accessories and an atomic particle. In securing the opening of the hole at a definite moment, an apparatus of this type might, for instance, be used for an accurate measurement of the time an electron or a photon takes to come from the diaphragm to some other place, but evidently, it would leave no possibility of controlling the energy transfer to

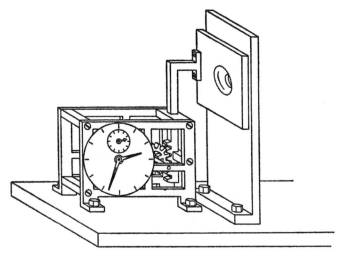

FIG. 6

the shutter with the aim of drawing conclusions as to the energy of the particle which has passed through the diaphragm. If we are interested in such conclusions we must, of course, use an arrangement where the shutter devices can no longer serve as accurate clocks, but where the knowledge of the moment when the hole in the diaphragm is open involves a latitude connected with the accuracy of the energy measurement by the general relation (4).

The contemplation of such more or less practical arrangements and their more or less fictitious use proved most instructive in directing attention to essential features of the problems. The main point here is the distinction between the *objects* under investigation and the *measuring instruments* which serve to define, in classical terms, the conditions under which the phenomena appear. Incidentally, we may remark that, for the illustration of the preceding considerations, it is not relevant that experiments involving an accurate control of the momentum or energy transfer from atomic particles to heavy bodies like diaphragms and shutters would be very difficult to perform, if practicable at all. It is only decisive that, in contrast to the proper measuring instruments, these bodies together with the particles would in such a case constitute the system to which the quantum-mechanical formalism has to be applied. As regards the specification of the conditions for any well-defined application of the formalism, it is moreover essential that the *whole experimental arrangement* be taken into account. In fact, the introduction of any further piece of apparatus, like a mirror, in the way of a particle might imply new interference effects essentially influencing the predictions as regards the results to be eventually recorded.

The extent to which renunciation of the visualization of atomic phenomena is imposed upon us by the impossibility of their subdivision is strikingly illustrated by the following example to which Einstein very early called attention and often has reverted. If a semi-reflecting mirror is placed in the way of a photon, leaving two possibilities for its direction of propagation, the photon may either be recorded on one, and only one, of two photographic plates situated at great distances in the two directions in question, or else we may, by replacing the plates by mirrors, observe effects exhibiting an interference between the two reflected wave-trains. In any attempt of a pictorial representation of the behaviour of the photon we would, thus, meet with the difficulty: to be obliged to say, on the one hand, that the photon always chooses *one* of the two ways and, on the other hand, that it behaves as if it had passed *both* ways.

It is just arguments of this kind which recall the impossibility of subdividing quantum phenomena and reveal the ambiguity in ascribing customary physical attributes to atomic objects. In particular, it must be realized that—besides in the account of the placing and timing of the instruments forming the experimental arrangement—all unambiguous use of space-time concepts in the description of atomic phenomena is confined to the recording of observations which refer to marks on a photographic plate or to similar practically irreversible amplification effects like the building of a water drop around an ion in a cloud-chamber. Although, of course, the existence of the quantum of action is ultimately responsible for the properties of the materials of which the measuring instruments are built and on which the functioning of the recording devices depends, this circumstance is not relevant for the problems of the adequacy and completeness of the quantum-mechanical description in its aspects here discussed.

These problems were instructively commented upon from different sides at the Solvay meeting,[10] in the same session where Einstein raised his general objections. On that occasion an interesting discussion arose also about how to speak of the appearance of phenomena for which only predictions of statistical character can be made. The question was whether, as to the occurrence of individual effects, we should adopt a terminology proposed by Dirac, that we were concerned with a choice on the part of "nature" or, as suggested by Heisenberg, we should say that we have to do with a choice on the part of the "observer" constructing the measuring instruments and reading their recording. Any such terminology would, however, appear dubious since, on the one hand, it is hardly reasonable to endow nature with volition in the ordinary sense, while, on the other hand, it is certainly not possible for the observer to influence the events which may appear under the conditions he has arranged. To my mind, there is no other alternative than to admit that, in this field of experience, we are dealing with individual phenomena and that our possibilities of handling the measuring instruments allow us only to make a choice between the different complementary types of phenomena we want to study.

The epistemological problems touched upon here were more explicitly dealt with in my contribution to the issue of *Naturwissenschaften* in celebration of Planck's 70th birthday in 1929. In this article, a comparison was also made between the lesson derived from the discovery of the universal quantum of action

[10] *Ibid.*, 248ff.

and the development which has followed the discovery of the finite velocity of light and which, through Einstein's pioneer work, has so greatly clarified basic principles of natural philosophy. In relativity theory, the emphasis on the dependence of all phenomena on the reference frame opened quite new ways of tracing general physical laws of unparalleled scope. In quantum theory, it was argued, the logical comprehension of hitherto unsuspected fundamental regularities governing atomic phenomena has demanded the recognition that no sharp separation can be made between an independent behaviour of the objects and their interaction with the measuring instruments which define the reference frame.

In this respect, quantum theory presents us with a novel situation in physical science, but attention was called to the very close analogy with the situation as regards analysis and synthesis of experience, which we meet in many other fields of human knowledge and interest. As is well known, many of the difficulties in psychology originate in the different placing of the separation lines between object and subject in the analysis of various aspects of psychical experience. Actually, words like "thoughts" and "sentiments," equally indispensable to illustrate the variety and scope of conscious life, are used in a similar complementary way as are space-time co-ordination and dynamical conservation laws in atomic physics. A precise formulation of such analogies involves, of course, intricacies of terminology, and the writer's position is perhaps best indicated in a passage in the article, hinting at the mutually exclusive relationship which will always exist between the practical use of any word and attempts at its strict definition. The principal aim, however, of these considerations, which were not least inspired by the hope of influencing Einstein's attitude, was to point to perspectives of bringing general epistemological problems into relief by means of a lesson derived from the study of new, but fundamentally simple physical experience.

At the next meeting with Einstein at the Solvay Conference in 1930, our discussions took quite a dramatic turn. As an objection to the view that a control of the interchange of momentum and energy between the objects and the measuring instruments was excluded if these instruments should serve their purpose of defining the space-time frame of the phenomena, Einstein brought forward the argument that such control should be possible when the exigencies of relativity theory were taken into consideration. In particular, the general relationship between energy and mass, expressed in Einstein's famous formula

$$E = mc^2 \qquad\qquad (5)$$

should allow, by means of simple weighing, to measure the total energy of any system and, thus, in principle to control the energy transferred to it when it interacts with an atomic object.

As an arrangement suited for such purpose, Einstein proposed the device indicated in Fig. 7, consisting of a box with

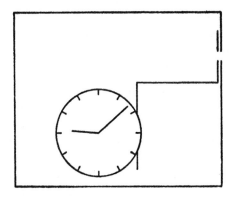

FIG. 7

a hole in its side, which could be opened or closed by a shutter moved by means of a clock-work within the box. If, in the beginning, the box contained a certain amount of radiation and the clock was set to open the shutter for a very short interval at a chosen time, it could be achieved that a single photon was released through the hole at a moment known with as great accuracy as desired. Moreover, it would apparently also be possible, by weighing the whole box before and after this event, to measure the energy of the photon with any accuracy wanted, in definite contradiction to the reciprocal indeterminacy of time and energy quantities in quantum mechanics.

This argument amounted to a serious challenge and gave rise to a thorough examination of the whole problem. At the outcome of the discussion, to which Einstein himself contributed effectively, it became clear, however, that this argument could not be upheld. In fact, in the consideration of the problem, it was found necessary to look closer into the consequences of the identification of inertial and gravitational mass implied in the application of relation (5). Especially, it was essential to take into account the relationship between the rate of a clock and its position in a gravitational field—well known from the red-shift of the lines in the sun's spectrum—following from Einstein's principle of equivalence between gravity effects and the phenomena observed in accelerated reference frames.

Our discussion concentrated on the possible application of an apparatus incorporating Einstein's device and drawn in Fig. 8 in the same pseudo-realistic style as some of the preceding figures. The box, of which a section is shown in order to exhibit its interior, is suspended in a spring-balance and is furnished with a pointer to read its position on a scale fixed to the balance support. The weighing of the box may thus be performed with any given accuracy Δm by adjusting the balance to its zero position by means of suitable loads. The essential point is now that any determination of this position with a given accuracy Δq will involve a minimum latitude Δp in the control of the momentum of the box connected with Δq by the relation (3). This latitude must obviously again be smaller than the total impulse which, during the whole interval T of the balancing procedure, can be given by the gravitational field to a body with a mass Δm, or

$$\Delta p \approx \frac{h}{\Delta q} < T \cdot g \cdot \Delta m, \qquad (6)$$

where g is the gravity constant. The greater the accuracy of the reading q of the pointer, the longer must, consequently, be the balancing interval T, if a given accuracy Δm of the weighing of the box with its content shall be obtained.

Now, according to general relativity theory, a clock, when displaced in the direction of the gravitational force by an amount of Δq, will change its rate in such a way that its reading

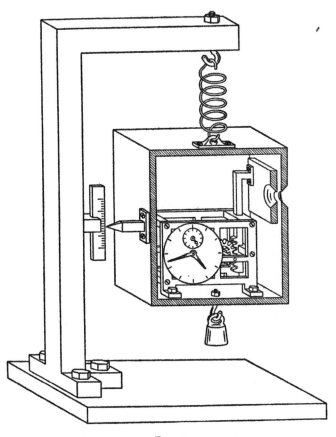

Fig. 8

in the course of a time interval T will differ by an amount ΔT given by the relation

$$\frac{\Delta T}{T} = \frac{1}{c^2}\, g\Delta q. \qquad (7)$$

By comparing (6) and (7) we see, therefore, that after the weighing procedure there will in our knowledge of the adjustment of the clock be a latitude

$$\Delta T > \frac{h}{c^2 \Delta m} \, .$$

Together with the formula (5), this relation again leads to

$$\Delta T \cdot \Delta E > h,$$

in accordance with the indeterminacy principle. Consequently, a use of the apparatus as a means of accurately measuring the energy of the photon will prevent us from controlling the moment of its escape.

The discussion, so illustrative of the power and consistency of relativistic arguments, thus emphasized once more the necessity of distinguishing, in the study of atomic phenomena, between the proper measuring instruments which serve to define the reference frame and those parts which are to be regarded as objects under investigation and in the account of which quantum effects cannot be disregarded. Notwithstanding the most suggestive confirmation of the soundness and wide scope of the quantum-mechanical way of description, Einstein nevertheless, in a following conversation with me, expressed a feeling of disquietude as regards the apparent lack of firmly laid down principles for the explanation of nature, in which all could agree. From my viewpoint, however, I could only answer that, in dealing with the task of bringing order into an entirely new field of experience, we could hardly trust in any accustomed principles, however broad, apart from the demand of avoiding logical inconsistencies and, in this respect, the mathematical formalism of quantum mechanics should surely meet all requirements.

The Solvay meeting in 1930 was the last occasion where, in common discussions with Einstein, we could benefit from the stimulating and mediating influence of Ehrenfest, but shortly before his deeply deplored death in 1933 he told me that Einstein was far from satisfied and with his usual acuteness had discerned new aspects of the situation which strengthened his critical attitude. In fact, by further examining the possibilities for the application of a balance arrangement, Einstein had perceived alternative procedures which, even if they did not allow the use he originally intended, might seem to enhance the paradoxes beyond the possibilities of logical solution. Thus, Einstein had pointed out that, after a preliminary weighing of the box with the clock and the subsequent escape of the photon, one was still left with the choice of either repeating the weighing or opening the box and comparing the reading of the clock with the standard time scale. Consequently, we are at this stage still free to choose whether we want to draw conclusions either about the energy of the photon or about the moment when it left the box. Without in any way interfering with the photon between its escape and its later interaction with other suitable measuring instruments, we are, thus, able to make accurate predictions pertaining *either* to the moment of its arrival *or* to the amount of energy liberated by its absorption. Since, however, according to the quantum-mechanical formalism, the specification of the state of an isolated particle cannot involve both a well-defined connection with the time scale and an accurate fixation of the energy, it might thus appear as if this formalism did not offer the means of an adequate description.

Once more Einstein's searching spirit had elicited a peculiar aspect of the situation in quantum theory, which in a most striking manner illustrated how far we have here transcended customary explanation of natural phenomena. Still, I could not agree with the trend of his remarks as reported by Ehrenfest. In my opinion, there could be no other way to deem a logically consistent mathematical formalism as inadequate than by demonstrating the departure of its consequences from experience or by proving that its predictions did not exhaust the possibilities of observation, and Einstein's argumentation could be directed to neither of these ends. In fact, we must realize that in the problem in question we are not dealing with a *single* specified experimental arrangement, but are referring to *two* different, mutually exclusive arrangements. In the one, the balance together with another piece of apparatus like a spectrometer is used for the study of the energy transfer by a photon; in the other, a shutter regulated by a standardized clock together with another apparatus of similar kind, accurately timed relatively to the clock, is used for the study of the time of propagation of a photon over a given distance. In both these cases, as also assumed by Einstein, the observable effects are expected to be in complete conformity with the predictions of the theory.

The problem again emphasizes the necessity of considering the *whole* experimental arrangement, the specification of which is imperative for any well-defined application of the quantum-mechanical formalism. Incidentally, it may be added that paradoxes of the kind contemplated by Einstein are encountered also in such simple arrangements as sketched in Fig. 5. In fact, after a preliminary measurement of the momentum of the diaphragm, we are in principle offered the choice, when an electron or photon has passed through the slit, either to repeat the momentum measurement or to control the position of the diaphragm and, thus, to make predictions pertaining to alternative subsequent observations. It may also be added that it obviously can make no difference as regards observable effects obtainable by a definite experimental arrangement, whether our plans of constructing or handling the instruments are fixed beforehand or whether we prefer to postpone the completion of our planning until a later moment when the particle is already on its way from one instrument to another.

In the quantum-mechanical description our freedom of constructing and handling the experimental arrangement finds its proper expression in the possibility of choosing the classically defined parameters entering in any proper application of the formalism. Indeed, in all such respects quantum mechanics exhibits a correspondence with the state of affairs familiar from classical physics, which is as close as possible when considering the individuality inherent in the quantum phenomena. Just in helping to bring out this point so clearly, Einstein's concern had therefore again been a most welcome incitement to explore the essential aspects of the situation.

The next Solvay meeting in 1933 was devoted to the problems of the structure and properties of atomic nuclei, in which field such great advances were made just in that period due to the experimental discoveries as well as to new fruitful applications of quantum mechanics. It need in this connection hardly be recalled that just the evidence obtained by the study of artificial nuclear transformations gave a most direct test of Einstein's fundamental law regarding the equivalence of mass and energy, which was to prove an evermore important guide for researches in nuclear physics. It may also be mentioned how Einstein's intuitive recognition of the intimate relationship between the law of radioactive transformations and the probability rules governing individual radiation effects (cf. p. 205) was confirmed by the quantum-mechanical explanation of spontaneous nuclear disintegrations. In fact, we are here dealing with a typical example of the statistical mode of description, and the complementary relationship between energy-momentum conservation and time-space co-ordination is most strikingly exhibited in the well-known paradox of particle penetration through potential barriers.

Einstein himself did not attend this meeting, which took place at a time darkened by the tragic developments in the political world which were to influence his fate so deeply and add so greatly to his burdens in the service of humanity. A few months earlier, on a visit to Princeton where Einstein was then guest of the newly founded Institute for Advanced Study to which he soon after became permanently attached, I had, however, opportunity to talk with him again about the epistemological aspects of atomic physics, but the difference between our ways of approach and expression still presented obstacles to mutual understanding. While, so far, relatively few persons had taken part in the discussions reported in this article, Einstein's critical attitude towards the views on quantum theory adhered to by many physicists was soon after brought to public attention through a paper[11] with the title "Can Quantum-Mechanical Description of Physical Reality Be Considered Complete?," published in 1935 by Einstein, Podolsky and Rosen.

The argumentation in this paper is based on a criterion which the authors express in the following sentence: "If, without in any way disturbing a system, we can predict with certainty (i.e., with probability equal to unity) the value of a physical quantity, then there exists an element of physical reality correspond-

---

[11] A. Einstein, B. Podolsky and N. Rosen, *Phys. Rev.*, 47, 777, (1935).

ing to this physical quantity." By an elegant exposition of the consequences of the quantum-mechanical formalism as regards the representation of a state of a system, consisting of two parts which have been in interaction for a limited time interval, it is next shown that different quantities, the fixation of which cannot be combined in the representation of one of the partial systems, can nevertheless be predicted by measurements pertaining to the other partial system. According to their criterion, the authors therefore conclude that quantum mechanics does not "provide a complete description of the physical reality," and they express their belief that it should be possible to develop a more adequate account of the phenomena.

Due to the lucidity and apparently incontestable character of the argument, the paper of Einstein, Podolsky and Rosen created a stir among physicists and has played a large rôle in general philosophical discussion. Certainly the issue is of a very subtle character and suited to emphasize how far, in quantum theory, we are beyond the reach of pictorial visualization. It will be seen, however, that we are here dealing with problems of just the same kind as those raised by Einstein in previous discussions, and, in an article which appeared a few months later,[12] I tried to show that from the point of view of complementarity the apparent inconsistencies were completely removed. The trend of the argumentation was in substance the same as that exposed in the foregoing pages, but the aim of recalling the way in which the situation was discussed at that time may be an apology for citing certain passages from my article.

Thus, after referring to the conclusions derived by Einstein, Podolsky and Rosen on the basis of their criterion, I wrote:

Such an argumentation, however, would hardly seem suited to affect the soundness of quantum-mechanical description, which is based on a coherent mathematical formalism covering automatically any procedure of measurement like that indicated. The apparent contradiction in fact discloses only an essential inadequacy of the customary viewpoint of natural philosophy for a rational account of physical phenomena of the type with which we are concerned in quantum mechanics. Indeed the *finite interaction between object and measuring agencies* conditioned

[12] N. Bohr, *Phys. Rev.*, *48*, 696, (1935).

by the very existence of the quantum of action entails—because of the impossibility of controlling the reaction of the object on the measuring instruments, if these are to serve their purpose—the necessity of a final renunciation of the classical ideal of causality and a radical revision of our attitude towards the problem of physical reality. In fact, as we shall see, a criterion of reality like that proposed by the named authors contains—however cautious its formulation may appear—an essential ambiguity when it is applied to the actual problems with which we are here concerned.

As regards the special problem treated by Einstein, Podolsky and Rosen, it was next shown that the consequences of the formalism as regards the representation of the state of a system consisting of two interacting atomic objects correspond to the simple arguments mentioned in the preceding in connection with the discussion of the experimental arrangements suited for the study of complementary phenomena. In fact, although any pair q and p, of conjugate space and momentum variables obeys the rule of non-commutative multiplication expressed by (2), and can thus only be fixed with reciprocal latitudes given by (3), the difference $q_1 - q_2$ between two space-co-ordinates referring to the constituents of the system will commute with the sum $p_1 + p_2$ of the corresponding momentum components, as follows directly from the commutability of q_1 with p_2 and q_2 with p_1. Both $q_1 - q_2$ and $p_1 + p_2$ can, therefore, be accurately fixed in a state of the complex system and, consequently, we can predict the values of either q_1 or p_1 if either q_2 or p_2, respectively, are determined by direct measurements. If, for the two parts of the system, we take a particle and a diaphragm, like that sketched in Fig. 5, we see that the possibilities of specifying the state of the particle by measurements on the diaphragm just correspond to the situation described on p. 220 and further discussed on p. 230, where it was mentioned that, after the particle has passed through the diaphragm, we have in principle the choice of measuring either the position of the diaphragm or its momentum and, in each case, to make predictions as to subsequent observations pertaining to the particle. As repeatedly stressed, the principal point is here that such measurements demand mutually exclusive experimental arrangements.

The argumentation of the article was summarized in the following passage:

From our point of view we now see that the wording of the above-mentioned criterion of physical reality proposed by Einstein, Podolsky, and Rosen contains an ambiguity as regards the meaning of the expression 'without in any way disturbing a system.' Of course there is in a case like that just considered no question of a mechanical disturbance of the system under investigation during the last critical stage of the measuring procedure. But even at this stage there is essentially the question of *an influence on the very conditions which define the possible types of predictions regarding the future behaviour of the system.* Since these conditions constitute an inherent element of the description of any phenomenon to which the term "physical reality" can be properly attached, we see that the argumentation of the mentioned authors does not justify their conclusion that quantum-mechanical description is essentially incomplete. On the contrary, this description, as appears from the preceding discussion, may be characterized as a rational utilization of all possibilities of unambiguous interpretation of measurements, compatible with the finite and uncontrollable interaction between the objects and the measuring instruments in the field of quantum theory. In fact, it is only the mutual exclusion of any two experimental procedures, permitting the unambiguous definition of complementary physical quantities, which provides room for new physical laws, the coexistence of which might at first sight appear irreconcilable with the basic principles of science. It is just this entirely new situation as regards the description of physical phenomena that the notion of *complementarity* aims at characterizing.

Rereading these passages, I am deeply aware of the inefficiency of expression which must have made it very difficult to appreciate the trend of the argumentation aiming to bring out the essential ambiguity involved in a reference to physical attributes of objects when dealing with phenomena where no sharp distinction can be made between the behaviour of the objects themselves and their interaction with the measuring instruments. I hope, however, that the present account of the discussions with Einstein in the foregoing years, which contributed so greatly to make us familiar with the situation in quantum physics, may give a clearer impression of the necessity of a radical revision of basic principles for physical explanation in order to restore logical order in this field of experience.

Einstein's own views at that time are presented in an article "Physics and Reality," published in 1936 in the *Journal of the Franklin Institute*.[13]." Starting from a most illuminating exposition of the gradual development of the fundamental principles in the theories of classical physics and their relation to the problem of physical reality, Einstein here argues that the quantum-mechanical description is to be considered merely as a means of accounting for the average behaviour of a large number of atomic systems and his attitude to the belief that it should offer an exhaustive description of the individual phenomena is expressed in the following words: "To believe this is logically possible without contradiction; but it is so very contrary to my scientific instinct that I cannot forego the search for a more complete conception."

Even if such an attitude might seem well-balanced in itself, it nevertheless implies a rejection of the whole argumentation exposed in the preceding, aiming to show that, in quantum mechanics, we are not dealing with an arbitrary renunciation of a more detailed analysis of atomic phenomena, but with a recognition that such an analysis is *in principle* excluded. The peculiar individuality of the quantum effects presents us, as regards the comprehension of well-defined evidence, with a novel situation unforeseen in classical physics and irreconcilable with conventional ideas suited for our orientation and adjustment to ordinary experience. It is in this respect that quantum theory has called for a renewed revision of the foundation for the unambiguous use of elementary concepts, as a further step in the development which, since the advent of relativity theory, has been so characteristic of modern science.

In the following years, the more philosophical aspects of the situation in atomic physics aroused the interest of ever larger circles and were, in particular, discussed at the Second International Congress for the Unity of Science in Copenhagen in July 1936. In a lecture on this occasion,[14] I tried especially to

[13] A. Einstein, *Journ. Frankl. Inst.*, *221*, 349, (1936).

[14] N. Bohr, *Erkenntnis*, *6*, 293, (1937), and *Philosophy of Science*, *4*, 289, (1937).

stress the analogy in epistemological respects between the limitation imposed on the causal description in atomic physics and situations met with in other fields of knowledge. A principal purpose of such parallels was to call attention to the necessity in many domains of general human interest to face problems of a similar kind as those which had arisen in quantum theory and thereby to give a more familiar background for the apparently extravagant way of expression which physicists have developed to cope with their acute difficulties.

Besides the complementary features conspicuous in psychology and already touched upon (cf. p. 224), examples of such relationships can also be traced in biology, especially as regards the comparison between mechanistic and vitalistic viewpoints. Just with respect to the observational problem, this last question had previously been the subject of an address to the International Congress on Light Therapy held in Copenhagen in 1932,[15] where it was incidentally pointed out that even the psycho-physical parallelism as envisaged by Leibniz and Spinoza has obtained a wider scope through the development of atomic physics, which forces us to an attitude towards the problem of explanation recalling ancient wisdom, that when searching for harmony in life one must never forget that in the drama of existence we are ourselves both actors and spectators.

Utterances of this kind would naturally in many minds evoke the impression of an underlying mysticism foreign to the spirit of science; at the above mentioned Congress in 1936 I therefore tried to clear up such misunderstandings and to explain that the only question was an endeavour to clarify the conditions, in each field of knowledge, for the analysis and synthesis of experience.[14] Yet, I am afraid that I had in this respect only little success in convincing my listeners, for whom the dissent among the physicists themselves was naturally a cause of scepticism as to the necessity of going so far in renouncing customary demands as regards the explanation of natural phenomena. Not least through a new discussion with Einstein in Princeton in 1937, where we did not get beyond a humourous contest con-

---

[15] II° Congrès international de la Lumière, Copenhague 1932 (reprinted in *Nature*, *131*, 421 and 457, 1933).

cerning which side Spinoza would have taken if he had lived
to see the development of our days, I was strongly reminded of
the importance of utmost caution in all questions of terminology
and dialectics.

These aspects of the situation were especially discussed at a
meeting in Warsaw in 1938, arranged by the International Institute of Intellectual Co-operation of the League of Nations.[16]
The preceding years had seen great progress in quantum physics due to a number of fundamental discoveries regarding the
constitution and properties of atomic nuclei as well as due to
important developments of the mathematical formalism taking
the requirements of relativity theory into account. In the last
respect, Dirac's ingenious quantum theory of the electron offered a most striking illustration of the power and fertility of
the general quantum-mechanical way of description. In the phenomena of creation and annihilation of electron pairs we have
in fact to do with new fundamental features of atomicity, which
are intimately connected with the non-classical aspects of quantum statistics expressed in the exclusion principle, and which
have demanded a still more far-reaching renunciation of explanation in terms of a pictorial representation.

Meanwhile, the discussion of the epistemological problems
in atomic physics attracted as much attention as ever and, in
commenting on Einstein's views as regards the incompleteness
of the quantum-mechanical mode of description, I entered more
directly on questions of terminology. In this connection I
warned especially against phrases, often found in the physical
literature, such as "disturbing of phenomena by observation" or
"creating physical attributes to atomic objects by measurements." Such phrases, which may serve to remind of the apparent paradoxes in quantum theory, are at the same time apt
to cause confusion, since words like "phenomena" and "observations," just as "attributes" and "measurements," are used in
a way hardly compatible with common language and practical
definition.

As a more appropriate way of expression I advocated the ap-

[16] *New Theories in Physics* (Paris 1938), 11.

plication of the word *phenomenon* exclusively to refer to the observations obtained under specified circumstances, including an account of the whole experimental arrangement. In such terminology, the observational problem is free of any special intricacy since, in actual experiments, all observations are expressed by unambiguous statements referring, for instance, to the registration of the point at which an electron arrives at a photographic plate. Moreover, speaking in such a way is just suited to emphasize that the appropriate physical interpretation of the symbolic quantum-mechanical formalism amounts only to predictions, of determinate or statistical character, pertaining to individual phenomena appearing under conditions defined by classical physical concepts.

Notwithstanding all differences between the physical problems which have given rise to the development of relativity theory and quantum theory, respectively, a comparison of purely logical aspects of relativistic and complementary argumentation reveals striking similarities as regards the renunciation of the absolute significance of conventional physical attributes of objects. Also, the neglect of the atomic constitution of the measuring instruments themselves, in the account of actual experience, is equally characteristic of the applications of relativity and quantum theory. Thus, the smallness of the quantum of action compared with the actions involved in usual experience, including the arranging and handling of physical apparatus, is as essential in atomic physics as is the enormous number of atoms composing the world in the general theory of relativity which, as often pointed out, demands that dimensions of apparatus for measuring angles can be made small compared with the radius of curvature of space.

In the Warsaw lecture, I commented upon the use of not directly visualizable symbolism in relativity and quantum theory in the following way:

Even the formalisms, which in both theories within their scope offer adequate means of comprehending all conceivable experience, exhibit deepgoing analogies. In fact, the astounding simplicity of the generalization of classical physical theories, which are obtained by the use of multidimensional geometry and non-commutative algebra, respectively, rests in both cases essentially on the introduction of the conventional symbol $\sqrt{-1}$. The abstract character of the formalisms concerned is indeed, on closer examination, as typical of relativity theory as it is of quantum mechanics, and it is in this respect purely a matter of tradition if the former theory is considered as a completion of classical physics rather than as a first fundamental step in the thoroughgoing revision of our conceptual means of comparing observations, which the modern development of physics has forced upon us.

It is, of course, true that in atomic physics we are confronted with a number of unsolved fundamental problems, especially as regards the intimate relationship between the elementary unit of electric charge and the universal quantum of action; but these problems are no more connected with the epistemological points here discussed than is the adequacy of relativistic argumentation with the issue of thus far unsolved problems of cosmology. Both in relativity and in quantum theory we are concerned with new aspects of scientific analysis and synthesis and, in this connection, it is interesting to note that, even in the great epoch of critical philosophy in the former century, there was only question to what extent *a priori* arguments could be given for the adequacy of space-time co-ordination and causal connection of experience, but never question of rational generalizations or inherent limitations of such categories of human thinking.

Although in more recent years I have had several occasions of meeting Einstein, the continued discussions, from which I always have received new impulses, have so far not led to a common view about the epistemological problems in atomic physics, and our opposing views are perhaps most clearly stated in a recent issue of *Dialectica*,[17] bringing a general discussion of these problems. Realizing, however, the many obstacles for mutual understanding as regards a matter where approach and background must influence everyone's attitude, I have welcomed this opportunity of a broader exposition of the development by which, to my mind, a veritable crisis in physical science has been overcome. The lesson we have hereby received would seem to have brought us a decisive step further in the never-

[17] N. Bohr, *Dialectica*, 1, 312 (1948).

ending struggle for harmony between content and form, and taught us once again that no content can be grasped without a formal frame and that any form, however useful it has hitherto proved, may be found to be too narrow to comprehend new experience.

Surely, in a situation like this, where it has been difficult to reach mutual understanding not only between philosophers and physicists but even between physicists of different schools, the difficulties have their root not seldom in the preference for a certain use of language suggesting itself from the different lines of approach. In the Institute in Copenhagen, where through those years a number of young physicists from various countries came together for discussions, we used, when in trouble, often to comfort ourselves with jokes, among them the old saying of the two kinds of truth. To the one kind belong statements so simple and clear that the opposite assertion obviously could not be defended. The other kind, the so-called "deep truths," are statements in which the opposite also contains deep truth. Now, the development in a new field will usually pass through stages in which chaos becomes gradually replaced by order; but it is not least in the intermediate stage where deep truth prevails that the work is really exciting and inspires the imagination to search for a firmer hold. For such endeavours of seeking the proper balance between seriousness and humour, Einstein's own personality stands as a great example and, when expressing my belief that through a singularly fruitful co-operation of a whole generation of physicists we are nearing the goal where logical order to a large extent allows us to avoid deep truth, I hope that it will be taken in his spirit and may serve as an apology for several utterances in the preceding pages.

———

The discussions with Einstein which have formed the theme of this article have extended over many years which have witnessed great progress in the field of atomic physics. Whether our actual meetings have been of short or long duration, they have always left a deep and lasting impression on my mind, and when writing this report I have, so-to-say, been arguing with Einstein all the time even when entering on topics apparently far removed from the special problems under debate at our meetings. As regards the account of the conversations I am, of course, aware that I am relying only on my own memory, just as I am prepared for the possibility that many features of the development of quantum theory, in which Einstein has played so large a part, may appear to himself in a different light. I trust, however, that I have not failed in conveying a proper impression of how much it has meant to me to be able to benefit from the inspiration which we all derive from every contact with Einstein.

NIELS BOHR

UNIVERSITETETS INSTITUT
FOR TEORETISK FYSIK
COPENHAGEN, DENMARK

VII. THE EPISTEMOLOGICAL PROBLEM OF NATURAL SCIENCE

NATURVIDENSKABENS ERKENDELSESPROBLEM
Overs. Dan. Vidensk. Selsk. Virks. Juni 1950 – Maj 1951, p. 39

Communication to the Royal Danish Academy on 19 January 1951

ABSTRACT

TEXT AND TRANSLATION

Niels Bohr gav en meddelelse: *Naturvidenskabens erkendelsesproblem.*

Efter en oversigt over den erkendelsesteoretiske indstilling, som dannede grundlaget for naturvidenskabens udvikling op til vort århundrede, blev der givet en fremstilling af de problemer, som atomfysikkens senere fremskridt har stillet os overfor, og som har belært os om en principiel begrænsning af årsagsbeskrivelsen. Den som komplementaritetsprincippet betegnede mere almindelige ramme for den logiske sammenfatning af erfaringerne er også egnet til at skabe baggrund for vor orientering med hensyn til andre af menneskelivets problemer.

TRANSLATION

Communication by Niels Bohr: *The Epistemological Problem of Natural Science.*

Following a survey of the epistemological attitude that provided the basis for the development of natural science until the turn of the century, an account was presented of the problems with which the later developments of atomic physics have confronted us, urging upon us a lesson about a limitation in the very principle of causal description. The more general framework for logical coordination of experience, referred to as the principle of complementarity, is also suited for establishing a background for our orientation with respect to other problems of human life.

VIII. QUANTUM PHYSICS AND PHILOSOPHY – CAUSALITY AND COMPLEMENTARITY

"Philosophy in the Mid-Century, A Survey" (ed. R. Klibansky),
La nuova Italia editrice, Firenze 1958, pp. 308–314

See Introduction to Part II, sect. 4.

QUANTUM PHYSICS AND PHILOSOPHY – CAUSALITY AND COMPLEMENTARITY (1958)

Versions published in English, German and Danish

English: Quantum Physics and Philosophy – Causality and Complementarity
A "Philosophy in the Mid-Century, A Survey" (ed. R. Klibansky), La nuova Italia editrice, Firenze 1958, pp. 308–314
B "Essays 1958–1962 on Atomic Physics and Human Knowledge", Interscience Publishers, New York 1963, pp. 1–7 (reprinted 1987 in: "Essays 1958–1962 on Atomic Physics and Human Knowledge, The Philosophical Writings of Niels Bohr, Vol. III", Ox Bow Press, Woodbridge, Connecticut 1987, pp. 1–7)

German: Über Erkenntnisfragen der Quantenphysik
C Max Planck Festschrift, VEB Verlag der Wissenschaften, Berlin 1958, pp. 169–175
D Naturwiss. Rundschau 1960, pp. 252–255
E *Atomphysik und Philosophie – Kausalität und Komplementarität* in: "Atomphysik und menschliche Erkenntnis II", Friedr. Vieweg & Sohn, Braunschweig 1966, pp. 1–7

Danish: Kvantefysik og filosofi – Kausalitet og komplementaritet
F "Atomfysik og menneskelig erkendelse II", J.H. Schultz Forlag, Copenhagen 1964, pp. 11–18

The German version (*C, D, E*) has an introduction paying homage to Planck. *C* also contains a summary. Otherwise, all these versions agree with each other.

[386]

Institut International de Philosophie

PHILOSOPHY IN THE MID-CENTURY
A Survey

LA PHILOSOPHIE
AU MILIEU DU VINGTIÈME SIÈCLE
Chroniques

edited by   par les soins de

RAYMOND KLIBANSKY

★

QUANTUM PHYSICS AND PHILOSOPHY
CAUSALITY AND COMPLEMENTARITY
by NIELS BOHR – *University of Copenhagen*

FIRENZE
LA NUOVA ITALIA EDITRICE
1958

[387]

QUANTUM PHYSICS AND PHILOSOPHY

CAUSALITY AND COMPLEMENTARITY

by NIELS BOHR

University of Copenhagen

THE significance of physical science for philosophy does not merely lie in the steady increase of our experience of inanimate matter, but above all in the opportunity of testing the foundation and scope of some of our most elementary concepts. Notwithstanding refinements of terminology due to accumulation of experimental evidence and developments of theoretical conceptions, all account of physical experience is, of course, ultimately based on common language, adapted to orientation in our surroundings and to tracing relationships between cause and effect. Indeed, Galileo's program—to base the description of physical phenomena on measurable quantities—has afforded a solid foundation for the ordering of an ever larger field of experience.

In Newtonian mechanics, where the state of a system of material bodies is defined by their instantaneous positions and velocities, it proved possible, by the well-known simple principles, to derive, solely from the knowledge of the state of the system at a given time and of the forces acting upon the bodies, the state of the system at any other time. A description of this kind, which evidently represents an ideal form of causal relationships, expressed by the notion of *determinism*, was found to have still wider scope. Thus, in the account of electromagnetic phenomena, in which we have to consider a propagation of forces with finite velocities, a deterministic description could be upheld by including in the definition of the state not only the positions and velocities of the charged bodies, but also the direction and intensity of the electric and magnetic forces at every point of space at a given time.

The situation in such respects was not essentially changed by the recognition, embodied in the notion of *relativity*, of the extent to which the description of physical phenomena depends on the reference frame chosen by the observer. We are here concerned with a most fruitful development which has made it possible to formulate physical laws common to all observers and to link phenomena which hitherto appeared uncorrelated. Although in this formulation use is made of mathematical abstractions such as a four-dimensional non-Euclidean metric, the physical interpretation for each observer rests on the usual separation between space and time, and maintains the deterministic character of the description. Since, moreover, as stressed by Einstein, the space-time coordination of different observers never implies reversal of what may be termed the causal sequence of events, relativity theory has not only widened the scope, but also strengthened the foundation of the deterministic account, characteristic of the imposing edifice generally referred to as classical physics.

A new epoch in physical science was inaugurated, however, by Planck's discovery of the *elementary quantum of action*, which revealed a feature of *wholeness* inherent in atomic processes, going far beyond the ancient idea of the limited divisibility of matter. Indeed, it became clear that the pictorial description of classical physical theories represents an idealization valid only for phenomena in the analysis of which all actions involved are sufficiently large to permit the neglect of the quantum. While this condition is amply fulfilled in phenomena on the ordinary scale, we meet in experimental evidence concerning atomic particles with regularities of a novel type, incompatible with deterministic analysis. These quantal laws determine the peculiar stability and reactions of atomic systems, and are thus ultimately responsible for the properties of matter on which our means of observation depend.

The problem with which physicists were confronted was therefore to develop a rational generalization of classical physics, which would permit the harmonious incorporation of the quantum of action. After a preliminary exploration of the experimental evidence by more primitive methods, this difficult task was eventually accomplished by the introduction of appropriate mathematical abstractions. Thus, in the quantal formalism, the quantities by which the state of a physical system is ordinarily defined are replaced by symbolic operators subjected to a non-commutative algorism involving Planck's constant. This procedure prevents a fixation of such quantities to the extent which would be required for the deterministic description of classical physics, but allows us to determine their spectral distribution as revealed by evidence about atomic processes. In conformity with the non-pictorial character of the formalism, its physical interpretation finds expression in laws, of an essentially statistical type, pertaining to observations obtained under given experimental conditions.

Notwithstanding the power of quantum mechanics as a means of ordering an immense amount of evidence regarding atomic phenomena, its departure from accustomed demands of causal explanation has naturally given rise to the question whether we are here concerned with an exhaustive description of experience. The answer to this question evidently calls for a closer examination of the conditions for the unambiguous use of the concepts of classical physics in the analysis of atomic phenomena. The decisive point is to recognize that the description of the experimental arrangement and the recording of observations must be given in plain language, suitably refined by the usual physical terminology. This is a simple logical demand, since by the word ' experiment ' we can only mean a procedure regarding which we are able to communicate to others what we have done and what we have learnt.

In actual experimental arrangements, the fulfilment of such requirements is secured by the use, as measuring instruments, of rigid bodies sufficiently heavy to allow a completely classical account of their relative positions and velocities. In this connection, it is also essential to remember that all unambiguous information concerning atomic objects is derived from the permanent marks —such as a spot on a photographic plate, caused by the impact of an electron—left on the bodies which define the experimental conditions. Far from involving any special intricacy, the irreversible amplification effects on which the recording of the presence of atomic objects rests rather remind us of the essential irreversibility inherent in the very concept of observation. The description of atomic phenomena has in these respects a perfectly objective character, in the sense that no explicit reference is made to any individual observer and that therefore, with proper regard to relativistic exigencies, no ambiguity is involved in the communication of information.

As regards all such points, the observation problem of quantum physics in no way differs from the classical physical approach. The essentially new feature in the analysis of quantum phenomena is, however, the introduction of a *fundamental distinction between the measuring apparatus and the objects under investigation*. This is a direct consequence of the necessity of accounting for the functions of the measuring instruments in purely classical terms, excluding in principle any regard to the quantum of action. On their side, the quantal features of the phenomenon are revealed in the information about the atomic objects derived from the observations. While, within the scope of classical physics, the interaction between object and apparatus can be neglected or, if necessary, compensated for, in quantum physics this interaction thus forms an inseparable part of the phenomenon. Accordingly, the unambiguous account of proper quantum phenomena must, in principle, include a description of all relevant features of the experimental arrangement.

The very fact that repetition of the same experiment, defined on the lines described, in general yields different recordings pertaining to the object, immediately implies that a comprehensive account of experience in this field must be expressed by statistical laws. It need hardly be stressed that we are not concerned here with an analogy to the familiar recourse to statistics in the description of physical systems of too complicated a structure to make practicable the complete definition of their state necessary for a deterministic account. In the case of quantum phenomena, the unlimited divisibility of events implied in such an account is, in principle, excluded by the requirement to specify the experimental conditions. Indeed, the feature of wholeness typical of proper quantum phenomena finds its logical expression in the circumstance that any attempt at a well-defined subdivision would demand a change in the experimental arrangement incompatible with the definition of the phenomena under investigation.

Within the scope of classical physics, all characteristic properties of a given object can in principle be ascertained by a single experimental arrangement, although in practice various arrangements are often convenient for the study of different aspects of the phenomena. In fact, data obtained in such a way simply supplement each other and can be combined into a consistent picture of the behaviour of the object under investigation. In quantum physics, however, evidence about atomic objects obtained by different experimental arrangements exhibits a novel kind of complementary relationship. Indeed, it must be recognized that such evidence which appears contradictory when combination into a single picture is attempted, exhausts all conceivable knowledge about the object. Far from restricting our efforts to put questions to nature in the form of experiments, the notion of *complementarity* simply characterizes the answers we can receive by such inquiry, whenever the interaction between the measuring instruments and the objects forms an integral part of the phenomena.

Although, of course, the classical description of the experimental arrangement and the irreversibility of the recordings concerning the atomic objects ensure a sequence of cause and effect conforming with elementary demands of causality, the irrevocable abandonment of the ideal of determinism finds striking expression in the complementary relationship governing the unambiguous use of the fundamental concepts on whose unrestricted combination the classical physical description rests. Indeed, the ascertaining of the presence of an atomic particle in a limited space-time domain demands an experimental arrangement involving a transfer of momentum and energy to bodies such as fixed scales and synchronized clocks, which cannot be included in the description of their functioning, if these bodies are to fulfil the role of defining the reference frame. Conversely, any strict application of the laws of conservation of momentum and energy to atomic processes implies, in principle, a renunciation of detailed space-time coordination of the particles.

These circumstances find quantitative expression in Heisenberg's indeterminacy relations which specify the reciprocal latitude for the fixation, in quantum mechanics, of kinematical and dynamical variables required for the definition of the state of a system in classical mechanics. In fact, the limited commutability of the symbols by which such variables are represented in the quantal formalism corresponds to the mutual exclusion of the experimental arrangements required for their unambiguous definition. In this context, we are of course not concerned with a restriction as to the accuracy of measurements, but with a limitation of the well-defined application of space-time concepts and dynamical conservation laws, entailed by the necessary distinction between measuring instruments and atomic objects.

In the treatment of atomic problems, actual calculations are most conveniently carried out with the help of a Schrödinger state function, from which the statistical laws governing observations obtainable under specified conditions can be deduced by definite mathematical operations. It must be recognized, however, that we are here dealing with a purely symbolic procedure, the unambiguous physical interpretation of which in the last resort requires a reference to a complete experimental arrangement. Disregard of this point has sometimes led to confusion, and in particular the use of phrases like " disturbance of phenomena by observation " or " creation of physical attributes of objects by measurements " is hardly compatible with common language and practical definition.

In this connection, the question has even been raised whether recourse to multivalued logics is needed for a more appropriate representation of the situation. From the preceding argumentation it will appear, however, that all departures from common language and ordinary logic are entirely avoided by reserving the word ' phenomenon ' solely for reference to unambiguously communicable information, in the account of which the word ' measurement ' is used in its plain meaning of standardized comparison. Such caution in the choice of terminology is especially important in the exploration of a new field of experience, where information cannot be comprehended in the familiar frame which in classical physics found such unrestricted applicability.

It is against this background that quantum mechanics may be seen to fulfil all demands on rational explanation with respect to consistency and completeness. Thus, the emphasis on permanent recordings under well-defined experimental conditions as the basis for a consistent interpretation of the quantal formalism corresponds to the presupposition, implicit in the classical physical account, that every step of the causal sequence of events in principle allows of verification. Moreover, a completeness of description like that aimed at in classical physics is provided by the possibility of taking every conceivable experimental arrangement into account.

Such argumentation does of course not imply that, in atomic physics, we have no more to learn as regards experimental evidence and the mathematical tools appropriate to its comprehension. In fact, it seems likely that the introduction of still further abstractions into the formalism will be required to account for the novel features revealed by the exploration of atomic processes of very high energy. The decisive point, however, is that in this connection there is no question of reverting to a mode of description which fulfils to a higher degree the accustomed demands regarding pictorial representation of the relationship between cause and effect.

The very fact that quantum regularities exclude analysis on classical lines necessitates, as we have seen, in the account of experience a logical distinction between measuring instruments and atomic objects, which in principle prevents comprehensive deterministic description. Summarizing, it may be stressed that, far from involving any arbitrary renunciation of the ideal of causality, the wider frame of complementarity directly expresses our position as regards the account of fundamental properties of matter presupposed in classical physical description, but outside its scope.

Notwithstanding all difference in the typical situations to which the notions of relativity and complementarity apply, they present in epistemological respects far-reaching similarities. Indeed, in both cases we are concerned with the exploration of harmonies which cannot be comprehended in the pictorial conceptions adapted to the account of more limited fields of physical experience. Still, the decisive point is that in neither case does the appropriate widening of our conceptual framework imply any appeal to the observing subject, which would hinder unambiguous communication of experience. In relativistic argumentation, such objectivity is secured by due regard to the dependence of the phenomena on the reference frame of the observer, while in complementary description all subjectivity is avoided by proper attention to the circumstances required for the well-defined use of elementary physical concepts.

In general philosophical perspective, it is significant that, as regards analysis and synthesis in other fields of knowledge, we are confronted with situations reminding us of the situation in quantum physics. Thus, the integrity of living organisms and the characteristics of conscious individuals and human cultures present features of wholeness, the account of which implies a typical complementary mode of description.[1] Owing to the diversified use of the rich vocabulary available for communication of experience in those wider fields, and above all to the varying interpretations, in philosophical literature, of the concept of causality, the aim of such comparisons has sometimes been misunderstood. However, the gradual development of an appropriate terminology for the description of the simpler situation in physical science indicates that we are not dealing with more or less vague analogies, but with clear examples of logical relations which, in different contexts, are met with in wider fields.

---

[1] Cf. N. BOHR, *Atomic Physics and Human Knowledge*, John Wiley and Sons, Inc., New York 1958.

.

IX. ON ATOMS AND HUMAN KNOWLEDGE

ATOMERNE OG DEN MENNESKELIGE ERKENDELSE
Overs. Dan. Vidensk. Selsk. Virks. Juni 1955 – Maj 1956, pp. 112–124

ON ATOMS AND HUMAN KNOWLEDGE
Dædalus **87** (1958) 164–175

See Introduction to Part II, sect. 4.

Niels Bohr's 70th birthday, 1955. Morning celebrations at the Institute.
Front row: Margrethe and Niels Bohr, their daughter-in-law and son – Ann and Hans Bohr – and
Jørgen Bøggild. The seventy matchboxes were a gift to the pipe smoker.

Bohr's 70th birthday. Pauli handing over a Festschrift (cf. Introduction to Part I, ref. 17).

[396]

ON ATOMS AND HUMAN KNOWLEDGE (1955)

Versions published in Danish, English and German

Danish: Atomerne og den menneskelige erkendelse
A   Overs. Dan. Vidensk. Selsk. Virks. Juni 1955 – Maj 1956, pp. 112–124
B   "Atomfysik og menneskelig erkendelse", J.H. Schultz Forlag, Copenhagen 1957, pp. 101–103

English: On Atoms and Human Knowledge
C   Dædalus **87** (1958) 164–175
D   *Atoms and Human Knowledge* in: "Atomic Physics and Human Knowledge", John Wiley & Sons, New York 1958, pp. 83–93 (reprinted 1987 in: "Essays 1933–1957 on Atomic Physics and Human Knowledge, The Philosophical Writings of Niels Bohr, Vol. II", Ox Bow Press, Woodbridge, Connecticut 1987, pp. 83–93)
E   "Centenary of the birth of Nikola Tesla 1856–1956", Nikola Tesla Museum, Belgrade 1959, pp. 106–115

German: Die Atome und die menschliche Erkenntnis
F   "Atomphysik und menschliche Erkenntnis", Friedr. Vieweg & Sohn, Braunschweig 1958, pp. 84–95

A and *B* agree.

C, *D* and *E* correspond to *A*; *C* and *E* contain footnotes referring to the occasions when the respective addresses were made. *C* agrees with *D*, whilst *E* is a different translation of *A*.

F corresponds to *A*.

ATOMERNE
OG DEN MENNESKELIGE ERKENDELSE

Foredrag i Det kgl. danske Videnskabernes Selskabs møde
den 14. oktober 1955.

Af Niels Bohr.

Den udforskning af atomernes verden, som vort århundrede
har bragt, har i videnskabens historie næppe noget sidestykke
hvad angår fremskridt i kendskab til og beherskelse af den natur, hvoraf vi selv er del. Med enhver forøgelse af viden og
kunnen er jo forbundet et større ansvar, og indfrielsen af de
rige løfter og overvindelsen af de nye farer, som atomtiden indebærer, stiller hele vor civilisation på den alvorligste prøve, der
kun kan bestås ved sammenhold af alle folkeslag hvilende på
indbyrdes forståelse af det menneskelige fællesskab. For forhåb-
ninger om at kunne bygge bro over skillelinier frembragt af forskelle i folkenes levevilkår og historie turde det være af betydning, at videnskaben, der ingen landegrænser kender og hvis
vindinger er fælleseje, gennem tiderne har forenet mennesker i
stræben efter at klarlægge grundlaget for vor erkendelse. Som
jeg skal forsøge at vise, har de studier af atomerne, der skulle
få så store følger og hvis fremgang har hvilet på et verdens-
omspændende videnskabeligt samarbejde, ikke alene uddybet
vor kundskab og forståelse på et nyt erfaringsområde, men tillige stillet almene erkendelsesproblemer i ny belysning.

Det kunne måske i første øjeblik synes overraskende, at
netop atomvidenskaben skulle rumme en belæring af almen
karakter, men vi må erindre, at den på alle stadier af sin udvikling har berørt dybtliggende erkendelsesproblemer. Allerede old-
tidens tænkere søgte jo i antagelsen af en grænse for stoffernes
delelighed at finde et grundlag for forståelsen af de træk af be-

standighed, som naturfænomenerne trods al mangfoldighed og
foranderlighed udviser. Omend atomforestillingerne under fysikkens og kemiens udvikling siden Renaissancen viste sig stadig
mere frugtbare, var det dog lige op til vort århundrede en ud-
bredt opfattelse, at det drejede sig om en hypotese, idet man
ansaa det for givet, at vore sanseorganer, selv opbyggede af praktisk talt utallige atomer, var altfor grove til iagttagelse af stoffernes mindste dele. Situationen skulle imidlertid ændres væsentligt
ved de store opdagelser omkring århundredeskiftet, og som bekendt blev det ved eksperimentalteknikkens fremskridt muligt
direkte at registrere virkninger af enkelte atomer og at vinde
oplysninger om selve atomernes opbygning af mere elementære
bestanddele.

Medens allerede den antike atomlære øvede dyb indflydelse
på udviklingen af den mekaniske naturopfattelse, var det dog studiet af de umidelbart tilgængelige astronomiske og fysiske erfaringer, der gav anledning til at efterspore de lovmæssigheder, som
finder udtryk i den såkaldte klassiske fysik. Med Galileis program, hvorefter redegørelsen for fænomenerne baseres på målelige størrelser, lykkedes det at frigøre sig for de sammenligninger
med oplevelser af legemlig anstrengelse under bevægelse og med
motiver for vore handlinger, der så længe havde hindret mekanikkens rationelle udformning. I Newtons principper fandt man således grundlaget for en deterministisk beskrivelse, der tillader ud
fra kendskabet til et fysisk systems tilstand til et givet tidspunkt
at forudsige tilstanden til enhver efterfølgende tid. Efter disse ret-
ningslinier lykkedes det også senere at indordne de elektromagnetiske fænomener, hvilket dog krævede, at man i beskrivelsen af
systemets tilstand foruden de elektriserede og magnetiserede lege-
mers plads og hastigheder medtager styrken og retningen af de
elektriske og magnetiske kræfter i hvert punkt af rummet til det
givne tidspunkt.

I den begrebsbygning, der kendetegner den klassiske fysik,
mente man jo længe at besidde det rette værktøj til beskrivelsen
af alle fysiske fænomener, og ikke mindst viste den sig egnet til
frugtbargørelse af de atomistiske forestillinger. Rigtignok kunne
der for systemer, der som de sædvanlige legemer er opbygget
af et uhyre stort antal smådele, ikke blive tale om en fuldstændig beskrivelse af systemernes tilstand. Uden opgivelse af

8

[399]

det deterministiske ideal viste det sig imidlertid muligt på grundlag af de klassiske mekaniske principper at udlede statistiske lovmæssigheder, der genspejlede mange af legemernes egenskaber. Til trods for, at de mekaniske bevægelseslove tillader en fuldstændig omvending af de enkelte processers forløb, fandt således varmefænomenernes karakteristiske irreversible træk vidtgående forklaring gennem den statistiske energiligevægt, som frembringes ved molekylernes vekselvirkninger. Denne store udvidelse af mekanikkens anvendelsesområde understregede yderligere atomforestillingernes uundværlighed i naturbeskrivelsen og åbnede de første muligheder for en tælling af stoffernes atomer.

Afklaringen af grundlaget for de termodynamiske love skulle imidlertid lede på sporet af et helhedstræk hos atomare processer, der går langt ud over den gamle lære om stoffernes begrænsede delelighed. Som bekendt var det den nærmere undersøgelse af varmestrålingsfænomenerne, der blev prøvestenen for rækkevidden af den klassiske fysiks forestillinger. Opdagelsen af de elektromagnetiske bølger havde jo givet et grundlag for forståelsen af lysets forplantning, der vidtgående tillod at forklare stoffernes optiske egenskaber, men bestræbelserne på at gøre rede for varmestrålingsligevægten stillede sådanne forestillinger over for uoverstigelige vanskeligheder. Det var netop den omstændighed, at man her havde at gøre med argumenter baserede på almindelige principper og ganske uafhængige af specielle antagelser om stoffernes byggestene, som i det første år af dette århundrede førte Planck til opdagelsen af det universelle virkningskvantum, der klart skulle vise, at man i den klassiske fysiks beskrivelsesmåde havde at gøre med en idealisation med begrænset anvendelighed. Ved fænomener i sædvanlig målestok er de optrædende virkninger så store i forhold til kvantet, at dette kan lades ude af betragtning. Derimod møder vi ved de egentlige kvanteprocesser lovmæssigheder, der er ganske fremmede for den mekaniske naturopfattelse og som unddrager sig billedlig deterministisk beskrivelse.

Den opgave, som Plancks opdagelse stillede fysikerne overfor, var intet mindre end gennem en indgående undersøgelse af forudsætningerne for anvendelsen af vore mest elementære be-

greber at skaffe plads for virkningskvantet i en rationel almindeliggørelse af den fysiske beskrivelse. Under kvantefysikkens udvikling, der har rummet så mange overraskelser, er vi gang på
gang blevet belært om vanskelighederne ved at orientere sig på
et erfaringsområde så fjernt fra det, til hvis beskrivelse vore udtryksmidler er tilpasset. De hastige fremskridt har været betinget
af et omfattende og intensivt samarbejde mellem fysikere fra
mange lande, hvorved forskellige indstillinger på mest frugtbar
måde har hjulpet til at bringe problemerne i stadig klarere belysning. Ved denne lejlighed vil det selvfølgelig ikke være muligt
nærmere at omtale de enkeltes bidrag, og som en baggrund for
de efterfølgende betragtninger skal jeg blot kort minde om nogle
af udviklingens hovedtræk.

Medens Planck forsigtigt holdt sig til statistiske argumenter
og understregede vanskelighederne ved i en mere detaljeret beskrivelse at forlade det klassiske grundlag, pegede Einstein i det
samme år, hvor han gennem skabelsen af relativitetsteorien gav
den klassiske fysiks begrebsbygning en så harmonisk afrunding,
med stor dristighed på nødvendigheden af at tage hensyn til
virkningskvantet ved individuelle atomare fænomener. Især på-
viste han, at beskrivelsen af iagttagelserne vedrørende fotoelektri-
ske effekter forlanger, at overførelsen af energi til hver enkelt
af de fra stofferne udjagede elektroner netop svarer til absorption
af et såkaldt strålingskvantum. Da der imidlertid ikke kunne
være tale om simpelthen at erstatte den for redegørelsen for lys-
virkningernes forplantning uundværlige bølgeforestilling med
noget anskueligt partikelbillede, stilledes man her over for et
ejendommeligt dilemma, hvis løsning skulle kræve indgående
undersøgelse af billedlige forestillingers rækkevidde.

Dette spørgsmål tilspidsedes som bekendt yderligere gennem
Rutherfords opdagelse af atomkernen, der trods sin lidenhed
indeholder næsten hele atomets masse og hvis elektriske ladning
svarer til antallet af elektroner i det neutrale atom. På den ene
side fik man herved et simpelt billede af atomet, der umiddelbart tilbød sig for anvendelse af mekaniske og elektromagnetiske forestillinger. På den anden side var det klart, at ingen
konfiguration af elektriske partikler efter den klassiske fysiks
principper vil besidde en stabilitet nødvendig til forklaring af
atomernes fysiske og kemiske egenskaber. Navnlig måtte enhver

8*

bevægelse af elektronerne omkring atomkernen efter den klassiske elektromagnetiske teori give anledning til en stadig udstråling af energi, der ville medføre en hurtig sammentrækning af systemet, indtil elektronerne forenede sig med kernen til en neutral partikel af dimensioner forsvindende små i forhold til dem, der må tillægges atomerne. I de hidtil ganske uforklarlige empiriske love for grundstoffernes liniespektre fik man imidlertid et fingerpeg om virkningskvantets afgørende betydning for atomernes stabilitet og strålingsreaktioner.

Udgangspunktet blev her det såkaldte kvantepostulat, hvorefter enhver ændring i et atoms energi fremkommer som resultat af en fuldstændig overgang mellem to af dettes stationære tilstande. Ved endvidere at antage, at det ved alle atomare strålingsreaktioner drejer sig om en emission eller absorption af et enkelt lyskvantum, blev det muligt ved hjælp af spektrene at bestemme de stationære tilstandes energiværdier. Selvfølgelig var det inden for en deterministisk beskrivelses rammer udelukket at forklare overgangsprocessernes udelelighed og nærmere at redegøre for deres forekomst under givne betingelser. Ved hjælp af det såkaldte korrespondensprincip, hvorefter man gennem en sammenligning med det på klassisk grundlag forventede forløb af processerne søgte direktiver for en med kvantepostulatet forenelig statistisk almindeliggørelse af beskrivelsen, lykkedes det imidlertid at opnå et overblik over elektronernes bindingsforhold i atomerne, der genspejlede mange af stoffernes egenskaber. Dog viste det sig mere og mere klart, at det for en modsigelsesfri redegørelse for atomfænomenerne var nødvendigt i endnu højere grad at give afkald på brugen af billeder og gennem en radikal omformning af hele beskrivelsen at skaffe plads til alle af virkningskvantet betingede træk.

Den løsning, som man nåede til gennem sindrige bidrag fra en række af vor tids mest fremragende teoretikere, var overraskende simpel. Ligesom ved formuleringen af relativitetsteorien fandt man i matematikkens højt udviklede abstraktioner de rette hjælpemidler. I den kvantemekaniske formalisme erstattes de størrelser, der i den klassiske fysik benyttes til beskrivelse af et systems tilstand, med symbolske operatorer, hvis ombyttelighed indskrænkes ved regneregler, hvori virkningskvantet indgår. Dette medfører, at størrelser som partiklernes stedkoordinater og de tilsvarende bevægelsesmængdekomponenter ikke på en gang kan tilskrives bestemte værdier, og netop herved fremtræder formalismens statistiske karakter som en utvungen generalisation af den klassiske fysiks deterministiske beskrivelsesmåde. Denne generalisation tillod endvidere på konsekvent måde at formulere lovmæssigheder, der begrænser ensartede partiklers individualitet og for hvilke der ligeså lidt som for virkningskvantet selv kan gives noget udtryk ved tilvante fysiske billeder.

Ved hjælp af de kvantemekaniske metoder lykkedes det i løbet af få år at gøre rede for et overordentlig stort erfaringsmateriale vedrørende stoffernes fysiske og kemiske egenskaber. Ikke alene blev det muligt i enkeltheder at klarlægge bindingsforholdene for elektronerne i atomer og molekyler, men tillige at opnå dybtgående indsigt i selve atomkernernes opbygning og reaktioner. I den sidste forbindelse har ikke mindst de tidligt opdagede sandsynlighedslove for de spontane radioaktive kernesønderdelinger harmonisk kunnet indlemmes i den statistiske kvantemekaniske beskrivelse. Også med hensyn til egenskaberne hos de nye elementarpartikler, som i de seneste år er blevet iagttaget ved studiet af atomkerneomdannelser ved høje energier, er der stadig opnået fremskridt ved formalismens tilpasning til relativitetsteoriens invariansfordringer. Her møder vi dog nye problemer, hvis løsning øjensynlig forlanger videregående abstraktioner, egnet til at sammenknytte virkningskvantet og den elektriske elementarladning.

Til trods for kvantemekanikkens frugtbarhed inden for så stort et erfaringsområde har afkaldet på tilvante krav til fysisk forklaring givet anledning til tvivl hos mange fysikere og filosoffer, om vi her har at gøre med en udtømmende beskrivelse af atomfænomenerne. Især er der givet udtryk for den opfattelse, at den statistiske beskrivelsesmåde må betragtes som en midlertidig udvej, der i hvert fald i princippet burde kunne erstattes med en deterministisk beskrivelse. Den indgående diskussion af dette spørgsmål har imidlertid medført en afklaring af vor stilling som iagttagere på atomfysikkens område, der netop har givet os den erkendelsesteoretiske belæring, som jeg i begyndelsen af foredraget hentydede til.

Da videnskabens mål er at forøge og ordne vore erfaringer, må enhver undersøgelse af vilkårene for menneskelig erkendelse hvile på overvejelser af vore meddelelsesmidlers karakter og rækkevidde. Grundlaget er jo her det for vor orientering i omgivelserne og for organisationen af det menneskelige samfund udviklede sprog. Ved erfaringernes forøgelse rejser sig imidlertid stadig spørgsmålet om tilstrækkeligheden af de begreber og forestillinger, der indgår i dagligsproget. På grund af de fysiske problemers forholdsvise simpelhed er de særlig egnet til undersøgelsen af meddelelsesmidlernes brug, og vi er netop gennem atomfysikkens udvikling blevet belært om, hvorledes det er muligt uden at forlade det fælles sprog at skabe en ramme tilstrækkelig vid for en udtømmende beskrivelse af nye erfaringer.

Det er i denne forbindelse afgørende at gøre sig klart, at man ved enhver redegørelse for fysiske erfarenheder må beskrive såvel forsøgsomstændighederne som iagttagelserne ved hjælp af de samme meddelelsesmidler som benyttes i den klassiske fysik. Ved undersøgelse af enkelte atomare partikler muliggøres dette ved irreversible forstærkningseffekter — som en plet på en fotografiplade efterladt ved en elektrons indtrængen i denne eller et strømstød frembragt i en tælleindretning — og iagttagelserne vedrører blot de derved opnåede oplysninger om, hvor og hvornår partiklen opfangedes på fotografipladen, eller om dens energi ved ankomsten til tælleren. Disse oplysninger forudsætter selvfølgelig kendskab til fotografipladens position i forhold til de andre i forsøgsopstillingen indgående dele, som regulerende blændere og lukkere der fastlægger rum-tidskoordinationen, eller elektriserede og magnetiserede legemer der bestemmer de på partiklen virkende ydre kraftfelter og derigennem afgiver grundlag for energimålinger. Forsøgsomstændighederne kan varieres på mangfoldige måder, men det afgørende er, at vi i hvert enkelt tilfælde må kunne meddele andre, hvad vi har gjort og hvad vi har lært, og at derfor måleinstrumenternes funktioner må beskrives inden for den klassiske fysiks forestillingskreds.

Da alle målinger således kun angår legemer, der er tilstrækkelig tunge til, at man ved deres beskrivelse kan se bort fra virkningskvantet, er der for så vidt intet nyt iagttagelsesproblem i atomfysikken. Den forstærkning af de atomare effekter, der tillader at basere redegørelsen på målelige størrelser og som giver fænomenerne en ejendommelig afsluttet karakter, understreger i denne forbindelse blot den irreversibilitet, der kendetegner selve iagttagelsesbegrebet. Medens der inden for den klassiske fysiks rammer ikke er nogen principiel forskel på målemidlernes og undersøgelsesobjekternes beskrivelse, ligger forholdene for kvantefænomenernes vedkommende væsentlig anderledes. Ikke alene peger spørgsmålet om objekternes egen stabilitet og individualitet på en begrænsning af forestillingen om veldefinerede atomare systemer, men allerede inden for det store erfaringsområde, hvor denne forestilling kan bevares, sætter virkningskvantet grænser for beskrivelsen af systemernes tilstand ved rum-tidskoordinater og bevægelsesmængde- og energistørrelser. Da den klassiske fysiks deterministiske beskrivelse netop hviler på antagelsen af en ubegrænset forenelighed af rum-tidskoordinationen og de dynamiske bevarelseslove, står vi åbenbart her over for problemet om en sådan beskrivelses uindskrænkede gennemførlighed for de atomare objekters vedkommende.

Til afklaringen af dette hovedpunkt bidrog især henvisningen til den rolle, som vekselvirkningen mellem objekterne og måleinstrumenterne spiller i kvantefænomenernes beskrivelse. Som fremhævet af Heisenberg medfører virkningskvantet, at konstateringen af et objekts tilstedeværelse i et begrænset rum-tidsområde vil være forbundet med en udveksling af bevægelsesmængde og energi imellem målemidlet og objektet, der er desto større jo mindre det omhandlede område vælges, og det måtte derfor blive af afgørende betydning at undersøge, i hvilken udstrækning sådan med iagttagelsen forbunden vekselvirkning kan tages særskilt i betragtning ved fænomenernes beskrivelse. Dette spørgsmål har været genstand for megen diskussion, og der er fremkommet mange forslag med sigte på en fuldstændig kontrol af alle vekselvirkninger. Ved sådanne betragtninger er der imidlertid ikke taget tilstrækkelig hensyn til, at selve redegørelsen for målemidlernes funktioner medfører, at enhver af virkningskvantet betinget vekselvirkning mellem disse og de atomare objekter på uadskillelig måde indgår i fænomenerne.

Ved en hvilken som helst forsøgsanordning, der tillader at konstatere en partikels tilstedeværelse i et begrænset rum-tidsområde, må der jo benyttes fast anbragte målestokke og synkroniserede

ure, der efter deres definition udelukker kontrol af en til dem
overført bevægelsesmængde eller energi. Omvendt kræver enhver
entydig anvendelse af dynamiske bevarelseslove i kvantefysikken,
at det omhandlede fænomens beskrivelse for de atomare objekters
vedkommende indebærer et principielt afkald med hensyn til
detaljeret rum-tidskoordination. Dette gensidige udelukkelsesforhold mellem de forsøgsbetingelser, som de elementære begrebers
brug forudsætter, medfører at hele forsøgsanordningen må tages
i betragtning ved fænomenernes veldefinerede beskrivelse. I
denne sammenhæng finder kvantefænomenernes udelelighed konsekvent udtryk deri, at enhver definerbar underdeling ville kræve
en ændring af forsøgsanordningen med optræden af nye individuelle fænomener. Selve grundlaget for en deterministisk beskrivelse
er således faldet bort, og forudsigelsernes statistiske karakter
fremtræder derved, at der under samme forsøgsomstændigheder
i almindelighed fremkommer iagttagelser svarende til forskellige
mulige individuelle processer.

Sådanne betragtninger har ikke alene opklaret det i det foregående omtalte dilemma vedrørende lysforplantningen, men også
bragt den fuldstændige løsning af de tilsvarende paradokser med
hensyn til billedlig fremstilling af materielle partiklers forhold,
som den senere udvikling har stillet os overfor. Der kan jo her
ikke blive tale om nogen fysisk forklaring i tilvant forstand, men
om den fjernelse af enhver tilsyneladende modsigelse, som er
alt, hvad vi på et nyt erfaringsområde kan forlange. Hvor store
kontraster en sammenligning mellem erfaringer vedrørende atomare objekter, vundet under forskellige forsøgsbetingelser, end
kan frembyde, må sådanne fænomener betegnes som komplementære i den forstand, at de hver især er veldefinerede og tilsammen
udtømmer al definerbart kendskab til de pågældende objekter.
Den kvantemekaniske formalisme, der tager direkte sigte på
sammenfatning af iagttagelser, der er opnået under forsøgsomstændigheder beskrevet ved elementære fysiske begreber, giver
netop en udtømmende komplementær redegørelse for et meget
stort erfaringsområde. Afkaldet på anskuelige billeder gælder
kun selve de atomare objekters tilstand, medens grundlaget
for beskrivelsen af forsøgsomstændighederne såvel som vor frihed til at vælge disse er fuldt ud bevaret. Hele formalismen,
hvis anvendelse kun er veldefineret for afsluttede fænomener, er

i alle sådanne henseender at betragte som en rationel almindeliggørelse af den klassiske fysik.

På baggrund af den indflydelse, som den mekaniske naturopfattelse har udøvet på filosofisk tænkning, er det forståeligt, at man fra mange sider har opfattet komplementaritetssynspunktet som rummende en med beskrivelsens objektivitet uforenelig henvisning til den subjektive iagttager. Selvfølgelig må vi på ethvert erfaringsområde opretholde en skarp adskillelse mellem iagttageren og indholdet af iagttagelserne, men vi må betænke, at virkningskvantets opdagelse har stillet selve grundlaget for naturbeskrivelsen i ny belysning og belært os om hidtil upåagtede forudsætninger for den rationelle anvendelse af de begreber, på hvilke meddelelserne om erfaringerne hviler. I kvantefysikken er, som vi har set, en redegørelse for måleinstrumenternes funktioner uundværlig for definitionen af fænomenerne, og vi må så at sige drage skillelinien mellem subjekt og objekt på en måde, der i hvert enkelt tilfælde sikrer den entydige anvendelse af de i meddelelserne benyttede elementære fysiske begreber. Langtfra at rumme en mod videnskabens ånd stridende mystik henviser betegnelsen komplementaritet blot til de med vor stilling ved beskrivelsen og sammenfatningen af erfaringerne på atomfysikkens område forbundne erkendelsesvilkår.

I lighed med tidligere fremskridt inden for fysikken har den belæring om almindelige vilkår for menneskelig erkendelse, som vi har fået gennem atomfysikkens udvikling, ikke kunnet undgå at give anledning til fornyet overvejelse af vore meddelelsesmidlers brug for objektiv beskrivelse på andre erfaringsområder. Ikke mindst rejser den særlige betoning af iagttagelsesproblemet spørgsmålet om de levende organismers plads i naturbeskrivelsen og om vor egen stilling som tænkende og handlende væsner. Omend det inden for den klassiske fysiks forestillingskreds var muligt i en vis udstrækning at sammenligne organismerne med maskiner, var det klart, at mange af livets karakteristiske træk ved sådan sammenligning ikke kom til deres ret, og den mekaniske naturopfattelses afmagt over for beskrivelsen af menneskets situation fandt særlig udtryk i de vanskeligheder, som den primitive adskillelse mellem sjæl og legeme frembyder.

122

De problemer, vi her møder, er øjensynlig forbundet med, at der til beskrivelsen af mange sider af den menneskelige tilværelse kræves en sprogbrug, der ikke umiddelbart hviler på simple fysiske billeder. Netop erkendelsen af den begrænsede anvendelighed af sådanne billeder ved redegørelsen for atomfænomenerne giver imidlertid et fingerpeg om, hvordan vi, stillet over for biologiens og psykologiens problemer, må søge plads for erfaringerne inden for en objektiv beskrivelses rammer. Ligesom i det foregående drejer det sig her om i hvert enkelt tilfælde at være opmærksom på skillelinien imellem iagttageren og indholdet af meddelelserne. Medens snittet mellem subjekt og objekt i den mekaniske naturopfattelse lå fast, er det netop den med betingelserne for udtryksmidlernes konsekvente brug forbundne forskellige placering af en sådan skillelinie, der giver den nødvendige plads for en mere omfattende beskrivelse.

Uden at forsøge nogen nærmere definition af organisk liv tør vi sige, at en levende organisme kendetegnes ved sin selvstændighed og tilpasningsevne, der bevirker, at vi ved beskrivelsen af organismens indre funktioner og dens reaktioner på ydre påvirkninger ofte må benytte det for fysik og kemi ganske fremmede ord formålstjenlighed. Omend atomfysikkens resultater har fundet mangfoldige anvendelser i biofysik og biokemi, er der jo ved de udelelige afsluttede kvantefænomener ikke tale om noget træk, der kalder på betegnelsen liv. Som vi har set, beroede den inden for et stort erfaringsområde udtømmende beskrivelse af atomfænomenerne på den frie brug af de for de elementære begrebers definitionsmæssige anvendelse nødvendige måleinstrumenter. I en levende organisme kan imidlertid en sådan skelnen mellem målemidler og undersøgelsesobjekter næppe fuldtud gennemføres, og vi må være forberedt på, at enhver forsøgsanordning med sigte på en i atomfysisk forstand veldefineret beskrivelse af organismens virkemåde ville være uforenelig med livets udfoldelse.

I den biologiske forskning benyttes henvisninger til de levende organismers helhedstræk og formålstjenlige reaktioner side om side med de stedse mere indgående oplysninger vedrørende organismernes struktur og regulationsprocesser, der ikke mindst har bragt så store fremskridt på lægekunstens område. Det drejer sig her om en praktisk stillingtagen til et emne, hvor de til redegørelsen for dets forskellige sider benyttede udtryksmidler henviser til gensidigt udelukkende iagttagelsesomstændigheder. I forbindelse med de som mekanistisk og finalistisk betegnede indstillinger må det erkendes, at vi ikke har at gøre med modstridende synspunkter, men med et med vor stilling som iagttagere af naturen forbundet komplementaritetsforhold. For at undgå misforståelser er det dog væsentligt at bemærke, at der ved beskrivelsen af det organiske liv og bedømmelsen af dets udviklingsmuligheder — i modsætning til redegørelsen for de af virkningskvantet betingede atomare lovmæssigheder — selvfølgelig ikke kan tages sigte på nogen fuldstændighed, men blot på begrebsrammens tilstrækkelige rummelighed.

Ved redegørelsen for de til livet knyttede psykiske oplevelser møder vi iagttagelsesomstændigheder og dertil svarende udtryksmidler, der fjerner sig endnu mere fra fysikkens sprogbrug. Ganske uanset spørgsmålet om, i hvilket omfang det er nødvendigt og berettiget at bruge udtryk som instinkt og fornuft ved beskrivelsen af dyrs adfærd, er det, når det drejer sig om menneskers tilværelse, uundgåeligt at gøre brug af ordet bevidsthed, ikke blot med henblik på en selv men også på ens medmennesker. Medens den til orienteringen i omverdenen tilpassede sprogbrug har kunnet tage udgangspunkt i simple fysiske billeder og årsagskrav, har redegørelsen for vore sindstilstande lige fra sprogets begyndelse krævet en typisk komplementær beskrivelsesmåde. Brugen af ord som tanker og følelser henviser jo ikke til en fast forbunden årsagskæde, men til oplevelser, der udelukker hverandre, idet de er betinget af forskellige snit imellem det bevidste indhold og den baggrund, som vi løst betegner som os selv.

Særlig lærerigt er forholdet mellem oplevelse af viljesfølelse og bevidst overvejelse af handlingsmotiver. Uundværligheden af sådanne tilsyneladende kontrasterende udtryksmidler ved beskrivelsen af bevidsthedslivets rigdom minder os slående om den måde, hvorpå elementære fysiske begreber benyttes i atomfysikken. Ved sådan sammenligning må vi dog ikke alene gøre os klart, at psykiske oplevelser ikke kan gøres til genstand for fysiske målinger, men tillige at selve viljesbegrebet, så langt fra at henvise til en almindeliggørelse af den deterministiske beskrivelse, på forhånd tager sigte på menneskelivets muligheder. Uden at gå nærmere ind på den gamle filosofiske diskussion om viljens frihed skal jeg blot minde om, at anvendelsen af ordet vilje i den objektive beskrivelse af vor situation ganske svarer til de for menneskelige meddelelser lige uundværlige ord som håb og ansvar.

Vi er her nået frem til problemer, der berører menneskers fællesskab, og hvor udtryksmidlernes mangfoldighed er betinget af umuligheden af ved nogen fast skillelinie at karakterisere individets rolle i samfundet. Med henblik på kontraster, som de under forskellige levevilkår udviklede menneskekulturer kan udvise hvad såvel hævdvundne traditioner som deres udtryksformer angår, kan man i en vis forstand betegne sådanne kulturer som komplementære. Det drejer sig dog her ingenlunde om definitive gensidige udelukkelsesforhold som dem, vi møder ved den objektive beskrivelse af fysikkens og psykologiens almindelige problemer, men om forskelle i indstilling, der kan værdsættes eller mildnes ved udvidet samkvem mellem folkeslagene. I vor tid, hvor forøgelsen af vore kundskaber og muligheder mere end nogensinde sammenknytter folkeslagenes skæbne, har samarbejdet på videnskabens udvikling fået vidtrækkende opgaver, for hvis fremme netop påmindelsen om almene vilkår for menneskelig erkendelse turde rumme nye håb.

Reprinted from *Dædalus: Proceedings of The American Academy of Arts and Sciences*, volume 87, number 2 (1958).

DÆDALUS

Journal of the American Academy of Arts and Sciences

EDITORIAL AND PUBLICATION OFFICE: American Academy of Arts and Sciences, 280 Newton Street, Brookline Station, Boston 46, Massachusetts. All communications should be sent to the Editor at that address.

SUBSCRIPTIONS: *Dædalus* is published quarterly. Subscription rate in the U. S. A. $6.50 a year, single copies $1.75. Subscription orders may be sent to the publication office.

NOTES FROM THE ACADEMY

On Atoms and Human Knowledge[*]

Niels Bohr

IN THE history of science, this century's exploration of the world of atoms has hardly any parallel in so far as the progress of knowledge and the mastery of that nature of which we ourselves are part are concerned. However, with every increase of knowledge and abilities is connected a greater responsibility; and the fulfilment of the rich promise and the elimination of the new dangers of the atomic age confront our whole civilization with a serious challenge which can be met only by cooperation of all peoples, resting on a mutual understanding of the human fellowship. In this situation, it is important to realize that science, which knows no national boundaries and whose achievements are the common possession of mankind, has through the ages united men in their efforts to elucidate the foundations of our knowledge. As I shall attempt to show, the study of atoms, which was to entail such far-reaching consequences and whose progress has been based on world-wide cooperation, not only has deepened our insight into a new domain of experience, but has thrown new light on general problems of knowledge.

At first, it might seem surprising that atomic science should contain a lesson of a general nature, but we must remember that it has in all stages of its development concerned profound problems of knowledge. Thus, thinkers of antiquity, by assuming a limit for the divisibility of substances, attempted to find a basis for understanding the features of permanency exhibited by natural phenomena, in spite of their multifariousness and variability. Although atomic ideas have contributed more and more fruitfully to the development of physics and chemistry since the Renaissance, they were regarded as a hy-

* The content of this article formed the basis of the author's introduction to the panel discussion at the 1401st meeting of the Academy, November 13, 1957. This article has appeared in Danish in the 1956 Yearbook of the Danish Academy of Science and Letters, and is included in a collection of articles by the author, to be published shortly by John Wiley and Sons, New York, under the title *Atomic Physics and Human Knowledge*.

pothesis right up to the beginning of this century. Indeed, it was taken for granted that our sense organs, themselves composed of innumerable atoms, were too coarse to observe the smallest parts of matter. This situation was, however, to become essentially changed by the great discoveries at the turn of the century and, as is well known, progress in experimental technique made it possible to record the effects of single atoms and to obtain information on the more elementary particles of which the atoms themselves were found to be composed.

In spite of the deep influence exerted by ancient atomism on the development of the mechanical conception of nature, it was the study of immediately accessible astronomical and physical experience which made it possible to trace the regularities expressed in the so-called classical physics. Galileo's dictum, according to which the account of phenomena should be based on measurable quantities, made it possible to eliminate such animistic views which had so long hindered the rational formulation of mechanics. In Newton's principles, the foundation was laid of a deterministic description permitting, from the knowledge of the state of a physical system at a given moment, prediction of its state at any subsequent time. On the same lines, it was possible to account for electromagnetic phenomena. This required, however, that the description of the state of the system should include, besides the positions and velocities of the electrified and magnetized bodies, the strength and direction of the electrical and magnetic forces at every point of space, at the given moment.

The conceptual framework which is characteristic of classical physics was long thought to provide the correct tool for the description of all physical phenomena, and not least was it suited to the utilization and development of atomic ideas. Of course, for systems such as ordinary bodies which are composed of an enormous number of constituent parts, there could be no question of an exhaustive description of the state of the system. Without abandoning the deterministic ideal, it became possible, however, on the basis of the principles of classical mechanics, to deduce statistical regularities reflecting many of the properties of material bodies. Even though the mechanical laws of motion permit a complete reversal of the course of single processes, full explanation of the characteristic feature of irreversibility in heat phenomena was found in the statistical energy equilibrium resulting from the interaction of the molecules. This great extension of the application of mechanics emphasized further the indispensability of atomic ideas to the description of nature and opened the first possibilities of counting the atoms of the substances.

However, clarification of the foundation of the laws of thermodynamics was to open the way for recognition of a feature of wholeness in atomic processes far beyond the old doctrine of the limited divisibility of matter. As is well known, the closer analysis of heat radiation became the test of the scope of classical physical ideas. The discovery of electromagnetic waves had already provided a basis for understanding the propagation of light, explaining many of the optical properties of substances; but endeavours to account for radiation equilibrium confronted such ideas with insurmountable difficulties. The circumstance that here one had to do with arguments based on general principles and quite independent of special assumptions regarding the constituents of the substances led Planck, in the first year of this century, to the discovery of the universal quantum of action, which showed clearly that the classical physical description is an idealization of limited applicability. In phenomena on the ordinary scale, the actions involved are so large compared to the quantum that it can be left out of consideration. However, in proper quantum processes, we meet regularities which are completely foreign to the mechanical conception of nature and which defy pictorial deterministic description.

The task with which Planck's discovery confronted physicists was nothing less than, by means of a thorough analysis of the presuppositions on which the application of our most elementary concepts are based, to provide room for the quantum of action in a rational generalization of the classical physical description. During the development of quantum physics, entailing so many surprises, we have time and again been reminded of the difficulties of orienting ourselves in a domain of experience far from that to the description of which our means of expression are adapted. Rapid progress has been made possible by a wide and intensive collaboration among physicists from many countries, whose diverse approaches have helped in a most fruitful way to focus the problem ever more sharply. On this occasion, of course, it will not be possible to deal in detail with individual contributions, but as a background for the following considerations I shall remind you briefly of some of the main features of the development.

While Planck cautiously limited himself to statistical arguments and emphasized the difficulties of abandoning the classical foundations in the detailed description of nature, Einstein daringly pointed to the necessity of taking the quantum of action into account in individual atomic phenomena. In the same year that he so harmoniously completed the framework of classical physics by establishing the theory of relativity, he showed that the description of observations on photoelectric effects requires that the transmission of energy to each of the electrons expelled from the substances corresponds to the absorption of a so-called quantum of radiation. Since the idea of waves is indispensable to the account of the propagation of light, there could be no question of simply replacing it with a corpuscular description, and one was therefore confronted with a peculiar dilemma whose solution was to require a thorough analysis of the scope of pictorial concepts.

As is well known, this question was further accentuated by Rutherford's discovery of the atomic nucleus which, despite its minuteness, contains almost the whole mass of the atom and whose electrical charge corresponds to the number of electrons in the neutral atom. This gave a simple picture of the atom which immediately suggested the application of mechanical and electromagnetic ideas. Yet, it was clear that, according to classical physical principles, no configuration of electrical particles could possess the stability necessary to the explanation of the physical and chemical properties of atoms. In particular, according to classical electromagnetic theory, every motion of the electrons around the atomic nucleus would produce a continual radiation of energy implying a rapid contraction of the system until the electrons became united with the nucleus into a neutral particle of dimensions vanishingly small relative to those which must be ascribed to atoms. However, in the hitherto entirely incomprehensible empirical laws for the line spectra of the elements was found a hint as to the decisive importance of the quantum of action for the stability and radiative reactions of the atom.

The point of departure became here the so-called quantum postulate, according to which every change in the energy of an atom is the result of a complete transition between two of its stationary states. By assuming further that all atomic radiative reactions involve the emission or absorption of a single light quantum, the energy values of the stationary states could be determined from the spectra. It was evident that no explanation of the indivisibility of the transition processes, or their appearance under given conditions, could be given within the framework of deterministic description. However, it proved possible to obtain a survey of the electron bindings in the atom, which reflected many of the properties of substances, with the aid of the so-called correspondence principle. On the basis of a comparison with the classically expected course of the processes, directives were sought for a statistical generalization of the description compatible with the quantum postulate. Still, it became more and more clear that in order to obtain a consistent account of atomic phenomena it was necessary to renounce even more the use of pictures, and that a radical reformulation of the whole description was needed to provide room for all features implied by the quantum of action.

The solution which was reached as a result of the ingenious contributions of many of the most eminent theoretical physicists of our time was surprisingly simple. As in the formulation of relativity theory, adequate tools were found in highly developed mathematical abstractions. The quantities which in classical physics are used to describe the state of a system are replaced in quantum mechanical formalism by symbolic operators whose commutability is limited by rules containing the quantum. This implies that quantities such as positional coordinates and corresponding momentum components of particles cannot simultaneously be ascribed definite values. In this way, the statistical character of the formalism is displayed as a natural generalization of the description of classical physics. In addition, this generalization permitted a consequent formulation of the regularities which limit the individuality of identical particles and which, like the quantum itself, cannot be expressed in terms of usual physical pictures.

By means of the methods of quantum mechanics it was possible to account for a very large amount of the experimental evidence on the physical and chemical properties of substances. Not only was the binding of electrons in atoms and molecules clarified in detail, but a deep insight was also obtained into the constitution and reactions of atomic nuclei. In this connection, we may mention that the probability laws for spontaneous radioactive transmutations have been harmoniously incorporated into the statistical quantum mechanical description. Also, the understanding of the properties of the new elementary particles, which have been observed in recent years in the study of transmutations of atomic nuclei at high energies, has been subject to continual progress resulting from the adaption of the formalism to the invariance requirements of relativity theory. Still, we are confronted with new problems whose solution obviously demands further abstractions suited to combine the quantum of action with the elementary electric charge.

In spite of the fruitfulness of quantum mechanics within such a wide domain of experience, the renunciation of accustomed demands on physical explanation has caused many physicists and philosophers to doubt that we are dealing with an exhaustive description of atomic phenomena. In particular, the view has been expressed that the statistical mode of description must be regarded as a temporary expedient which, in principle, ought to be replaceable by a deterministic description. The thorough discussion of this question has, however, led to that clarification of our position as observers in atomic physics which has given us the epistemological lesson referred to in the beginning of this lecture.

As the goal of science is to augment and order our experience, every analysis of the conditions of human knowledge must rest on considerations of the character and scope of our means of communication. Our basis is, of course, the language developed for orientation in our surroundings and for the organization of human communities. However, the increase of experience has repeatedly raised questions as to the sufficiency of the concepts and ideas incorporated in daily language. Because of the relative simplicity of physical problems, they are especially suited to investigate the use of our means of communication. Indeed, the development of atomic physics has taught us how, without leaving common language, it is possible to create a framework sufficiently wide for an exhaustive description of new experience.

In this connection, it is imperative to realize that in every account of physical experience one must describe both experimental conditions and observations by the same means of communication as one used in classical physics. In the analysis of single atomic particles, this is made possible by irreversible amplification effects — such as a spot on a photographic plate left by the impact of an electron, or an electric discharge created in a counter device — and the observations concern only where and when the particle is registered on the plate or its energy on arrival at the counter. Of course, this information presupposes knowledge of the position of the photographic plate relative to the other parts of the experimental arrangement, such as regulating diaphragms and shutters defining space-time coordination or electrified and magnetized bodies which determine the external force fields acting on the particle and permit energy measurements. The experimental conditions can be varied in many ways, but the point is that in each case we must be able to communicate to others what we have done and what we have learned, and that therefore the functioning of the measuring instruments must be described within the framework of classical physical ideas.

As all measurements thus concern bodies sufficiently heavy to permit the quantum to be neglected in their description, there is, strictly speaking, no new observational problem in atomic physics. The amplification of atomic effects, which makes it possible to base the account on measurable quantities and which gives the phenomena a peculiar closed character, only emphasizes the irreversibility characteristic of the very concept of observation. While, within the frame of classical physics, there is no difference in principle between the description of the measuring instruments and the objects under investigation, the situation is essentially different when we study quantum phenomena, since the quantum of action imposes restrictions on the description of the state of the systems by means of space-time coordinates and momentum-energy quantities. Since the deterministic description of classical physics rests on the assumption of an unrestricted compatability of space-time coordination and the dynamical conservation laws, we are obviously confronted here with the problem of whether, as regards atomic objects, such a description can be fully retained.

The role of the interaction between objects and measuring instruments in the description of quantum phenomena was found to be especially important for the clarification of this main point. Thus, as stressed by Heisenberg, the locating of an object in a limited space-time domain involves, according to quantum mechanics, an exchange of momentum and energy between instrument and object which is the greater the smaller the domain chosen. It was therefore of the utmost importance to investigate the extent to which the interaction entailed in observation can be taken into account separately in the description of phenomena. This question has been the focus of much discussion, and there have appeared many proposals which aim at the complete control of all interactions. In such considerations however, due regard is not taken to the fact that the very account of the functioning of measuring instruments involves that any interaction implied by the quantum, between these and the atomic objects, be inseparably entailed in the phenomena.

Indeed, every experimental arrangement permitting the registration of an atomic particle in a limited space-time domain demands fixed measuring rods and synchronized clocks which, from their very definition, exclude the control of momentum and energy transmitted to them. Conversely, any unambiguous application of the dynamical conservation laws in quantum physics requires that the description of the phenomena involve a renunciation in principle of detailed space-time coordination. This mutual exclusiveness of the experimental conditions implies that the whole experimental arrangement must be taken into account in a well-defined description of the phenomena. The indivisibility of quantum phenomena finds its consequent expression in the circumstance that every definable subdivision would require a change of the experimental arrangement with the appearance of new individual phenomena. Thus, the very foundation of a deterministic description has disappeared and the statistical character of the predictions is evidenced by the fact that in one and the same experimental arrangement there will in general appear observations corresponding to different individual processes.

Such considerations not only have clarified the above-mentioned dilemma with respect to the propagation of light, but have also completely solved the corresponding paradoxes confronting pictorial representation of the behavior of material particles. Here, of course, we cannot seek a physical explanation in the customary sense; all we can demand in a new field of experience is the removal of any apparent contradiction. However great the contrasts exhibited by atomic phenomena under different experimental conditions, such phenomena must be termed complementary in the sense that each is well defined and that together they exhaust all definable knowledge about the objects concerned. The quantum-mechanical formalism, the sole aim of which is the comprehension of observations obtained under experimental conditions described by simple physical concepts, gives just such an exhaustive complementary account of a very large domain of experience. The renunciation of pictorial representation involves only the state of atomic objects, while the foundation of the description of the experimental condition, as well as our freedom to choose them, is fully retained. The whole formalism which can be applied only to closed phenomena must in all such respects be considered a rational generalization of classical physics.

In view of the influence of the mechanical conception of nature on philosophical thinking, it is understandable that one has sometimes seen in the notion of complementarity a reference to the subjective observer, incompatible with the objectivity of scientific description. Of course, in every field of experience we must retain a sharp distinction between the observer and the content of the observations, but we must realize that the discovery of the quantum of action has thrown new light on the very foundation of the description of nature, and revealed hitherto unnoticed presuppositions to the rational use of the concepts on which the communication of experience rests. In quantum physics, as we have seen, an account of the functioning of the measuring instruments is indispensable to the definition of phenomena and we must, so-to-say, distinguish between subject and object in such a way that each single case secures the unambiguous application of the elementary physical concepts used in the description. Far from containing any mysticism foreign to the spirit of science, the notion of complementarity points to the logical conditions for description and comprehension of experience in atomic physics.

The epistemological lesson of atomic physics has naturally, just as have earlier advances in physical science, given rise to renewed consideration of the use of our means of communication for objective description in other fields of knowledge. Not least the emphasis placed on the observational problem raises the questions of the position of living organisms in the description of nature and of our own situation as thinking and acting beings. Even though it was, to some extent, possible within the frame of classical physics to compare organisms with machines, it was clear that such comparisons did not take sufficient account of many of the characteristics of life. The inadequacy of the mechanical concept of nature for the description of man's situation is particularly evident in the difficulties entailed in the primitive distinction between soul and body.

The problems with which we are confronted here are obviously connected with the fact that the description of many aspects of human existence demands a terminology which is not immediately founded on simple physical pictures. However, recognition of the limited applicability of such pictures in the account of atomic phenomena gives a hint as to how biological and psychological phenomena may be comprehended within the frame of objective description. As before, it is here important to be aware of the separation between the observer and the content of the communications. While in the mechanical conception of nature the subject-object distinction was fixed, room is provided for a wider description through the recognition that the consequent use of our concepts requires different placings of such a separation.

Without attempting any exhaustive definition of organic life, we may say that a living organism is characterized by its integrity and adaptability, which implies that a description of the internal functions of an organism and its reaction to external stimuli often requires the word *purposeful*, which is foreign to physics and chemistry. Although the results of atomic physics have found a multitude of applications in biophysics and biochemistry, the closed individual quantum phenomena exhibit, of course, no feature suggesting the notion of life. As we have seen, the description of atomic phenomena, exhaustive within a wide domain of experience, is based on the free use of such measuring instruments as are necessary to the proper application of the elementary concepts. In a living organism, however, such a distinction between the measuring instruments and the objects under investigation can hardly be fully carried through, and we must be prepared that every experimental arrangement whose aim is a description of the functioning of the organism, which is well defined in the sense of atomic physics, will be incompatible with the display of life.

In biological research, references to features of wholeness and purposeful reactions of organisms are used together with the increasingly detailed information on structure and regulatory processes that has resulted in such great progress, not least in medicine. Here, we have to do with a practical approach to a field where the means of expression used for the description of its various aspects refer to mutually exclusive conditions of observation. In this connection, it must be realized that the attitudes termed mechanistic and finalistic are not contradictory points of view, but rather exhibit a complementary relationship which is connected with our position as observers of nature. To avoid misunderstanding, however, it is essential to note that — in contrast to the account of atomic regularities — a description of organic life and an evaluation of its possibilities of development cannot aim at completeness, but only at sufficient width of the conceptual framework.

In the account of psychical experiences, we meet conditions of observation and corresponding means of expression still further removed from the terminology of physics. Quite apart from the extent to which the use of words like "instinct" and "reason" in the description of animal behavior is necessary and justifiable, the word " consciousness," applied to oneself as well as to others, is indispensable when describing the human situation. While the terminology adapted to orientation in the environment could take as its starting point simple physical pictures and ideas of causality, the account of our states of mind required a typical complementary mode of description. Indeed, the use of words like "thought" and "feeling" does not refer to a firmly connected causal chain, but to experiences which exclude each other because of different distinctions between the conscious content and the background which we loosely term ourselves.

The relation between the experience of a feeling of volition and conscious pondering on motives for action is especially instructive. The indispensability of such apparently contrasting means of expression to the description of the richness of conscious life strikingly reminds us of the way in which elementary physical concepts are used in atomic physics. In such a comparison, however, we must recognize that psychical experience cannot be subjected to physical measurements and that the very concept of volition does not refer to a generalization of a deterministic description, but from the outset points to characteristics of human life. Without entering into the old philosophical discussion of freedom of the will, I shall only mention that in an objective description of our situation the use of the word "volition" corresponds closely to that of words like "hope" and " responsibility," which are equally indispensable to human communication.

We have here reached problems which touch human fellowship and where the variety of means of expression originates from the impossibility of characterizing by any fixed distinction the role of the individual in the society. The fact that human cultures, developed under different conditions of living, exhibit such contrasts with respect to established traditions and social patterns allows one, in a certain sense, to call such cultures complementary. However, we are here in no way dealing with definite mutually exclusive features, such as those we meet in the objective description of general problems of physics and psychology, but with differences in attitude which can be appreciated or ameliorated by extended intercourse between peoples. In our time, when increasing knowledge and ability more than ever link the fate of all peoples, international collaboration in science has far-reaching tasks which may be furthered not least by an awareness of the general conditions for human knowledge.

APPENDIX

A. EINSTEIN, B. PODOLSKY AND N. ROSEN

CAN QUANTUM-MECHANICAL DESCRIPTION OF PHYSICAL REALITY BE CONSIDERED COMPLETE?

Phys. Rev. **47** (1935) 777–780

See Introduction to Part II, sect. 2.

MAY 15, 1935 PHYSICAL REVIEW VOLUME 47

Can Quantum-Mechanical Description of Physical Reality Be Considered Complete?

A. Einstein, B. Podolsky and N. Rosen, *Institute for Advanced Study, Princeton, New Jersey*
(Received March 25, 1935)

In a complete theory there is an element corresponding to each element of reality. A sufficient condition for the reality of a physical quantity is the possibility of predicting it with certainty, without disturbing the system. In quantum mechanics in the case of two physical quantities described by non-commuting operators, the knowledge of one precludes the knowledge of the other. Then either (1) the description of reality given by the wave function in quantum mechanics is not complete or (2) these two quantities cannot have simultaneous reality. Consideration of the problem of making predictions concerning a system on the basis of measurements made on another system that had previously interacted with it leads to the result that if (1) is false then (2) is also false. One is thus led to conclude that the description of reality as given by a wave function is not complete.

1.

ANY serious consideration of a physical theory must take into account the distinction between the objective reality, which is independent of any theory, and the physical concepts with which the theory operates. These concepts are intended to correspond with the objective reality, and by means of these concepts we picture this reality to ourselves.

In attempting to judge the success of a physical theory, we may ask ourselves two questions: (1) "Is the theory correct?" and (2) "Is the description given by the theory complete?" It is only in the case in which positive answers may be given to both of these questions, that the concepts of the theory may be said to be satisfactory. The correctness of the theory is judged by the degree of agreement between the conclusions of the theory and human experience. This experience, which alone enables us to make inferences about reality, in physics takes the form of experiment and measurement. It is the second question that we wish to consider here, as applied to quantum mechanics.

Whatever the meaning assigned to the term *complete*, the following requirement for a complete theory seems to be a necessary one: *every element of the physical reality must have a counterpart in the physical theory*. We shall call this the condition of completeness. The second question is thus easily answered, as soon as we are able to decide what are the elements of the physical reality.

The elements of the physical reality cannot be determined by *a priori* philosophical considerations, but must be found by an appeal to results of experiments and measurements. A comprehensive definition of reality is, however, unnecessary for our purpose. We shall be satisfied with the following criterion, which we regard as reasonable. *If, without in any way disturbing a system, we can predict with certainty (i.e., with probability equal to unity) the value of a physical quantity, then there exists an element of physical reality corresponding to this physical quantity.* It seems to us that this criterion, while far from exhausting all possible ways of recognizing a physical reality, at least provides us with one such way, whenever the conditions set down in it occur. Regarded not as a necessary, but merely as a sufficient, condition of reality, this criterion is in agreement with classical as well as quantum-mechanical ideas of reality.

To illustrate the ideas involved let us consider the quantum-mechanical description of the behavior of a particle having a single degree of freedom. The fundamental concept of the theory is the concept of *state*, which is supposed to be completely characterized by the wave function ψ, which is a function of the variables chosen to describe the particle's behavior. Corresponding to each physically observable quantity A there is an operator, which may be designated by the same letter.

If ψ is an eigenfunction of the operator A, that is, if

$$\psi' \equiv A\psi = a\psi, \qquad (1)$$

where a is a number, then the physical quantity A has with certainty the value a whenever the particle is in the state given by ψ. In accordance with our criterion of reality, for a particle in the state given by ψ for which Eq. (1) holds, there is an element of physical reality corresponding to the physical quantity A. Let, for example,

$$\psi = e^{(2\pi i/h)p_0 x}, \qquad (2)$$

where h is Planck's constant, p_0 is some constant number, and x the independent variable. Since the operator corresponding to the momentum of the particle is

$$p = (h/2\pi i)\partial/\partial x, \qquad (3)$$

we obtain

$$\psi' = p\psi = (h/2\pi i)\partial\psi/\partial x = p_0\psi. \qquad (4)$$

Thus, in the state given by Eq. (2), the momentum has certainly the value p_0. It thus has meaning to say that the momentum of the particle in the state given by Eq. (2) is real.

On the other hand if Eq. (1) does not hold, we can no longer speak of the physical quantity A having a particular value. This is the case, for example, with the coordinate of the particle. The operator corresponding to it, say q, is the operator of multiplication by the independent variable. Thus,

$$q\psi = x\psi \neq a\psi. \qquad (5)$$

In accordance with quantum mechanics we can only say that the relative probability that a measurement of the coordinate will give a result lying between a and b is

$$P(a, b) = \int_a^b \bar{\psi}\psi dx = \int_a^b dx = b - a. \qquad (6)$$

Since this probability is independent of a, but depends only upon the difference $b - a$, we see that all values of the coordinate are equally probable.

A definite value of the coordinate, for a particle in the state given by Eq. (2), is thus not predictable, but may be obtained only by a direct measurement. Such a measurement however disturbs the particle and thus alters its state. After the coordinate is determined, the particle will no longer be in the state given by Eq. (2). The usual conclusion from this in quantum mechanics is that *when the momentum of a particle is known, its coordinate has no physical reality.*

More generally, it is shown in quantum mechanics that, if the operators corresponding to two physical quantities, say A and B, do not commute, that is, if $AB \neq BA$, then the precise knowledge of one of them precludes such a knowledge of the other. Furthermore, any attempt to determine the latter experimentally will alter the state of the system in such a way as to destroy the knowledge of the first.

From this follows that either (1) *the quantum-mechanical description of reality given by the wave function is not complete* or (2) *when the operators corresponding to two physical quantities do not commute the two quantities cannot have simultaneous reality.* For if both of them had simultaneous reality—and thus definite values—these values would enter into the complete description, according to the condition of completeness. If then the wave function provided such a complete description of reality, it would contain these values; these would then be predictable. This not being the case, we are left with the alternatives stated.

In quantum mechanics it is usually assumed that the wave function *does* contain a complete description of the physical reality of the system in the state to which it corresponds. At first sight this assumption is entirely reasonable, for the information obtainable from a wave function seems to correspond exactly to what can be measured without altering the state of the system. We shall show, however, that this assumption, together with the criterion of reality given above, leads to a contradiction.

2.

For this purpose let us suppose that we have two systems, I and II, which we permit to interact from the time $t=0$ to $t=T$, after which time we suppose that there is no longer any interaction between the two parts. We suppose further that the states of the two systems before $t=0$ were known. We can then calculate with the help of Schrödinger's equation the state of the combined system I+II at any subsequent time; in particular, for any $t>T$. Let us designate the corresponding wave function by Ψ. We cannot, however, calculate the state in which either one of the two systems is left after the interaction. This, according to quantum mechanics, can be done only with the help of further measurements, by a process known as the *reduction of the wave packet*. Let us consider the essentials of this process.

Let a_1, a_2, a_3, $\cdots$ be the eigenvalues of some physical quantity A pertaining to system I and $u_1(x_1)$, $u_2(x_1)$, $u_3(x_1)$, $\cdots$ the corresponding eigenfunctions, where x_1 stands for the variables used to describe the first system. Then Ψ, considered as a function of x_1, can be expressed as

$$\Psi(x_1, x_2) = \sum_{n=1}^{\infty} \psi_n(x_2) u_n(x_1), \qquad (7)$$

where x_2 stands for the variables used to describe the second system. Here $\psi_n(x_2)$ are to be regarded merely as the coefficients of the expansion of Ψ into a series of orthogonal functions $u_n(x_1)$. Suppose now that the quantity A is measured and it is found that it has the value a_k. It is then concluded that after the measurement the first system is left in the state given by the wave function $u_k(x_1)$, and that the second system is left in the state given by the wave function $\psi_k(x_2)$. This is the process of reduction of the wave packet; the wave packet given by the

infinite series (7) is reduced to a single term $\psi_k(x_2) u_k(x_1)$.

The set of functions $u_n(x_1)$ is determined by the choice of the physical quantity A. If, instead of this, we had chosen another quantity, say B, having the eigenvalues b_1, b_2, b_3, $\cdots$ and eigenfunctions $v_1(x_1)$, $v_2(x_1)$, $v_3(x_1)$, $\cdots$ we should have obtained, instead of Eq. (7), the expansion

$$\Psi(x_1, x_2) = \sum_{s=1}^{\infty} \varphi_s(x_2) v_s(x_1), \qquad (8)$$

where φ_s's are the new coefficients. If now the quantity B is measured and is found to have the value b_r, we conclude that after the measurement the first system is left in the state given by $v_r(x_1)$ and the second system is left in the state given by $\varphi_r(x_2)$.

We see therefore that, as a consequence of two different measurements performed upon the first system, the second system may be left in states with two different wave functions. On the other hand, since at the time of measurement the two systems no longer interact, no real change can take place in the second system in consequence of anything that may be done to the first system. This is, of course, merely a statement of what is meant by the absence of an interaction between the two systems. Thus, *it is possible to assign two different wave functions* (in our example ψ_k and φ_r) *to the same reality* (the second system after the interaction with the first).

Now, it may happen that the two wave functions, ψ_k and φ_r, are eigenfunctions of two noncommuting operators corresponding to some physical quantities P and Q, respectively. That this may actually be the case can best be shown by an example. Let us suppose that the two systems are two particles, and that

$$\Psi(x_1, x_2) = \int_{-\infty}^{\infty} e^{(2\pi i/h)(x_1-x_2+x_0)p} dp, \qquad (9)$$

where x_0 is some constant. Let A be the momentum of the first particle; then, as we have seen in Eq. (4), its eigenfunctions will be

$$u_p(x_1) = e^{(2\pi i/h) p x_1} \qquad (10)$$

corresponding to the eigenvalue p. Since we have here the case of a continuous spectrum, Eq. (7) will now be written

$$\Psi(x_1, x_2) = \int_{-\infty}^{\infty} \psi_p(x_2) u_p(x_1) dp, \quad (11)$$

where

$$\psi_p(x_2) = e^{-(2\pi i/h)(x_2-x_0)p}. \quad (12)$$

This ψ_p however is the eigenfunction of the operator

$$P = (h/2\pi i)\partial/\partial x_2, \quad (13)$$

corresponding to the eigenvalue $-p$ of the momentum of the second particle. On the other hand, if B is the coordinate of the first particle, it has for eigenfunctions

$$v_x(x_1) = \delta(x_1 - x), \quad (14)$$

corresponding to the eigenvalue x, where $\delta(x_1-x)$ is the well-known Dirac delta-function. Eq. (8) in this case becomes

$$\Psi(x_1, x_2) = \int_{-\infty}^{\infty} \varphi_x(x_2) v_x(x_1) dx, \quad (15)$$

where

$$\varphi_x(x_2) = \int_{-\infty}^{\infty} e^{(2\pi i/h)(x-x_2+x_0)p} dp$$

$$= h\delta(x-x_2+x_0). \quad (16)$$

This φ_x, however, is the eigenfunction of the operator

$$Q = x_2 \quad (17)$$

corresponding to the eigenvalue $x+x_0$ of the coordinate of the second particle. Since

$$PQ - QP = h/2\pi i, \quad (18)$$

we have shown that it is in general possible for ψ_k and φ_r to be eigenfunctions of two noncommuting operators, corresponding to physical quantities.

Returning now to the general case contemplated in Eqs. (7) and (8), we assume that ψ_k and φ_r are indeed eigenfunctions of some noncommuting operators P and Q, corresponding to the eigenvalues p_k and q_r, respectively. Thus, by measuring either A or B we are in a position to predict with certainty, and without in any way

disturbing the second system, either the value of the quantity P (that is p_k) or the value of quantity Q (that is q_r). In accordance with our criterion of reality, in the first case we must consider the quantity P as being an element of reality, in the second case the quantity Q is an element of reality. But, as we have seen, both wave functions ψ_k and φ_r belong to the same reality.

Previously we proved that either (1) the quantum-mechanical description of reality given by the wave function is not complete or (2) when the operators corresponding to two physical quantities do not commute the two quantities cannot have simultaneous reality. Starting then with the assumption that the wave function does give a complete description of the physical reality, we arrived at the conclusion that two physical quantities, with noncommuting operators, can have simultaneous reality. Thus the negation of (1) leads to the negation of the only other alternative (2). We are thus forced to conclude that the quantum-mechanical description of physical reality given by wave functions is not complete.

One could object to this conclusion on the grounds that our criterion of reality is not sufficiently restrictive. Indeed, one would not arrive at our conclusion if one insisted that two or more physical quantities can be regarded as simultaneous elements of reality *only when they can be simultaneously measured or predicted*. On this point of view, since either one or the other, but not both simultaneously, of the quantities P and Q can be predicted, they are not simultaneously real. This makes the reality of P and Q depend upon the process of measurement carried out on the first system, which does not disturb the second system in any way. No reasonable definition of reality could be expected to permit this.

While we have thus shown that the wave function does not provide a complete description of the physical reality, we left open the question of whether or not such a description exists. We believe, however, that such a theory is possible.

PART III

SELECTED CORRESPONDENCE

INTRODUCTION

The letters to and from Bohr, quoted or referred to in the Introductions to Parts I and II, are reproduced here in the original language, in alphabetical order according to correspondents. Relevant letters to and from other correspondents are also included. In the few cases where passages of a purely private nature have been omitted, this is indicated in the text. Letters in Danish are followed by a full translation. This also applies to the German letters with a few exceptions.

The editors have used their discretion tacitly to correct "trivial" mistakes, e.g. in spelling and punctuation. We have tried, however, to preserve "characteristic" mistakes. Italics are used to indicate underlining in the original text. Underlined italics represent double underlining in the original text.

In the reproduction of the letters we have attempted to make the lay-out of letterheads etc. correspond as closely as possible to that of the original letters.

The list preceding the letters provides references to the pages in the Introductions where the letters appear so that the reader can readily find the context in which a particular letter is quoted. A page number in parentheses indicates that the letter is merely referred to.

Unless otherwise indicated, the letters are taken from BSC, AHQP.

CORRESPONDENCE INCLUDED

[1] Permission granted by the Albert Einstein Archives, the Hebrew University of Jerusalem, Israel.

		Reproduced p.	Translation p.	Quoted p.
WOLFGANG PAULI				
Bohr to Pauli,	25 January 1933	463	465	(14)
Pauli to Peierls,	22 May 1933[2]	–	468	(14)
Bohr to Pauli,	15 February 1934	469	31	31
Bohr to Pauli,	24 February 1934	471	472	(31)
Bohr to Pauli,	15 March 1934	474	477	31; 35
Pauli to Heisenberg,	15 June 1935[2]	480	251	251
Bohr to Pauli,	16 May 1947	Vol. 6, p. [451]	–	267
Pauli to Bohr,	29 May 1947	Vol. 6, p. [455]	–	268
Pauli to Bohr,	18 February 1948	277	–	277
Pauli to Bohr,	17 August 1948	268	–	268
Bohr to Pauli,	15 September 1948	270	–	270
Pauli to Rosenfeld,	5 July 1949[3]	482	–	(50)
Pauli to Rosenfeld,	11 July 1949[3]	485	–	(50)
Bohr to Pauli,	15 August 1949	486	490	Vol. 9, p. [79]
Pauli to Bohr,	21 August 1949	493	–	(50)
Bohr to Pauli,	29 August 1949[4]	496	–	(50)
Pauli to Bohr,	29 August 1949	498	–	(50)
Pauli to Bohr,	25 October 1949[5]	500	–	(50)
LÉON ROSENFELD				
Bohr to Rosenfeld, 2 September 1949		501	502	(50)
ERWIN SCHRÖDINGER				
Schrödinger to Bohr, 13 October 1935[6]		503	507	(255)
Bohr to Schrödinger, 26 October 1935[6]		510	511	(255)
VICTOR WEISSKOPF				
Weisskopf to Rosenfeld, 2 December [1933]		512	14	14
Bohr to Weisskopf, 5 December 1933		513	15	15

[2] PWB II.
[3] Rosenfeld Papers, NBA.
[4] With enclosure.
[5] NBA, not microfilmed.
[6] Schrödinger Archiv, Alpbach; AHQP, microfilm no. 92.

ALBERT EINSTEIN

EINSTEIN TO BOHR, 4 April 1949
[Typewritten]

THE INSTITUTE FOR ADVANCED STUDY
SCHOOL OF MATHEMATICS
PRINCETON, NEW JERSEY
den 4. April 1949

Lieber Bohr:

Ich danke Ihnen herzlich für alles, was Sie gelegentlich eines an sich so unwesentlichen Anlasses an freundlichen Bemühungen für mich aufgebracht haben. Auch für die freundliche Gratulation seitens der Mitglieder des Kopenhagener Institutes meinen herzlichen Dank.

Jedenfalls ist dies eine der Gelegenheiten, die nicht von der bangen Frage abhängt, ob Gott wirklich würfelt und ob wir an einer der physikalischen Beschreibung zugänglichen Realität festhalten sollen oder nicht. In meiner Antwort auf die im Schilpp'schen Buche erscheinenden Arbeiten habe ich wieder mein einsames altes Liedchen gesungen, das mich selber an den Refrain jenes alten Büchleins erinnert:

Ueber diese Rede des Kandidaten Jobses
Allgemeines Schütteln des Kopses.

Mit herzlichen Grüssen
Ihr
Albert Einstein

Translation, see p. [281]

BOHR TO EINSTEIN, 11 April 1949
[Carbon copy]

[København,] 11. April 1949.

Lieber Einstein,

Vielen Dank für Ihre freundlichen Zeilen. Es war für uns alle eine grosse Freude, anlässlich Ihres Geburtstages unseren Gefühlen Ausdruck zu geben. Um in demselben scherzhaften Tone zu sprechen, kann ich nicht umhin, über die bangen Fragen zu sagen, dass es sich meines Erachtens nicht darum handelt, ob wir an einer der physikalischen Beschreibung zugänglichen Realität festhalten sollen oder nicht, sondern darum, den von Ihnen gewiesenen Weg weiter zu verfolgen und die logischen Voraussetzungen für die Beschreibung der Realitäten zu erkennen. In meiner frechen Weise möchte ich sogar sagen, dass niemand – und nicht mal der liebe Gott selber – wissen kann, was ein Wort wie würfeln in diesem Zusammenhang heissen soll.

Mit herzlichen Grüssen
Ihr
[Niels Bohr]

Translation, see p. [282]

WERNER HEISENBERG

HEISENBERG TO BOHR, 16 June 1929
[Handwritten]

Chicago 16. Juni [19]29.

Lieber Bohr!

Zunächst will ich Dir und den anderen Physikern noch herzlich für Eure Karte aus Kopenhagen danken, ich wäre gern bei Eurer Tagung dabei gewesen und hoffe jedenfalls, bei der nächsten derartigen Zusammenkunft wieder mittun zu dürfen. Hier in der neuen Welt führ ich ein unruhiges Leben und reise fast jede Woche 1000 km hin und her; das hat den Nachteil, dass man nicht zur Arbeit kommt, aber den Vorteil, viel von der Welt zu sehen. Neulich war ich in Californien (Los Angeles) für eine Woche und auf dem Weg dorthin hab ich einige interessante Ausflüge unternommen. Zuerst hab ich am Rande der Rocky Mountains Halt gemacht und bin dort auf einige Berge gestiegen; die Berge sehen dort ein wenig wie in Norwegen aus, weniger steil und weniger kompliziert, als die Alpen. Einige sind aber ziemlich hoch (4300 m); da das Klima sehr trocken ist, gibt es keine Gletscher. Dann bin ich in Arizona und Colorado mehrere Tage in der Wüste herumgefahren, besonders eine Mondnacht in der Sandwüste war herrlich, und bin in den Colorado-Canyon hinuntergestiegen; die Ebene liegt dort etwa 2300 m hoch, der Fluss fliesst unten in etwa 500 m Höhe; die Wände des Canyon leuchten in allen Farben zwischen braunrot und violett. Auch von Pasadena aus hab ich mit Zwicky und mit Millikans Sohn eine Bergtour unternommen die uns auf eine Spitze (3500 m) am Rand der Wüste führte; man sieht von dort unmittelbar hinunter auf einen Teil der Mojave-Wüste, der noch unter dem Meeresniveau liegt und kann in der Ferne die ersten Bergketten jenseits der mexikanischen Grenze erkennen. – Hier in Chicago ist mein Leben allerdings weniger romantisch, immerhin bin ich gestern Nachmittag mit Bekannten auf dem Lake Michigan gesegelt und hab dabei Kopenhagener Erinnerungen aufgefrischt. Das Boot hier ist ganz ähnlich, wie Euer Kopenhagener Boot, nur hat es noch einen zweiten kleinen Mast und ein Besan-segel. Zur Ehre von Bjerrum muss ich aber hervorheben, dass die Diziplin u. Ordnung auf Eurem Boot viel besser war, wie hier; wenn meine Freunde hier einmal bei etwas "besserem Wind" segeln würden, müsste sich ein schauderhaftes Durcheinander einstellen.

Damit Du aber nicht denkst, dass ich die Physik völlig vergesse, will ich noch ein wenig davon erzählen. Ich hab hauptsächlich an der Arbeit von Pauli und mir weitergearbeitet und über die Frage von nichtkombinierenden Termsystemen in der Quantentheorie der Wellen nachgedacht. Das gibt ganz interessante Resultate, insbesondere glaub ich jetzt, ein Verfahren gefunden zu haben, wie man auch ohne die ε-Zusatzglieder von Pauli u. mir (vielleicht erinnerst Du Dich daran) auskommen kann. Auch stellte sich heraus, dass man in dieser Wellentheorie ziemlich ungezwungen Prozesse einführen kann, bei denen ein Proton u. ein Elektron sich zusammentun und ein Lichtquant ergeben – was in der Konfigurationsraumtheorie ja nicht möglich ist. Dagegen weiss ich über die Diracübergänge garnichts neues.

Auch hab ich gelegentlich meiner Vorlesung hier darüber nachgedacht, wie man die Unsicherheitsrelationen der Wellenamplituden erläutern könnte; mir ist auch ein mögliches Schema eingefallen, aber ich bin noch nicht recht zufrieden und glaub, dass es erst halb richtig ist; aber mich würde interessieren, gelegentlich Deine Meinung über diese Frage im allgemeinen zu hören: Der Gedanke ist etwa dieser:

Irgend eine Messung wird von vornherein nicht E und H an einem exakten *Punkt* des Raumes liefern, sondern Mittelwerte über eventuell sehr kleine Raumstücke. Sei das Volumen des Raumstückes Δv, so lauten die V[ertauschungs-]R[elationen] für E und Potentiale Φ

$$E_i \Phi_k - \Phi_k E_i = \delta_{ik}\, 2hci\frac{1}{\Delta v}, \tag{1}$$

wobei unter E_i und Φ_k jetzt Mittelwerte über das Raumstück Δv verstanden sind. Also würde man Unbestimmtheitsrelationen der Form

$$\Delta E_i \Delta \Phi_k \gtrsim \delta_{ik}\frac{hc}{\Delta v} \quad \text{erwarten;} \quad \text{oder} \quad \Delta E_x \Delta H_y \gtrsim \frac{hc}{\Delta v \Delta \ell} \quad \text{u. zykl.} \tag{2}$$

wenn $\Delta v = (\Delta \ell)^3$ als kubisch angenommen wird.

Eine Methode der Herleitung wäre die: Die Energie in Δv muss sein

$$\text{Energie} = \frac{\Delta v}{8\pi}(E^2 + H^2).\tag{3}$$

Wären E und H genau bekannt, so ergäbe sich bei kleinem Δv ein Widerspruch zum Partikelbild, da die Energie stets aus diskreten Lichtquanten besteht. Die grössten, überhaupt nachweisbaren Lichtquanten sind solche, deren Wellenlänge nicht viel kleiner als $\Delta\ell$ ist; die Ungenauigkeit der linken Seite von (3) müsste also, um den Widerspruch zu beseitigen, etwa gleich diesem grössten nachweisbaren hv sein, also

$$\Delta E_x \Delta H_y \gtrsim \frac{hc}{\Delta v \Delta\ell}.$$

Eine andere, noch weniger saubere Ableitung dachte ich mir folgendermassen:

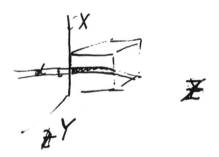

Zur Messung der Felder E_x u. H_y schicke man zwei Elektronenstrahlen der Breite d durch den Kubus $\Delta\ell^3$, den einen in der positiven, den andern in der negativen Z-Richtung. Der Ablenkungswinkel ist

$$\frac{1}{p_z} e(E_x \pm \frac{p_z}{\mu c} H_y) \cdot \frac{\mu\Delta\ell}{p_z},$$

also die Genauigkeit der H_y und E_x-Bestimmung

$$\Delta E_x = \frac{\lambda}{d} \cdot \frac{p_z}{e} \cdot \frac{p_z}{\mu\Delta\ell} = \frac{h}{de} \cdot \frac{p_z}{\mu\Delta\ell}; \quad \Delta H_y = \frac{h}{de} \cdot \frac{p_z}{\mu\Delta\ell} \cdot \frac{\mu c}{p_z}.\tag{4}$$

Dabei ist aber bisher vergessen, dass die in den Strahlen laufenden Elektronen selbst wieder das Feld merklich modifizieren u. zwar um einen teilweise unbekannten Betrag, da man nicht weiss, *wo* im Strahl das Elektron läuft. Die so hervorgerufene Unsicherheit in den Mittelwerten von E_x und H_y über den Kubus $(\Delta\ell)^3$ ist[7]

$$\Delta E_x \sim \frac{ed}{(\Delta\ell)^3}; \quad \Delta H_y \sim \frac{ed}{(\Delta\ell)^3} \cdot \left[\frac{p_z}{\mu c}\right].\tag{5}$$

Aus (4) und (5) folgt dann:

$$\Delta E_x \Delta H_y \gtrsim \frac{hc}{\Delta v \Delta\ell}.$$

---

[7] Introduction to Part I, ref. 5.

Ich glaube, dass wesentliche Züge hieran richtig sind, aber die Benützung der Mittelwerte von E und H in dem Kubus ist noch nicht ganz sauber. Aber würdest Du grundsätzlich so eine Diskussion für vernünftig halten?

Nun wünsch ich Dir noch einen recht schönen Sommer! Viele herzliche Grüsse!

Dein
Werner Heisenberg

Translation, see p. [5]

HEISENBERG TO PAULI, 12 March 1931
[Handwritten]

[Leipzig,] 12.3.[19]31.

Lieber Pauli!

Vielen Dank für Deinen Brief. Deine Kritik der Kopenhagener Physik ist insofern berechtigt, als natürlich niemand etwas Positives weiss. Immerhin ist die Bohrsche Kritik der Landau–Peierls Arbeit ganz interessant. Bohr ist mit den Ungenauigkeitsrelationen von L.–P. einverstanden. Die Ableitung derselben hält er an einigen Stellen für schlampig, aber dieser Punkt ist nicht wesentlich. Die Hauptkritik richtet sich vielmehr gegen die Schlüsse, die L.–P. aus den U[ngenauigkeits] R[elationen] ziehen. Bohr meint, dass die U.R. keineswegs bedeuten, dass die relativ. Wellenmechanik zu eng sei und einem allgemeineren Formalismus weichen müsse. Vielmehr sagt Bohr: auch in der nichtrelativ. Wellenmech. ist nur ein kleiner Teil aller Operatoren messbar. In der relativ. W.M. sind gewisse Operatoren, z.B. der Ort des Elektrons, nicht mehr unmittelbar messbar. Trotzdem könnte man das Resultat vieler Ortsmessungen etwa an einem $1S$-Zustand statistisch vorhersagen und die Form der Wellenfunktionen ist daher exakt nachprüfbar. Es ist daher unrichtig, wenn L.P. behaupten, die Wellenfunktion sei in der relativ. W.M. prinzipiell unbeobachtbar. Bohr meint also, alles in allem, dass man in der jetzigen relativ. W.M. (Dirac + Qu. Elektr. Dyn.) ein *befriedigendes* Schema habe, das erst verlassen werden müsste und könnte, wenn man die Grösse h/mc diskutiert. Bohr glaubt also *nicht* an die Möglichkeit des Grenzüberganges $m \to 0$. Er sagt etwa so: Man versteht bisher die Theorie im Limes $c \to \infty$ oder auch $m \to \infty$. Dagegen wird Lim $m \to 0$ ebensowenig Sinn haben, wie Lim $c \to 0$.

Er kann ja Recht haben, aber "nichts gewisses weiss man nicht". Im Einzelnen machte Bohr einige ganz hübsche Bemerkungen. L. u. P. sind besonders stolz darauf, dass man in der nichtrelativ. Theorie die Energie in beliebig kurzer Zeit messen kann. Bohr meint, diese Möglichkeit helfe jedenfalls nicht weiter, als eine langdauernde Energiemessung, da man *vor* der E. Messg. lange Zeit braucht. Man kann z.B. einen Spektralapparat aufstellen, eine Platte einlegen und den Spalt lange offen lassen; danach kann man dann *plötzlich* die Platte entwickeln und hat so eine kurzdauernde Energiemessung vorgenommen. – Dass nicht alle Operatoren in der W.M. messbar sind, geht auch besonders aus folgendem hervor: Man messe etwa den Impuls eines Elektrons *im* Gamowberg. Er würde sich imaginär ergeben müssen, wenn er messbar wäre. – Also: die Nichtmessbarkeit eines Operators ist *kein* gutes Kriterium für die Nicht-Geschlossenheit der Theorie.

Der Kopenhagener Film war sehr gut. Von den Reklamebildern auf der Tafel im Auditorium muss ich einige hier wiedergeben, da sie sehr gut waren z.B.: Einige unverständliche Zeichen mit der Unterschrift: Dirac's holes, nedsat pris.

Ein Bild von Landau mit dem Text:

Automatisk Grammophon.

Stort Udvalg. Hvert Stykke kun 5 Øre.

Politik

Kaerlighed (10 Øre)

Kunst

Religion

Specialitet: *Pligt*

Teoretisk Fysik (Let beskadiget, 2 Øre).

Sonst wars in jeder Beziehung sehr gemütlich.
Also viele Grüsse

Dein

W. Heisenberg

Translation

[Leipzig,] March 12, [19]31

Dear Pauli,

Many thanks for your letter. Your criticism of the Copenhagen physics is justified in so far as, of course, nobody knows anything definite. Still, Bohr's criticism of Landau and Peierls's work is quite interesting. Bohr agrees with the uncertainty relations of Landau and Peierls. He considers their derivations to be sloppy in some places, but this point is not essential. The main criticism is rather directed at the conclusions that Landau and Peierls draw from

[441]

the uncertainty relations. Bohr thinks that the uncertainty relations do not at all mean that relativistic wave mechanics is too narrow and has to give way for a more general formalism. Bohr says rather: also in non-relativistic wave mechanics only a small part of all operators is measurable. In relativistic wave mechanics certain operators, e.g. the position of the electrons, are no longer directly measurable. Still, one could predict statistically the result of many position measurements, say in a $1S$-state, and the form of the wave function may thus be tested exactly. Hence it is wrong when Landau–Peierls assert that the wave function in relativistic wave mechanics is unobservable in principle. Thus Bohr thinks, all in all, that in the present relativistic wave mechanics (Dirac + quantum electrodynamics) one has got a *satisfactory* scheme, which should and could only be abandoned when you discuss the quantity h/mc. Thus Bohr does *not* believe in the possibility of the limiting process $m \to 0$. He says something like this: We understand so far the theory in the limit $c \to \infty$ or $m \to \infty$. In contrast, the limit $m \to 0$ will make just as little sense as the limit $c \to 0$.

He may of course be right, but "nichts gewisses weisst man nicht"[8]. Concerning the details Bohr made some quite nice remarks. Landau and Peierls are particularly proud that in the non-relativistic theory energy may be measured in an arbitrarily short time. Bohr thinks that this possibility does not help us any more than an energy measurement of long duration, since one needs a long time *before* the energy measurement. One may, for example, set up a spectrograph, put in a plate and let the slit be open for a long time, then one may *suddenly* develop the plate, and one has thus performed a fast energy measurement. That not all operators in wave mechanics are measurable is especially apparent from the following consideration: Imagine that one measures something like the momentum of an electron *inside* the Gamow barrier. It had to turn out imaginary if it were measurable. Thus: Non-measurability of an operator is *not* a good criterion for incompleteness of the theory.

The Copenhagen film was very good. I must here reproduce some of the pictures from the advertisements on the blackboard in the lecture room which were very good, e.g. some incomprehensible signs with the text: Dirac holes, reduced price.

A picture of Landau with the text:

Automatic grammophone
Large Selection. Only 5 Øre a piece
Politics

---

[8] A Bavarian saying which can be translated as "one can know nothing for certain". Cf. PWB II, p. 68.

Love (10 Øre)
Art
Religion
Specialty: *Duty*

Theoretical Physics (slightly damaged, 2 Øre).

Also otherwise it was in every respect great fun.
Now many greetings
Yours,
W. Heisenberg

BOHR TO HEISENBERG, 13 March 1933
[Carbon copy]

[København,] 13. Marts [19]33.

Kære Heisenberg,

Jeg er mere ked, end jeg kan sige, af at maatte telegrafere, at Christian og jeg i sidste Øjeblik alligevel er blevet forhindret i at komme ned til Dig. Vi havde begge glædet os saa meget til Rejsen og talt saa meget om den vidunderlige Tid, jeg tilbragte i Din Skihytte i Fjor, og jeg vil i de følgende Dage stadig følge Dig og Bloch og Weizsäcker i Tankerne paa jeres Ture og i jeres Hytteliv. Jeg haaber meget, at vi snart igen alle skal kunne være sammen paa et saadant Æventyr, som Besøget hos Dig i Fjor var og altid vil staa for mig; men som jeg telegraferede, gør Arbejdet her det denne Gang umuligt for mig at komme.

Det, som har krydset mine Planer, er de uforudsete Vanskeligheder, der stadig har forhindret Færdiggørelsen af Rosenfelds og min Undersøgelse over de elektrodynamiske Problemer, der, hver Gang vi troede, at vi var naaet til Bunds, igen aabenbarede nye Afgrunde. Selv om Grundlaget er det samme, er hele Stemningen om Arbejdet undergaaet flere fuldstændige Drejninger siden Blochs sidste Besøg i København. Diskussionen har navnlig drejet sig om Realiteten af de uforudsigelige Variationer i Feltkomponenterne, som efter Kvanteelektrodynamikken er uadskilleligt forbundne med ethvert Kendskab til Felternes Lyskvantesammensætning, og som derfor ogsaa optræder i det tomme Rum. Ikke mindst under et Besøg i forrige Uge af Pauli, der efter en – selv efter hans Maalestok at regne – meget kraftig Skældsordsbyge pr. Brev, i København viste sig som en ren Engel i Mildhed og Medfølelse, blev jeg klar over, at Forstaaelsen af dette Problem endnu lod meget tilbage at ønske.

Vi er efterhaanden kommet saa langt ud i Komplementaritetsbetragtninger og Kompensationsmekanismer, at selv Pauli ikke mere tør tage Stilling dertil. Alligevel tror jeg, at det i de allersidste Dage virkelig er lykkedes mig at puffe Sandheden ud i den bundløse Afgrund, hvor den hører hjemme, og hvis ingen ny Katastrofe sker, skal der ligge et færdigt Manuskript og vente paa Dig i München, naar Du kommer tilbage fra Hytten.

Selv om dette Arbejde har været Anledningen til mine Vanskeligheder, er det dog ikke den egentlige Grund til, at jeg har maattet opgive at komme med paa Skituren; ja, det vilde jo endda være en stor Glæde og Hjælp for mig at kunne diskutere Situationen med Dig. Paa Grund af Arbejdets Trækken i Langdrag og min sædvanlige Optimisme med Hensyn til dets Afslutning har jeg imidlertid i mange Uger opsat alle mine andre Forpligtelser, som hviler særlig tungt paa mig, fordi jeg om en Maanedstid sammen med min Kone rejser til Amerika for et Fjerdingaar. Jeg havde stadig haabet, at det altsammen kunde naas, men efter at jeg i de sidste Dage har været lidt træt og er kommet yderligere tilbage, er det blevet mig klart, at jeg ikke foreløbig kan forlade København, og netop da Dit venlige Telegram kom, stod jeg i Begreb med selv at telegrafere til Dig for at bringe dette sørgelige Budskab, som ikke alene betyder en stor Skuffelse for mig, men som jeg endda er bange for vil give Anledning til Ulejlighed og Besvær for Dig.

Paa Grund af den hele for mig lidt indviklede Situation i Aar har vi endda maattet opgive vor sædvanlige Paaskekonferens og udskyde den til September. Jeg haaber alligevel, at vi skal ses, før jeg rejser til Amerika, og skal skrive nærmere derom, naar jeg sender Manuskriptet. Hvis det ikke lader sig gøre, haaber jeg i det mindste, at Du kan komme hertil allerede i August, saa at vi baade kan gøre en Sejltur med Bjerrum og Drengene og faa truffet Forberedelser til Solvaymødet før Konferensen her. Men som sagt herom skriver jeg snart nærmere og sender denne Gang endnu kun de venligste Hilsner til Din Moder, Bloch og Weizsäcker og Dig selv fra os alle.

Din

[Niels Bohr]

Translation

[Copenhagen,] March 13, [19]33

Dear Heisenberg,

I am more sorry than I can say because I had to wire you that at the last moment Christian and I were prevented from visiting you after all. We had both been looking forward very much to the trip and had talked so much about the wonderful time I spent in your skiing cabin last year, and in the next few days I shall be following you and Bloch and Weizsäcker in my thoughts on your excursions and in your cabin life. I hope very much that we may soon join each other again in such an adventure as the visit to you last year was, and always will remain, to me. But, as I wired you, this time the work here makes it impossible for me to come.

What has thwarted my plans are the unforeseen difficulties that have still prevented the completion of the investigation by Rosenfeld and me of the electrodynamic problems which, whenever we believed that we had reached the bottom, revealed new abysses. Although the basis is the same, the entire mood of the work has undergone several complete turnabouts since Bloch's latest visit to Copenhagen. The discussion has especially been about the reality of the unpredictable variations in the field components, which according to quantum electrodynamics are inextricably connected with any knowledge of the light quantum composition of the fields, and which therefore also appear in empty space. Especially during a visit last week from Pauli, who after an epistolatory shower of invective – violent even according to his standards – showed himself in Copenhagen to be a pure angel of gentleness and compassion, did it become clear to me that our understanding of this problem left much to be desired. We have gradually become so deeply immersed in complementarity considerations and compensation mechanisms that even Pauli dare no longer commit himself. Yet I believe that during the very last few days I have really succeeded in pushing the truth down into the bottomless pit where it belongs and, if no further catastrophe occurs, a finished manuscript will be awaiting you in Munich on your return from the cabin.

Even though this work has been the origin of my difficulties, it is not the real reason why I had to give up joining you on the skiing trip. Indeed, it would have been a great pleasure and help for me to be able to discuss the situation with you. Because of the slow progress of the work and my usual optimism with respect to its completion, I have, however, for many weeks postponed all other obligations which are resting especially heavily on me because in a month I shall be going to the USA together with my wife for three months. I kept hoping that I could manage all of it, but after having been a little tired in the last few days and falling still further in arrears it has become clear to me that I cannot leave Copenhagen for the time being, and just when your kind telegram arrived, I was myself on the point of wiring you this sad message which not only means a disappointment to me, but which I even fear will cause you inconvenience and trouble.

Because of this whole, for me somewhat tangled, situation this year, we have even had to give up our usual Easter conference and have postponed it till September. Nevertheless I hope that we shall meet before I leave for the USA and I am going to write you more definitely about this when I send the manuscript. If it is not possible, I hope at least that you can come here already in August so that we may go for a sailing trip with Bjerrum and the boys and also make some preparations for the Solvay meeting before the conference here. But, as already mentioned, I am soon going to write you more definitely about that, and this time I am merely sending the kindest regards to your mother, Bloch, Weizsäcker and yourself from us all.

Yours
[Niels Bohr]

HEISENBERG TO BOHR, 12 March 1934[9]
[Handwritten]

Klosters 12.3.[19]34

Lieber Bohr!

Hab vielen Dank für Deinen Brief und das Manuskript Deines Solvay-berichts. Über Deine Absage war ich traurig, es wäre schön gewesen, auf der Schihütte in Ruhe über die physikalischen und die nichtphysikalischen Seiten des Lebens nachzudenken und zu sprechen. Andererseits war ich in den letzten Wochen durch den vergangenen Winter ziemlich erledigt, deshalb hab ich mich entschlossen, diesmal garnicht auf die Schihütte zu gehen, sondern in einem Berghotel der Schweiz einen sehr viel bürgerlicheren und langweiligeren Schiurlaub zu verbringen. So kam es, dass ich auch in den letzten Monaten so selten geschrieben hab. Aber ich will heute "keine Müdigkeit vortäuschen" und versuchen, auch über Physik etwas ausführliches zu schreiben.

Der Inhalt Deines Manuskripts und Deines Briefs an Pauli gefällt mir dem Grundton nach sehr gut. Insbesondere die von Dir angedeutete Vermutung, dass die Bestimmung von e^2/hc einerseits, die Gesetze des Kernaufbaus andererseits unabhängige Probleme seien, die vielleicht getrennt gelöst werden können, scheint mir sehr verlockend. Dagegen ist meine Stimmung gegenüber dem Problem der unendlichen Selbstenergie etwas verschieden von dem, was Du darüber schreibst. Ich bin zwar ganz mit Dir darin einig, dass wir

---

[9] Note that this letter from Heisenberg is *not* a reply to the previous letter from Bohr, as a comparison of the dates will show.

bisher keine Theorie besitzen, die über eine konsequente Anwendung des Korrespondenzprinzips auf die Maxwellsche Theorie wesentlich hinausgeht; daraus folgt auch, dass Punktladungen eine unendliche Selbstenergie haben müssen – schon ohne jeden Hinweis auf den Formalismus der Qu. El. Dyn. Aber mir scheint, dass eine vernünftige Quantisierung des Feldes – über die Korrespondenz hinausgehend – diese unendliche Selbstenergie beseitigen müsste, und zwar genau in der Weise, wie Klein u. Jordan den elektrostatischen Teil der Selbstenergie vermieden haben. Ich halte diesen "Trick" von Klein und Jordan keineswegs für einen oberflächlichen Kunstgriff, sondern für die folgerichtige Ausnützung eines bestimmten formalen Zuges in der Quantentheorie der Wellenfelder, den man in Zukunft noch allgemeiner ausnützen muss. In der Quantentheorie der Wellenfelder scheint nämlich das Problem der *einzelnen* Punktladung analog zur Behandlung eines quantenmechanischen Systems im *tiefsten* Quantenzustand; ebenso, wie nun alle Übergangswahrscheinlichkeiten vom tiefsten Quantenzustand eines Oszillators nach "abwärts" automatisch verschwinden müssen (das ist eine Forderung, die aus dem Korrespondenzprinzip nicht folgt, ihm aber auch nicht widerspricht, und die für die Anfänge der Qu. M. wichtig war), so muss man in der Quantenth. der Wellen fordern, dass die einzelne Punktladung eine triviale Lösung der Bewegungsgleichungen ist, d.h. dass alle Übergangswahrscheinlichkeiten, die zur Entstehung anderer Teilchen Anlass gäben, verschwinden. Diese Forderung widerspricht der Korrespondenz nicht, denn das Korrespondenzprinzip liefert nur Aussagen über Systeme, die viele Elektronen oder Lichtquanten enthalten. Ich glaube also, wenn man aus Korrespondenzargumenten auf die unendliche Selbstenergie schliesst (so wie z.B. Pauli u. ich dies in der Qu. El. Dyn. tun mussten), so ist das in Wirklichkeit genau so falsch, wie wenn man aus Korrespondenzgründen behaupten wollte, das Wasserstoffatom im Grundzustand müsse wegen der Bahnbewegung des Elektrons noch strahlen. Dass es in der Qu. El. Dyn. bisher nicht möglich war, die unendl. Selbstenergie zu beseitigen, bedeutet also wohl nur, dass wir eine falsche Hamiltonfunktion benützen und wahrscheinlich kann man die richtige Hamiltonfunktion erst finden, wenn e^2/hc bestimmt ist u. wenn die Löchertheorie ausgearbeitet ist. – Übrigens bin ich mit der letzten Diracschen Arbeit[10] nicht sehr zufrieden, insbesonders halte ich seine Schlussfolgerungen über den Teil der unendlichen Ladungsdichte, der abgezogen werden muss, für direkt falsch.

Darf ich in diesem Zusammenhang noch gegen einen Satz etwas einwenden,

---

[10] Probably P.A.M. Dirac, *Discussion of the Infinite Distribution of Electrons in the Theory of the Positron,* Proc. Camb. Phil. Soc. **30** (1934) 150–163.

der auf S. 13 Deines Manuskripts steht. "Mais la question, qui se pose, est de savoir si … ces actions de gravitation soient liées à des particules atomiques." Ich glaube, dass man – sobald man eine konsequente Feldtheorie zulässt und die Abweichungen vom Energiesatz sozusagen durch ein neues dazukommendes Feld interpretiert – auch zulassen muss, dass dieses neue dazukommende Feld "gequantelt" ist, d.h. gewissermassen aus Korpuskeln besteht, ähnlich wie das Licht aus Lichtquanten, die negative Ladung aus Elektronen. Die Neutrinos werden also vielleicht nur den Realitätsgrad der Lichtquanten besitzen, aber das reicht doch hin, um mit ihnen eine Beschreibung im Sinne der Erhaltungssätze durchzuführen. – Vielleicht empfindest Du meine neuentstandene Liebe zu den Neutrinos als eine seelische Verirrung, aber ich muss gestehen, dass mich insbesondere das Experiment von Joliot sehr in dieser Liebe bestärkt hat.

Übrigens muss ich allem, was ich bisher geschrieben habe, die General-Entschuldigung beifügen, dass es sich meist um Stimmungen, und nicht um bestimmte Meinungen gehandelt hat. – Pauli's und meine Arbeit geht langsam vorwärts, vielleicht können wir Dir nächstens ein Manuskript schicken. Meine Pläne für die kommenden Wochen sind noch ganz unbestimmt, ich würde natürlich gern nach Kopenhagen kommen, weiss aber noch nicht, ob es gehen wird.

Viele Grüsse Euch allen!

Dein Werner Heisenberg.

Translation[9]

Klosters, March 12, [19]34

Dear Bohr,

Many thanks for your letter and the manuscript of your Solvay report. I was sorry that you could not come, it would have been nice to think and talk quietly in the skiing cabin about physical and non-physical aspects of life. On the other hand, I have felt quite miserable during the last weeks of the past winter, and I therefore decided not to go to the skiing cabin at all, but to spend a much more bourgeois and boring skiing holiday in a Swiss mountain hotel. That is the reason why I have written so rarely during the last months. Today, however, I am not going to "pretend to be tired" and I will try to write also about physics in more detail.

I like very much the content of your manuscript and your letter to Pauli as far as the keynote is concerned. In particular the conjecture at which you hint that the determination of e^2/hc on the one hand, and the laws for the nuclear constitution on the other hand represent two independent problems, which perhaps could be solved separately, appears most alluring to me. However, my own mood regarding the infinite self-energy is somewhat different from what you write about it. True enough, I quite agree with you that we possess so far no theory which goes substantially beyond a systematic application of the correspondence principle to the Maxwell theory; this also implies that point charges must have an infinite self-energy – even without any reference to the formalism of quantum electrodynamics. But it seems to me that a reasonable quantization of the field – going beyond the correspondence – must get rid of this infinite self-energy, and actually just in the way in which Klein and Jordan can avoid the electrostatic part of the self-energy. I do not at all consider this "trick" of Klein and Jordan as a superficial artifice, but as a consistent exploitation of a definite formal feature of the quantum theory of wave fields, which one has to exploit in the future in an even more general manner. Indeed it seems to me that in the quantum theory of wave fields the problem of a *single* point charge is analogous to the treatment of a quantum mechanical system in the *lowest* quantum state; similarly, just as all transition probabilities "downwards" from the lowest quantum state of an oscillator must vanish automatically (this is a requirement that does not follow from the correspondence principle, but which does not contradict it either, and which was important for the beginnings of quantum mechanics), one must in the quantum theory of waves [wave fields] require that a single point charge represents a trivial solution of the equations of motion, i.e. that all transition probabilities which give rise to the creation of other particles, vanish. This requirement does not contradict correspondence since the correspondence principle only yields statements about systems containing many electrons or light quanta. Thus I believe that when one from correspondence arguments infers an infinite self-energy (such as, e.g. Pauli and I had to do in quantum electrodynamics), then this is really just as erroneous as if one, on the basis of correspondence, would maintain that a hydrogen atom in its ground state must still radiate on account of the orbital motion of the electron. The fact that it has so far not been possible to get rid of the infinite self-energy in quantum electrodynamics may probably only mean that we are using a wrong Hamiltonian and presumably we will only be able to find the correct Hamiltonian when e^2/hc is determined and when the hole theory is further elaborated. By the way, I am not very satisfied with Dirac's latest paper[10]; in particular I consider his conclusions as regards that part of the infinite charge density that has to be subtracted downright wrong.

May I in this connection still raise an objection against a sentence which appears on p. 13 of your manuscript. "Mais la question, qui se pose, est de savoir si ... ces actions de gravitation soient liées à des particules atomiques." I believe that as soon as one allows a consistent field theory and, so to speak, interprets the deviations from energy conservation in terms of a new additional field – then one must also allow that this new additional field be "quantized", i.e. that in a certain sense it consists of corpuscles in a similar way as light consists of light quanta and the negative charge of electrons. Thus, neutrinos may only possess the same degree of reality as light quanta, but still that suffices in these terms to accomplish a description within the framework of the conservation laws. Perhaps you will regard my newborn affection for the neutrino as an aberration of the mind, but I must confess that in particular the experiment by Joliot has very much strengthened this affection of mine.

Incidentally I must add a general apology for all that I have written so far; it has more to do with mood than with definite opinions. The paper by Pauli and me progresses slowly, perhaps we may send you a manuscript soon. My plans for the coming weeks are still very uncertain, I would of course like to come to Copenhagen, but I don't know yet whether it will be possible.

Many greetings to you all.

Yours, Werner Heisenberg

BOHR TO HEISENBERG, 22 May 1935
[Carbon copy]

[København,] 22. Mai [19]35.

Kære Heisenberg,

Jeg er meget ked af ikke før at have takket Dig for Dit rare Brev og Fremstillingen af Dine Tanker om et nyt Angreb paa Elektronproblemet. I disse Uger har jeg imidlertid haft saa mange praktiske Sager at ordne med nye Udvidelser af Instituttet, især med Henblik paa et forhaabentlig frugtbart Samarbejde med biologiske Institutioner, at jeg først i de allersidste Dage har haft Tid og Ro til igen at tænke nærmere over Elektronproblemerne og drøfte dem med Rosenfeld.

Før jeg gaar ind paa Dit Arbejdsprogram, for hvilket jeg i første Øjeblik var meget begejstret, men senere er blevet mere tvivlende, hvad den hele Tendens angaar, skal jeg sige et Par Ord om Maalingsproblemet, som Rosenfeld og jeg netop nu er kommet til mere Klarhed over. Jeg ved ikke, om Du har set et Arbejde af Oppenheimer i Physical Review[11] for kort Tid siden, hvori han ud fra den i og for sig rigtige, ja selvfølgelige Tanke, at alle Tæthedsmaalinger efter Deres Art maa føres tilbage til Feltmaalinger, har ment at kunne paavise en Modsigelse imellem Dine Beregninger af Tæthedsfluktuationer og Rosenfelds og mine Beregninger af Feltfluktuationerne. Han har imidlertid ganske misforstaaet Hovedpunktet i vort Arbejde, hvorefter Feltfluktuationerne ikke i og for sig sætter nogen principiel Grænse for Feltmaalinger, men blot for Resultaternes Tilbageførelse til Feltkilderne. Fluktuationer betyder derfor ingenlunde en Begrænsning for Tæthedsmaalinger, idet de fluktuerende Felter er principielt divergensfrie, i det mindste saa længe man holder sig paa den sædvanlige Kvanteelektrodynamiks Grund. Sammenhængen mellem Dine Tæthedsfluktuationer og Problemet om Tæthedsmaalinger er derimod paa det intimste forbundet med de af Euler og Kockel[12] behandlede Modifikationer af Feltligningerne i Positronteorien. En nærmere Undersøgelse viser da ogsaa her en fuldkommen Modsigelsesfrihed af Formalismen, hvad Sammenhængen mellem Definitions- og Maalingsmuligheder angaar. Som Rosenfeld og jeg med Smerte har overbevist os om, maa man for at eftervise en saadan Sammenhæng være ligesaa agtpaagivende overfor Faldgruber som i vor gamle Strid med Feltmaalingsproblemet. Vi skriver i Øjeblikket paa en lille Artikel til Physical Review om disse Spørgsmaal, hvoraf jeg skal sende Dig Manuskriptet en af de første Dage.

I alt dette er vi sikkert ganske enige. Men jeg er endnu som sagt slet ikke klar over, om jeg har forstaaet Tendensen i Dit Arbejdsprogram, og navnlig indser jeg ikke, hvorledes Du i Felt- og Tæthedsfluktuationerne finder Argumenter for primært at betragte Middelværdier over endelige mere eller mindre skarpt begrænsede Rum–Tidsomraader. Resultatet af Rosenfelds og mit Arbejde over Feltmaalingerne var jo netop, at den hidtidige Form for Kvanteelektrodynamikken, bortset fra Selvenergispørgsmaalet, var mere tilfredsstillende end man efter tidligere Undersøgelser over Maalingsspørgsmaalet maaske kunde være tilbøjelig til at mene, og at navnlig Betragtningen af Middelværdier var en integrerende Del af Formalismen og derfor ingenlunde pegede paa Nødvendigheden af en Ændring af de hidtidige Grundlag for den korrespondensmæssige Beskrivelse. Efter Dine og Dine Medarbejderes seneste

---

[11] R. Oppenheimer, *Note on Charge and Field Fluctuations*, Phys. Rev. **47** (1935) 144–145.

[12] H. Euler and B. Kockel, *Über die Streuung von Licht an Licht nach der Diracschen Theorie*, Naturwiss. **23** (1935) 246–247.

Arbejder og vort lille nye Bidrag til Maalingsproblemerne, forekommer det mig, at en meget lignende Situation foreligger i Elektronteorien. Naturligvis er jeg ganske enig med Dig i, at hele Grundlaget maa revideres paa væsentlig Maade før en virkelig tilfredsstillende og sammenhængende Teori kan opbygges; men jeg synes samtidig, at der er meget, der tyder paa, at en saadan Indførelse af Middelværdier, som Du foreslaar, næppe vil hjælpe tilstrækkeligt. Det karakteristiske Træk ved de velkendte Paradokser vedrørende Straalingen ved hurtige Elektroners Sammenstød er jo netop, at det ved de omhandlede Problemer allerede drejer sig om Rum–Tidsmiddelværdier, og at man derfor ikke indser, hvordan en explicit Indførelse af Middelværdien paa det nuværende Grundlag kan hjælpe.

Jeg er derfor stadig mere tilbøjelig til at tro, at hele Vanskeligheden i Elektronteorien beror derpaa, at saavel Definitions- som Maalingsproblemerne er opfattet paa alt for klassisk Maade, og at en Løsning af Gaaderne, først kan naas ved en helt ny Drejning af Grundlaget, hvori Eksistensen af Ladningskvantet bliver ganske anderledes intimt forbundet med Eksistensen af Virkningskvantet end i de hidtige Metoder. Medens alle saadanne Antydninger jo for Øjeblikket nødvendigvis maa være dunkel Tale, forekommer det mig, at det er muligt, i det mindste, naar man vil tillade sig at spøge lidt, at gøre Dine Argumenter vedrørende Nødvendigheden af at tage mere explicit Hensyn til de i Betragtning kommende Maalemetoder til Genstand for en mere utilsløret Kritik. Jeg forstaar i hvert Fald ikke, hvad Du mener dermed, at man allerede ved Hjælp af Formalismen skal udskyde Muligheden for Fænomenernes Forstyrrelse ved kosmiske Straaler og deslige. Hvis saadanne Virkninger hører med til det foreliggende fysiske Problem maa der jo nødvendigvis tages fornøden Hensyn til dem i Teorien, og hvis de kan elimineres fra de teoretiske Problemer maa det jo betyde, at de ogsaa fysisk kan holdes borte i tilstrækkelig Grad. En teoretisk Fysiker, der lever i en dyb Grube maa i det mindste, hvis han ikke er meget klogere end vi andre, dog have ligesaa meget Hovedbrud af Selvenergiparadokserne, som en der boede paa Jungfraujoch. Jeg forstaar selvfølgelig noget bedre, hvor Du vil hen end en saadan Sætning kunde give det Indtryk af, men jeg er dog ikke sikker paa, at Spøgen alligevel ikke i nogen Grad træffer Sagens Kærne. Men jeg er som sagt forberedt paa, at jeg ganske har misforstaaet Tendensen i Dine Argumenter, og skal være allermest glad for en Tilrettevisning.

Med mange venlige Hilsner fra os alle til hele Leipziger Kredsen,

Din
[Niels Bohr]

Translation

[Copenhagen,] May 22, [19]35

Dear Heisenberg,

I am very sorry not to have thanked you earlier for your nice letter and the presentation of your ideas for a new attack on the electron problem. However, during these weeks I have had so many practical matters to settle in connection with new extensions of the institute – especially with a view to a hopefully fruitful collaboration with biological institutions – that only during the very last days have I found time and peace to think more closely about the electron problems again and to discuss them with Rosenfeld.

Before commenting on your working plan, about which I was quite enthusiastic at first, but have later become more doubtful of, as far as the entire trend is concerned, I shall say a few words about the measuring problem, of which Rosenfeld and I have just now arrived at a better understanding. I don't know whether you have seen a paper by Oppenheimer in the Physical Review[11] a short while ago, in which – on the basis of the in itself correct, not to say obvious, idea that all density measurements according to their very nature have to be traced back to field measurements – he has thought himself able to demonstrate a contradiction between your evaluation of the density fluctuations and Rosenfeld's and my evaluation of the field fluctuations. However, he has completely missed the main point of our paper, according to which the field fluctuations in themselves do not entail any fundamental limitation to field measurements, but only to the [possibility of] tracing the results back to the field sources. Thus, fluctuations do not at all imply a limitation of density measurements, since the fluctuating fields as a matter of principle are divergence free, at least as long as we remain within the domain of the usual quantum electrodynamics. The relationship between your density fluctuations and the problem of density measurements is, however, most intimately connected with the modifications of the field equations in the positron theory as treated by Euler and Kockel[12]. Indeed, a closer investigation reveals here a complete consistency of the formalism as regards the connection between the possibilities of definition and measurement. As Rosenfeld and I have painfully convinced ourselves, in order to demonstrate such a connection one has to be just as alert against pitfalls as in our old struggle with the problem of field measurements. For the moment we are working on a small paper for the Physical Review about these questions and I am going to send you the manuscript in a few days.

About all this I am sure we quite agree. But it is not at all clear to me whether I have understood the trend of your working plan, and in particular

I do not see how you find arguments in the field and density fluctuations for primarily considering average values over finite, more or less sharply limited space–time regions. Indeed, the result of the paper by Rosenfeld and me was just that apart from the self-energy question, the present form of quantum electrodynamics was more satisfactory than one might be inclined to believe on the basis of earlier investigations of the question of measurement, and in particular that the consideration of average values was an integral part of the formalism and therefore by no means pointed to the necessity of changing the present basis for the correspondence description. After the latest papers by you and your collaborators and our own small new contribution to the measuring problems, it seems to me that we are facing a very similar situation in electron theory. Of course I quite agree with you that the entire basis has to be revised substantially before one can build a really satisfactory and coherent theory; but at the same time it appears to me that there is considerable evidence that an introduction of average values, such as you suggest, will hardly suffice. The characteristic feature of the well-known paradoxes concerning the radiation [emitted] in collisions between swift electrons is precisely the fact that in these problems we are already concerned with space–time averages and that therefore we fail to see how an explicit introduction of average values on the present basis could prove very useful.

I am therefore still more inclined to believe that the whole difficulty of electron theory is due to the fact that the problems of definition as well as of measurement are conceived in a much too classical manner, and that a solution to the riddles is only to be achieved by revising the basis entirely so that the existence of the quantum of charge becomes far more intimately connected with the existence of the quantum of action than in the methods used so far. While all such suggestions at present, of course, are bound to be obscure talk, it appears possible to me – at least when we allow ourselves to joke a little – to subject to more undisguised criticism your arguments concerning the necessity of considering more explicitly the relevant measuring methods. In any case, I do not understand what you have in mind when you say that already by means of the formalism one should eliminate the possibility of the phenomena being disturbed by cosmic rays and the like. If such effects are part of the given physical problem it is of course necessary that they are duly considered in the theory; and if they can be eliminated from the theoretical problems, this must of course imply that they may also physically be excluded to a sufficient degree. A theoretical physicist living in a deep pit must after all – unless he is much wiser than we are – be just as puzzled by the self-energy paradoxes as a physicist living on the Jungfraujoch. Of course I understand what you have in mind somewhat better than it appears from this sentence; yet I am not sure whether the joke does not after all to some extent hit the heart of the matter. But, as already said, I am prepared to accept that I may have quite misunderstood the trend of your arguments and I should be most happy for a rebuke.

With many kind regards from us all to the entire Leipzig circle,

Yours,
[Niels Bohr]

HEISENBERG TO BOHR, 30 May [1935]
[Handwritten]

Leipzig 30.5.[1935]

Lieber Bohr!

Hab vielen Dank für Deinen an Diskussionsstoff reichen Brief, der hier insbesondere durch das Argument mit dem Physiker auf dem Jungfraujoch grosse Begeisterung ausgelöst hat.

Um auf die Einzelheiten einzugehen: Zunächst hab ich Deine Kritik an der Oppenheimerschen Arbeit nicht ganz verstanden. Oppenheimer hat doch nicht viel anderes getan, als durch Rechnung gezeigt, dass es zwei zunächst unabhängige Ursachen für Feldschwankungen gibt: die Quantelung des Feldes und die Paarerzeugung; und dass diese beiden Arten der Schwankung sich im Resultat um den Faktor $e^2/\hbar c$ unterscheiden. Ich hatte beim Lesen der Oppenheimerschen Arbeit den Eindruck, als sei O. garnicht so unvorsichtig gewesen, aus seinen Rechnungen physikalische Konsequenzen zu ziehen. Die wenigen Andeutungen, die er hier gibt, können freilich unrichtig sein. Aber es wird mich natürlich ausserordentlich interessieren, zu hören, was Du und Rosenfeld geschrieben hast.

Was nun meine eigenen Pläne angeht, so gebe ich zunächst gern zu, dass ich Deine und Rosenfelds Arbeit in meinem letzten Brief nicht ganz korrekt angewendet habe. Was ich meinte, war etwa folgendes: Deine u. Rosenfelds Arbeit zeigte, dass der bisherige Formalismus seiner Natur nach nur Aussagen über Mittelwerte über beliebige Raum–Zeit-Gebiete zulässt und insoweit auch widerspruchsfrei ist. Da also vernünftigerweise in der mathematischen Formulierung der Theorie die durch die Messapparate vorgeschriebene Mittelung explizit auftritt (wie z.B. bei Dir u. Rosenfeld), so ist es vielleicht möglich, ohne die Korrespondenz zu verletzen, diese Mittelung auch noch an anderen

[455]

Stellen im Formalismus vorzunehmen, wo sie bisher nicht zulässig schien. Ein starkes Argument hierfür schien mir die Tatsache, dass eigentlich in allen bisherigen Rechnungen eine solche Mittelung gewissermassen unbewusst durchgeführt wurde. Denn wenn man die Selbstenergie streicht, oder eine Störungsrechnung abbricht, weil die höheren Glieder divergieren, so tut man eigentlich nichts anderes, als was ich mit "Mittelung" bezeichnete: man verschmiert etwa das Elektron auf $\hbar/mc$ und bekommt dann keine Selbstenergie. Was ich suchte, war also zunächst nur eine Art von Legitimation des bisher üblichen Verfahrens. Meine Vermutung war also: Für alle die Experimente, in denen eine Lokalisation genauer als z.B. $\hbar/mc$ prinzipiell unmöglich ist, ist das Elektron "wirklich" über $\hbar/mc$ verschmiert und besitzt keine merkliche Selbstenergie. Wenn man jedoch mit Lichtquanten der Ordnung $137mc^2$ experimentiert, so darf man eben *nicht* mehr über so grosse Gebiete ausschmieren. Dies bedeutet, dass man eine Störungsrechnung nicht mehr, wie bisher, beim zweiten Glied abbrechen darf, sondern dass man z.B. für die Bremsstrahlung auch die höheren Glieder des Störungsverfahrens ernst nehmen muss (z.B. die Euler–Kochelschen Glieder). Zu dem zweiten Satz auf S. 3 Deines Briefs möchte ich also sagen: Die Bremsstrahlung sollte nach meiner Ansicht dadurch in Ordnung kommen, dass man – im Gegensatz zu den bisherigen Rechnungen – *nicht* mehr über grosse Gebiete mittelt (d.h. nicht mehr die höheren Näherungen streicht), sondern nur noch über entsprechend kleinere Gebiete. Sollte sich herausstellen, dass die höheren Näherungen (Euler–Kochel u.s.w.) überhaupt alle Divergenzschwierigkeiten beseitigen, so wäre die explizite Einführung von Mittelwerten, so wie ich es vorschlug, nur eine mathematische Vereinfachung zur Herleitung der gesuchten Resultate, man könnte dann durch kleinermachen der Gebiete zur bisherigen Theorie übergehen. Ich halte es aber für unwahrscheinlich, dass die Divergenzschwierigkeiten durch die höheren Näherungen von selbst wegfallen. Vielmehr sieht es bisher so aus, als ob z.B. der Zustand tiefster Energie, der zur Ladung e gehört, nun dann so ähnlich aussieht, wie: ein einzelnes Elektron, wenn über Gebiete $> e^2/mc^2$ ausgeschmiert wird. Wenn man über viel kleinere Gebiete mittelt, so sieht der Zustand tiefster Energie eher aus wie: ein Elektron u. eine grosse Anzahl von Paaren.

Hoffentlich hab ich mich jetzt klarer ausgedrückt, als in meinem letzten Brief. Noch eins möchte ich hinzufügen: Es ist klar, dass ein Formalismus vom Typus der Gleichung (8) in meinem letzten Brief, wenn er überhaupt richtig ist, den Wert von $e^2/\hbar c$ festlegt. Denn e kommt in ihm garnicht vor. Es wird sich aus einem solchen Formalismus also z.B. eine Coulombsche Anziehung der Elektronen von der Form $const \cdot \hbar c/r$ ergeben, aus der dann $e^2/\hbar c$ folgt. Die spezielle Form (8) gibt übrigens, soviel ich bisher sehen konnte, einen viel zu grossen Wert für $e^2/\hbar c$ ($e^2/\hbar c \sim 3/2$, bis auf Faktoren π) und gibt überhaupt keine vernünftige Korrespondenz.

Hoffentlich kommt nächstens Dein und Rosenfelds Manuskript! Mit vielen Grüssen an alle gemeinsamen Freunde.

Dein
Werner Heisenberg

BOHR TO HEISENBERG, 21 January 1937
[Photostat, Léon Rosenfeld's handwriting]

Paris, 21. I. [19]37

Kære Heisenberg,

Jeg var meget ked af, at jeg, fordi jeg havde lovet at holde et Par Forelæsninger i Paris paa Vej til Amerika, maatte rejse fra København umiddelbart efter at have faaet Dit venlige Brev. Det vilde jo ogsaa for mig have været en meget stor Glæde om vi, som Du foreslog, kunde have haft Lejlighed til at træffes før den lange Rejse og talt rigtig sammen om Fysikkens Problemer og mange andre Spørgsmaal. Ikke mindst i Forbindelse med Bemærkningen i Dit Brev om Vanskelighederne ved Feltbegrebets Anvendelse i Elektronteorien vil det maaske interessere Dig, at Rosenfeld, der er saa venlig at skrive dette Brev paa min Diktat her i Paris, og jeg i den sidste Tid i Forbindelse med et dog mislykket Forsøg paa at gøre vores Arbejde om Ladningsmaalingerne færdigt til Offentliggørelse før min Rejse har taget hele Spørgsmaalet om Anvendelsen af klassiske Begreber i Kvanteelektrodynamikken og Elektronteori op til ny Diskussion. Medens vi i vort gamle og i vort nye Arbejde kunde vise, at disse Teorier, hvad Maalingsmulighederne angaar, i rent logisk Henseende danner en konsekvent Idealisation, har jeg nemlig i allersidste Tid faaet Mistanke om, at de uhyrlige Konsekvenser af en saadan Idealisation gaar langt ud over Virkelighedens Grænser allerede hvad den tilgrundlagte Identifikation af Feltstørrelser med deres Virkning paa Prøvelegemer af endelige Dimensioner angaar. Den i Rosenfelds og mit Arbejde paaviste fuldstændige Uadskillelighed imellem de teoretiske Felt- og Ladningsfluktuationer og den paa Grund af den individuelle Karakter af Straalings- og Pardannelsesprocesserne ikke kompenserbare Del af Tilbagevirkningen paa Prøvelegemet tillader aabenbart Muligheden af, at frakende alle saadanne Fluktuationer enhver simpel fysisk Realitet. Hvis man tager dette Standpunkt, der i Øjeblikket forekommer mig det eneste fornuftige, vilde det betyde, at vi atter paa dette Omraade er henvist til simple Korrespondens- og Komplementaritetsbetragtninger. Jeg har i disse

Dage diskuteret dette Perspektiv nærmere med Rosenfeld, og har faaet det Indtryk, at det selv for en lærdere Mand end jeg maaske dog ikke er saa afskrækkende som det i første Øjeblik kunde synes. Jeg vil imidlertid være meget taknemmelig for at høre lidt fra Dig om, hvordan Du selv vilde stille Dig til en saadan Situation. Jeg vil være i Princeton (Adresse: Prof. O. Veblen) i den første Uge af Februar og et Brev fra Dig kunde maaske naa mig der, hvor jeg haaber at faa Lejlighed til rigtig at diskutere alle Slags Sager med Einstein og Neumann. Der var meget mere jeg gerne vilde skrive om, men da jeg nu maa skynde mig Afsted for at naa Toget til Cherbourg, hvor jeg skal møde Skibet fra Southampton med min Kone og Hans, maa jeg slutte med mange venlige Hilsner og gode Ønsker, især om et glædeligt Gensyn til Sommer.

Din
Niels Bohr

Translation

Paris, January 21, [19]37

Dear Heisenberg,

I was very sad that because of a promise to give a couple of lectures in Paris on my way to America, I had to leave Copenhagen immediately after I received your friendly letter. It would also have given me very great pleasure if we could have met before the long journey as you suggested, and had a good discussion about physics problems and many other topics. Not least in connection with the remark in your letter about the difficulties met in applying the field concept in electron theory it might interest you that Rosenfeld, who is so kind as to write this letter according to my dictation here in Paris, and I have recently – in connection with a vain attempt to make our work on charge measurements ready for publication before my journey – taken up for renewed discussion the whole question of application of classical concepts in quantum electrodynamics and electron theory. It is a fact that both in our old and in our new work we were able to show that as far as measuring possibilities are concerned, these theories represent in a purely logical respect a consistent idealization. However, quite recently I have begun to suspect that the preposterous consequences of such an idealization lead far beyond the borders of reality already with respect to the basic presupposition of identifying field quantities with their actions on test bodies of finite dimensions. As Rosenfeld and I demonstrated in our paper, the theoretical field and charge fluctuations are absolutely inseparable from that part of the reaction on the test body which cannot be compensated due to the individuality of the processes of radiation and pair creation. Obviously this circumstance allows the possibility of depriving all such fluctuations of any simple physical reality. If one accepts this point of view – which for the moment appears to me as the only reasonable one – it would mean that we are forced again, in this domain, to rely on simple correspondence and complementarity arguments. These days I have discussed this perspective in more detail with Rosenfeld and got the impression that, even for a man more erudite than I, it may perhaps be less deterring than it might appear at first glance. However, I would be very pleased to hear about your attitude to this. I will be in Princeton (Address: Prof. O. Veblen) in the first week of February, and a letter from you might reach me there, where I hope to get the opportunity of really discussing all kinds of things with Einstein and Neumann. I would like to write about much more, but as I now must rush to catch the train to Cherbourg where I am meeting the ship from Southampton bringing my wife and Hans, I must close with many kind regards and good wishes, especially for a happy reunion in the summer.

Yours,

Niels Bohr

OSKAR KLEIN

KLEIN TO BOHR, 20 July 1935
[Handwritten]

Söderbyle, Väddöbacka 20.7.1935.

Kære Bohr!

Mange Tak for Dit rare Brev og for Din Venlighed at sende mig et Exemplar af Dit Svar til Einstein, som jeg har læst med stor Interesse og Belæring. Det glædede mig meget at se hvordan Du har benyttet denne Lejlighed til at fremstille Dit komplementære Synspunkt, som saa længe har ligget Dig paa Hjertet, paa en mere indtrængende og lettere forstaaelig Maade end tidligere. Samtidigt morede det mig at se hvor kraftigt Du bemöder Einsteins Tvivl paa Kvantemekanikken, idet Du ikke blot hvad angaar det fysiske og filosofiske men ogsaa med Hensyn til det matematiske har fundet et overlegent Standpunkt. Jeg synes at den Maade hvorpaa Du omtaler Virkningen af en enkelt Blænder er særlig lærerig; at man efter den förste Impulsbestemmelse endnu har Mulighed til Valg mellem at kende Partikelns Sted ved Gennemgangen gennem Skærmen eller dens Impuls. Forfærdelig morsomt ogsaa at lære hvordan Einsteins Tankeexperiment lader sig gennemføre i Enkeltheder ved den Forsøgsanordning Du angiver. Det skal bli morsomt at se om Einstein besvarer Din Artikel. Han har naturligvis en særlig Evne til at sige Ting paa en Maade, der virker overbevisende baade paa ham selv og andre, men denne Gang synes jeg han maa tage Indtryk af den Maade hvorpaa Du belyser hans Paradoks udfra Dine almene Synspunkter.

...

Din Oskar Klein

Translation

Söderbyle, July 20, 1935

Dear Bohr,

Many thanks for your nice letter and for your kindness in sending me a copy of your answer to Einstein which I have read with great interest and from which I have learned much. It was a great pleasure for me to see how you have grasped this opportunity to give an exposition – more emphatic and easier to comprehend than before – of your complementarity point of view, which has been on your mind for so long. At the same time I enjoyed seeing how forcefully you repudiate Einstein's doubts about quantum mechanics – in so far as you have found a superior standpoint not only as regards the physical and philosophical aspects, but also with respect to the mathematics. I find the way you explain the effect of a single diaphragm particularly instructive; the fact that after the initial momentum measurement we still have the possibility of a choice between a determination of the particle's position during its passage through the screen or its momentum. Also, it was a special joy to learn how Einstein's thought experiment may be performed in detail through the experimental arrangement that you describe. It will be interesting to see whether Einstein is going to answer your article. He has of course a special talent to express things in a way that appears convincing to himself as well as to others; but this time I think he must be impressed by the way you illuminate his paradox on the basis of your general points of view.

...

Yours, Oskar Klein

HENDRIK A. KRAMERS

KRAMERS TO BOHR, 9 August 1935
[Handwritten]

OEGSTGEEST. 9 Aug. 1935
POELGEESTERWEG 2

Kære Bohr,

...

Phys. Rev. Manuskriptet har jeg læst flere Gange igennem, og det var morsomt at se hvordan det egentlig ligger med Einstein's Paradox. Jeg har nogle enkelte Bemærkninger, som ikke har med selve Sagen at gøre, – for den er klar – men som nærmest er af pædagogisk Natur. Den ene er næsten filosofisk, det er, at man nu kan se hvor subtilt Aarsagsbegrebet er blevet, samtidigt med at det ikke har tabt noget af sin gamle Styrke. Det kommer saa morsomt frem, naar Du et Sted i Begyndelsen siger at (den fysiske) "Aarsag" til Impulsusikkerheden ligger i Spaltens Tilstedeværelse, der kan bevirke, at der finder Impulsudbytning Sted.

Anden Bemærkning er følgende. Det er Synd, at det enkle Eksempel med to Partikler, som du diskuterer, ikke er indrettet saaledes, at der eventuelt kunde være Tale om en *fuldkommen nøjagtig* Bestemmelse af 2^{en} Partikels Sted, paa Grund af en Maaling, der beskæftiger sig kun med 1ste Partikel. For Sagens almindelige Sammenhæng gør det ikke noget, og du siger ogsaa noget om det (de to primære Spalter skal ikke være alt for snævre). Jeg nævner det kun, fordi saadant et lille Punkt bidrager til at gøre hele Stilen saa forskellig fra Einstein's, at jeg ikke er sikker paa at han overhovedet vil komme ind paa din Tankegang, og anerkende dens Styrke. Det samme gælder, naar du – i Argumenterne – fra Begyndelsen af – opererer med Ubestemthedsrelationernes fysiske Anvendelse, saaledes at det kunde se ud, som om man ræsonnerer i en viciøs Kreds. Dette er ikke sandt, men hvad der ligger bagved, er at Einstein ogsaa burde være utilfreds med Ubest. relationerne i deres enkleste Form, lige saa stærkt som han er det med de teknisk mere indviklede Konsekvenser, som han stadigt stirrer paa. Men paa vis Maade er det helt overflødigt at jeg fremhæver dette, fordi dette netop har været Grunden til, at du først talte om Forsøgene med en enkel Partikel.

...

Med de mest hengivne Hilsner forbliver jeg din
H.A. Kramers

Translation

Oegstgeest, August 9, 1935

Dear Bohr,

...

I have gone through the Phys. Rev. manuscript several times and it was a pleasure to see how it really stands with Einstein's paradox. I have a few remarks that have nothing to do with this matter itself – because that is clear – but are rather of a pedagogical nature. The first, almost philosophical, is this: We can now recognize how subtle the concept of causality has become although, at the same time, it has lost nothing of its old strength. It is so pleasantly revealed when, somewhere in the beginning, you say that the (physical) "cause" of the uncertainty in momentum is the presence of the slit, which may give rise to an exchange of momentum.

The second remark is the following: It is a pity that the simple example with two particles which you discuss is not so devised that it might yield a *completely precise* determination of the position of the 2nd particle on the basis of a measurement that only concerns the 1st particle. As far as the general context is concerned, it does not matter, and you do say something about it (the two primary slits should not be too narrow). I mention this only because such a small point contributes to making the style so different from Einstein's that I am not sure whether he at all will appreciate your way of thinking and recognize its strength. This also applies in your argumentation when – from the beginning – you operate with the physical application of the uncertainty relations, so that it might appear that your reasoning is a vicious circle. This is not true, but basically Einstein ought to be just as strongly dissatisfied with the uncertainty relations in their most simple form as he is as regards the more complicated consequences on which he is still focusing. However, in a certain sense it is quite superfluous that I emphasize this point because it has been the very reason for you to discuss the experiments with a single particle in the first place.

...

With my most affectionate greetings, I remain yours,
H.A. Kramers

WOLFGANG PAULI

BOHR TO PAULI, 25 January 1933
[Typewritten]

UNIVERSITETETS INSTITUT
FOR
TEORETISK FYSIK

BLEGDAMSVEJ 15, KØBENHAVN Ø.
DEN 25. Januar 1933.

Kære Pauli,

Hvad tænker Du dog om mig, at jeg aldrig fik besvaret Dine rare Breve fra i Sommer? Jeg behøver jo ikke at sige, hvor stor en Glæde det var for mig at forstaa, at Du er nogenlunde tilfreds med Livet i Almindelighed og med Dit eget Arbejde. At Du skulde se ukritisk paa Dine gamle Venners beskedne Bestræbelser vilde jo ogsaa forbavse mig meget. For at blive i Faust-Parodiens Aand, følte jeg dog, at Du vist har læst min stakkels Faraday-Lecture og Rom-Diskussionen paa en Maade, der lidt for meget ligner den Maade, som man i det mindste paa dansk siger, at Fanden læser Biblen. Ved begge Lejligheder var det først og fremmest mit Formaal at understrege, at vi ved Kerneproblemet ikke længere staar paa Mekanikkens faste Grund, og at vi derfor trods alle Dine gamle Advarsler og Trusler maaske alligevel bliver tvungne til at slaa ind paa helt nye Veje. Naar jeg i denne Forbindelse har fremhævet den Begrænsning af Partikelbegrebet, som vi allerede møder i den klassiske Elektronradius, er det jo dermed ikke Meningen, at denne Radius alene skulde danne Grænsen for Anvendelsen af den relativistiske Kvantemekanik. Ja, som jeg søgte at fremhæve i Rom-Beretningen ved Omtalen af Klein–Nishina-Formlen, kan man ikke engang tale om en absolut Grænse i den Henseende. Jeg tror heller ikke, at vort Syn paa Diracs Teori er saa forskelligt, som Du synes at mene. Med min Betoning af det Klein'ske Paradoks' principielle Betydning, navnlig paa Grund af den Grænse det sætter for Muligheden af at se bort fra Maaleinstrumenternes atomistiske Struktur, var det jo altid Hensigten at understrege Umuligheden af at opbygge en sammenhængende Teori, saa længe Elementarpartiklerne og Virkningskvantet betragtes som uafhængige Elementer. Naar jeg derimod ikke kunde dele Din Begejstring for Landau og Peierls Arbejde, laa det ingenlunde i, at jeg mente, at de paa alle Punkter havde Uret i deres Bemærkninger om den relativistisk begrundede Begrænsning af Rum–Tidsmaalinger, som jo i Hovedsagen heller ikke var mig ukendt, men fordi jeg stadig mener, at de skød over Maalet og misforstod de Fordringer, man kan stille til en fysisk Teori. Jeg tror jo ogsaa af vore senere Samtaler at

[463]

forstaa, at vi er enige i, at det er umuligt at modbevise en relativistisk invariant Formalisme ved simple Relativitetsargumenter. Efter min Mening maa enhver Bedømmelse af Formalismens Begrænsning explicit tage sit Udgangspunkt i Urigtigheden af dens Konsekvenser. Jeg har derfor altid understreget, at Begrænsningen i Diracs Teori kun kan søges i Forekomsten af de ved de negative Energiværdier symboliserede ufysiske Konsekvenser. Selvfølgelig var jeg ikke dengang, ligesaa lidt som Peierls og Landau, i Stand til paa alle Punkter at gennemskue den deraf følgende Begrænsning af Diracs Teori, og jeg er stadig yderst interesseret i at lære den nøjagtige Gennemregning af det af Dig fremdragne, vigtige Dispersionsproblem at kende. Jeg forstaar derfor slet ikke Din Fanatisme i den hele Sag, og navnlig synes jeg, at Du rent forglemmer den komplementære Karakter af Dit Yndlingsudtryk "Beschwich-tigungsphilosophie". I Øjeblikket er der imidlertid næppe Anledning til at skændes om Enkeltheder vedrørende Partikelmekanikkens Begrænsning, thi vi er vist alle enige om, at vi staar foran en ny Udviklingsfase, der kræver helt nye Metoder. Hvad der ligger mig paa Sinde i denne Situation er blot, at vi ikke foregriber denne Udvikling ved Misbrug af tilsyneladende logiske Argumenter. En Advarsel om den Forsigtighed, der her udkræves, og den Fare som Landau og Peierls Indstilling rummer, har da ogsaa en nærmere Undersøgelse af Begrænsningen af den kvanteelektrodynamiske Formalisme givet. Selv om jeg ikke straks kunde finde nogen Fejl i Peierls og Landaus Argumenter vedrørende elektromagnetiske Feltstørrelsers Maalelighed, var jeg altid uhyggelig ved deres Kritik af Formalismens Grundlag, fordi det jo her til syvende og sidst drejede sig om rene Korrespondensbetragtninger, hvori Partikelproblemet ikke explicit indgik. Sammen med Rosenfeld har jeg i Efteraaret undersøgt Sagen nærmere, og vi er kommet til det Resultat, at der er fuld Overensstemmelse mellem den principielle Begrænsning af elektromagnetiske Kræfters Maalelighed og Formalismens Ombytningsrelationer for Feltkomponenter. Først og fremmest viste det sig, at den Forstyrrelse af Maalingerne, som Prøvelegemernes Ud-straaling efter Landau og Peierls Mening skulde foraarsage, fuldstændig kan elimineres. Selv efter at vi var klare over dette Punkt, var der imidlertid endnu mange Paradoksier tilbage at løse. Det viste sig bl.a., at Middelværdierne af alle elektriske og magnetiske Feltkomponenter, taget over det samme Rum–Tidsomraade, var fuldstændig ombyttelige, og at derfor det i Heisenbergs Bog fremdragne Eksempel slet ikke var egnet til en nærmere Prøvelse af Formalismen. Hovedgrunden til, at jeg ikke før har besvaret Dine Breve, var, at jeg stadig opsatte at skrive, indtil jeg kunde fortælle nærmere om dette Arbejde, som jeg tænkte maaske vilde interessere Dig mere end den platoniske Diskussion af de uløste Problemer. Paa Grund af de stadig opdukkende, nye Vanskeligheder trak Arbejdet imidlertid ud, og vi naaede ikke at faa det færdig

[464]

redigeret, førend Rosenfeld lidt før Jul maatte forlade København paa Grund af sine Pligter i Liège. Han kommer imidlertid meget snart tilbage, og jeg haaber, at vi inden faa Uger kan sende Dig et fuldstændigt Manuskript eller en Korrektur. Jeg ved ikke, hvor langt Du er gaaet ind paa disse Spørgsmaal i Din store Artikel til Handbuch der Physik. Bloch, der netop er paa et lille Besøg i København har lige vist mig Korrekturen af de første 150 Sider, og jeg har allerede deraf faaet et meget stærkt Indtryk af det store Arbejde, Du har nedlagt deri, og den Omhu hvormed det hele er udarbejdet. Jeg glæder mig meget til at studere det hele nærmere og til at lære deraf. For at Du kan se, at jeg ikke selv har ligget paa den lade Side under Rosenfelds Fraværelse, sender jeg som en lille Nytaarshilsen Korrekturen af et Foredrag, som jeg holdt i Sommer, og som jeg i disse Uger har udarbejdet og oversat til dansk[13]. Jeg skal være meget glad for at høre, hvad Du synes derom, og Du skal ikke være bange for at bruge lige saa mange Skældsord, som Du kan finde paa. Vor sædvanlige lille Konferens vil i Aar finde Sted indenfor de første Uger af April, og jeg regner bestemt med, at Du denne Gang kan modtage en Indbydelse fra Instituttet, og at vi ikke ligesom sidst skal nøjes med Din Optræden i en Parodi, selv om den var meget morsom.

Med mange venlige Hilsener fra os alle og de bedste Ønsker for det nye Aar,

Din

Niels Bohr

Translation

Copenhagen, January 25, 1933

Dear Pauli,

What *do* you think of me that I never got around to answering your nice letters from last summer? Of course, I need not say how happy I was to realize that you are reasonably satisfied with life in general and with your own work. That you should look uncritically upon the modest endeavours of your old friends, would of course also greatly surprise me. To remain in the spirit of the Faust parody I felt, however, that you had perhaps read my poor Faraday

---

[13] N. Bohr, *Lys og Liv*, Naturens Verden **17** (1933) 49–59. Translation of *Light and Life*, an address delivered at the opening meeting of the International Congress on Light Therapy in Copenhagen, 15 August 1932. Reproduced in Vol. 10.

lecture and the Rome discussion in a way which is a little too much like the way the devil reads the Bible, as we say in Danish at least. On both occasions my purpose was primarily to emphasize that in the problem of the nucleus we no longer stand on the firm ground of mechanics, and that we therefore, in spite of all your old warnings and threats, might perhaps still be compelled to choose quite new paths. When in this connection I have emphasized the limitation of the particle concept which we already encounter in the classical electron radius, it was of course not the idea that this radius by itself should set the limit for the application of relativistic quantum mechanics. Indeed, as I tried to emphasize in the Rome report when commenting on the Klein–Nishina formula, one cannot even talk about an absolute limit in this respect. Also I do not believe that our views on Dirac's theory are so different as you seem to think. When I stressed the fundamental importance of the Klein paradox – especially because of the limit that it sets for the possibility of disregarding the atomic structure of the measuring instruments – the intention was of course always to emphasize the impossibility of constructing a coherent theory as long as the elementary particles and the quantum of action are regarded as independent elements. When, on the other hand, I was unable to share your enthusiasm for the paper by Landau and Peierls, it was not at all because I thought that they were wrong on all points in their remarks about the relativistic limitations in space–time measurements, which of course were also in the main not unfamiliar to me. But it was because I still think that they overshot the mark and misunderstood what one can demand from a physical theory. Also I do believe to understand from our later conversations that we agree that it is impossible to disprove a relativistically invariant formalism by means of simple relativity arguments. In my opinion, any judgment of the limitations of the formalism must explicitly take its point of departure in the incorrectness of its consequences. I have therefore always stressed that the limitations in Dirac's theory only had to be sought in the appearance of the unphysical consequences symbolized by the negative energy values. I was of course at the time no more able than Peierls and Landau to see through the resulting limitations of Dirac's theory at all points, and I am still very eager to learn about the exact calculation of the important dispersion problem to which you have called attention. Therefore, I do not at all understand your fanaticism in this whole matter, and above all I think that you quite forget the complementary character of your favourite expression "soothing philosophy". At the moment, however, there is hardly any reason to quarrel about details concerning the limitations of particle mechanics, since we probably all agree that we are entering a new phase of the development which calls for quite new methods. In this situation I am only anxious that we do not forestall this development by misuse of apparently logical arguments. A closer in-

[466]

vestigation of the limitations of the formalism of quantum electrodynamics has indeed also given a warning as to the care that is required here, and the danger contained in Landau and Peierls's attitude. Although I could not immediately find any error in Peierls and Landau's arguments concerning the measurability of electromagnetic field quantities, I always felt uneasy about their criticism of the foundations of the formalism, because after all here we had to do with pure correspondence considerations in which the particle problem did not enter explicitly. During the autumn I have investigated the matter more closely together with Rosenfeld, and we have reached the conclusion that there is full agreement between the fundamental limitation in the measurability of electromagnetic forces and the commutation relations for field components given by the formalism. Above all it turned out that the disturbance of the measurements which, according to the opinion of Landau and Peierls, should be caused by the radiation from the test bodies, can be completely eliminated. Even after we had clarified this point, however, many paradoxes still remained to be solved. Among other things it turned out that the average values of all electric and magnetic field components, taken over the same space–time domain, commuted completely and that therefore the example discussed in Heisenberg's book was not at all suited for a closer examination of the formalism. The main reason why I did not answer your letters before was that I kept postponing to write until I could tell you something more definite about this work, which I thought perhaps would interest you more than the Platonic discussion of the unsolved problems. Because of new difficulties that kept turning up, the work dragged on, however, and we did not manage to finish editing it before Rosenfeld had to leave Copenhagen, shortly before Christmas, because of his duties in Liège. However, he will be back very soon and I hope that we can send you a complete manuscript or a proof within a few weeks. I do not know how far you entered into these questions in your long article for the Handbuch der Physik. Bloch, who is now paying a short visit to Copenhagen, has just shown me the proofs of the first 150 pages, and I have already got a very strong impression of the great effort you have devoted to it and the care with which all of it has been worked out. I am looking very much forward to studying all of it more closely and to learn from it. To show you that I have not been idle myself during Rosenfeld's absence, I am forwarding as a little New Year's greeting the proofs of a lecture that I gave last summer and which I have prepared during these weeks and translated into Danish[13]. I would be very happy to learn what you think about it and you should not be afraid of using all the invectives you can think of. This year our usual small conference is going to take place during the first weeks of April and I definitely reckon that this time you will be able to accept the invitation from the Institute and that we shall not have to content ourselves, like the last time, with your appearance in a parody, even though it was very funny.

With many kind regards from us all and best wishes for the new year,

Yours,

Niels Bohr

PAULI TO PEIERLS, 22 May 1933
[Translated extract of German text in PWB II, letter [310]]

May 22, 1933

Dear Peierls,

...

Thus, while I believe that Bohr is quite mistaken in his fight against energy conservation in nuclear physics,* I do believe, on the other hand, that he is absolutely correct as far as the question of field measurements is concerned. I do not believe this out of a general belief in Bohr's authority (I never entertained such a belief), and not even only because I have not found any errors in his proof. But I believe it because, conversely, not a single watertight argument exists for the inequalities of the field strengths in the paper by you and Landau. (The discussion of a particular measuring device *never* provides an argument!) If there is an inequality for limitations in the measuring accuracy that is not based directly on the foundations of the theoretical formalism in the form of commutation relations, *then such a relation must have a definite physical origin.* In the case of the uncertainty relations in your paper for the momentum and position measurements of particles, this origin is the actual absence of negative energy states (a fact contradicted by the theory). But the negative energy states do not play any rôle in the field measurements, and there is no argument whatsoever for an inequality which does not follow from the commutation relations and which *does not contain the charge, mass or dimensions of the test bodies.* (The apparent paradoxes concerning the "zero point fluctuations" of the field strengths were resolved by Bohr.)

---

* By the way, when I saw him in March, he absolutely rejected the positive electron and thought that Blackett had just produced some "pathological photographs".

On the other hand, something could of course happen when the linear dimensions of the space–time region, over which the fields are averaged in the measurement, becomes of the order of magnitude h/mc (respectively h/mc^2) (m = the mass of the test body), that is, more generally, when the atomic structure of the test bodies plays a rôle. This is also Bohr's opinion. (By the way, the promised proofs of the paper by Rosenfeld and [him][14] have so far not arrived.)

...

Yours,
Pauli

BOHR TO PAULI, 15 February 1934
[Typewritten]

UNIVERSITETETS INSTITUT   BLEGDAMSVEJ 15, KØBENHAVN Ø.
            FOR   DEN 15. Februar 1934.
     TEORETISK FYSIK

Kære Pauli,

Tak for Dit rare Brev, der kom som en meget velkommen Anledning til en lille Hjerteudgydelse. Jeg er nemlig i de sidste Dage med stor Begejstring vendt tilbage til mit gamle primitive Syn paa Elektronteoriens Problemer, efter at jeg i nogen Tid har været meget foruroliget over den af Dig med saa stor Styrke betonede Vanskelighed med Elektronernes uendelige Selvenergi. Det har nemlig slaaet mig, at den kvanteelektrodynamiske Formalisme ikke alene, saadan som Rosenfeld og jeg har prøvet at vise, er en fuldstændig konsekvent, af Atomteorien uafhængig Idealisation, men at enhver Anvendelse af den staar i et rent Udelukkelsesforhold til Elementarpartikelproblemerne. Enhver Feltmaaling indenfor Kvanteteoriens Omraade fordrer jo, som vi har vist, Prøvelegemer med en elektrisk Ladning $E \gg \sqrt{hc}$, og som er kontinuert fordelt over et Omraade, hvis lineære Udstrækning er $L \gg h/Mc$, hvor M er Prøvelegements Masse. Det vil imidlertid sige, at Teoriens Anvendelsesomraade falder helt udenfor de egentlige Atomproblemer og er indskrænket til Problemer, hvor der i Modsætning til disse er en tilstrækkelig stor Kobling mellem de ladede Legemer og Feltet til at sikre det i Formalismen forudsatte komplementære Forhold mellem Felt- og Photonbegrebet. I de egentlige Atomproblemer, hvis

[14] A slip of the pen corrected: "me", in the original, should read "him" (Bohr).

[469]

karakteristiske Træk netop er den overordentlig ringe Kobling mellem Partiklerne og Straalingsfelterne, er Betingelserne for Feltbegrebets Anvendelse en helt anden. Netop den ringe Værdi af ε^2/hc og Elektronradien deraf følgende ringe Størrelse i Forhold til h/mc betinger, dels at man i de egentlige kvantemekaniske Problemer inklusive Hulteorien med stor Tilnærmelse kan benytte Feltbegrebet paa ren klassisk Maade, dels at man i de egentlige Straalingsproblemer med tilsvarende Tilnærmelse kan indskrænke sig til de gamle beskedne Korrespondensargumenter paa konsekvent Maade.

Jeg er bange for, at en saadan Indstilling i første Øjeblik maaske vil forekomme Dig altfor reaktionær, men hvis jeg ikke tager rent fejl, er det virkelig det eneste nøgterne Syn paa Atomproblemerne, der for Øjeblikket er muligt. Pointen er nemlig, at ligesom den klassiske Elektronteori var en Idealisation, der gælder for alle Virkninger, der er store i Forhold til h, saaledes er Kvanteelektrodynamikken en Idealisation, der kun gælder, dersom alle Ladninger er store i Forhold til ε. Dette betyder jo ingenlunde, at den sidste Teori ikke kan finde Anvendelse ved Diskussionen af saadanne universelle Problemer som Varmestraalingsloven, men blot at Situationen indenfor Atomteorien ligger væsentlig anderledes end antaget i enhver af de nævnte Idealisationer.

At Feltbegrebet maa benyttes med saa stor Forsigtighed, er ogsaa en nærliggende Tanke, naar man erindrer, at alle Feltvirkninger til syvende og sidst kun kan iagttages ved deres Virkning paa Materien. Som Rosenfeld og jeg viste, er det saaledes slet ikke muligt at afgøre, hvorvidt Feltfluctuationerne allerede er til Stede i det tomme Rum eller først skabes af Prøvelegemerne. Selv om Fluctuationerne er et uundgaaeligt Led af Kvanteelektrodynamikken, er man derfor berettiget til i Elektronteorien paa modsigelsesfri Maade ganske at se bort fra disse Fluctuationer og alle andre Konsekvenser af Feltkvantiseringen. Dette fremgaar jo ogsaa af Einsteins simple Udledning af Varmestraalingsloven paa korrespondensmæssigt Grundlag.

Denne Indstilling har navnlig været mig en stor Befrielse, hvad de relativistiske Maaleproblemer angaar. Som Du ved, har jeg længe ment, at man ikke ved Undersøgelse af disse Problemer kan finde noget som helst Argument mod at benytte Rum–Tidsbegreberne i Elektronteorien ogsaa indenfor Omraader, der er smaa i Forhold til h/mc. Jeg var dog alligevel begyndt at blive meget foruroliget ved Mistanken om, at Nødvendigheden af en Feltkvantisering kunde gøre saavel Behandlingen af Maaleproblemerne som Forsøgene paa den videre Udbygning af Hulteorien ganske illusorisk. Som Sagerne nu ligger, er der imidlertid efter min Mening ingensomhelst Anledning til at tro, at vi i Maaleproblemerne kan møde nogensomhelst Overraskelse, der ikke implicite er skjult i Hulteoriens matematiske Formulering, hvis blot denne opfylder de tilstrækkelige Fordringer

[470]

til Invarians. Den detaillerede Behandling af mulige Maaleanordninger er vel i Almindelighed meget kompliceret, men vi har jo ogsaa alle nu forstaaet, at de teoretiske Forudsigelser, som det ved saadanne Maalinger vil kunne dreje sig om at prøve, ogsaa oftest vil være af meget kompliceret Natur. Kun i det særlige Tilfælde hvor Problemet med tilstrækkelig Tilnærmelse kan betragtes som et 1-Partikelproblem, ligger Forholdene simplere, og her kan man, som jeg antydede i Bryssel, let bevise, at man ved at gøre tilstrækkelig mange Forsøg med tilstrækkelig varierede Straalingsbundter, kan prøve alle Teoriens principielt statistiske Forudsigelser, hvad Tæthedsfunktionerne angaar.

Jeg forsøger netop i disse Dage ved Rosenfelds Hjælp at udarbejde en Fremstilling af Korrespondensargumentets konsekvente Anvendelse i Atomteorien, og skulde derfor være Dig meget taknemmelig for, om Du med at par Linier vil lade mig vide, om Du paa Forhaand er helt uenig i en saadan Indstilling som den, jeg her har antydet.

Med mange venlige Hilsener fra os alle,

Din ældgamle
Niels Bohr

Translation, see p. [31]

BOHR TO PAULI, 24 February 1934
[Typewritten]

UNIVERSITETETS INSTITUT
FOR
TEORETISK FYSIK

BLEGDAMSVEJ 15, KØBENHAVN Ø.
DEN 24. Februar 1934.

Kære Pauli,

Mange Tak for Dine rare Breve med det saavel hurtige som indgaaende Svar paa min lille Hjerteudgydelse. Baade Rosenfeld og jeg har studeret Dine Breve meget nøje og er naturligvis enige med det meste deri, men jeg tror alligevel, at Du undervurderer Konsekvensen i det helt igennem paa den ringe Værdi af Finstrukturkonstanten baserede Korrespondensangreb paa Atomproblemerne. Med den forbigaaende Foruroligelse, som jeg hentydede til i mit Brev, tænkte jeg paa et forvildet (for ogsaa at benytte Din Sprogbrug) Forsøg ud fra Kvanteelektrodynamikken at finde en nedre Grænse for Finstrukturkonstanten, og med Befrielsen paa den af Naturen eller vel rettere Fornuften tilbudte gamle Redningsplanke fra det bundløse Dyb, som vi her ser ned i. Denne Sætning vil jo sikkert give Dig ny Anledning til at bruge Dine store Evner som Psykoanalytiker; men min Indstilling har alligevel ikke alene en psykisk, men ogsaa en meget teknisk Baggrund. Jo mere jeg fordyber mig i Maaleproblemet, ser jeg Finstrukturkonstanten dukke op paa alle Punkter, og Umuligheden af en i Kvanteelektrodynamikkens Aand rationel Behandling af Straalings- og Gravitationsproblemerne forekommer mig mere og mere indlysende. Hvad de første angaar, mener jeg, at Du ganske overdriver Betydningen af den formelle Forbindelse mellem Selvenergiparadokset og Behandlingen af Spektralliniebreddeproblemet, der finder en utvungen, omend maaske ikke elegant Forklaring ud fra Atommekanikkens simpleste Forudsætninger. Hvad Gravitationsproblemerne angaar, har Du vel allerede forstaaet, at Svaret paa de af Rosenfeld behandlede Paradoxier efter min Indstilling vil blive en lignende Henvisning til Forskellen mellem makro- og mikroskopiske Beskrivelsesmuligheder som i Elektronteorien. Jeg har netop i disse Dage faaet Korrektur fra Langevin om Diskussionen i Bryssel og er netop i Færd med at gøre mine egne Bemærkninger til Indførelse deri færdig. Om nogle faa Dage skal jeg skrive nærmere til Dig derom. I dag vil jeg kun tilføje, at jeg nu, da Foraaret nærmer sig, som sædvanlig tænker paa Muligheden af et mere end velkomment lille Besøg af Dig her oppe, og at jeg derfor skulde være meget glad for at høre et Par Ord om Dine Planer for den nærmeste Tid.

Din

Niels Bohr

P.S. Hvis Du har en Kopi eller Korrektur af Dit og Heisenbergs nye Arbejde, vil vi alle her være meget interesserede i at se den.

Translation

Copenhagen, February 24, 1934

Dear Pauli,

Many thanks for your nice letters with the quick as well as thorough answer to the little outpouring from my heart. Both Rosenfeld and I have studied your letters very carefully and of course we agree with most of the contents. However, I still believe that you underestimate the consequences of the correspondence attack on the atomic problems, which is based throughout on the smallness of the fine structure constant. By the passing uneasiness to which I referred in my letter, I had in mind a bewildered (to use also your mode of expression) attempt to find a lower limit for the fine structure constant on the basis of quantum electrodynamics, and by the relief, I had in mind the old life-saver offered by nature, or perhaps rather by reason, from the bottomless pit we are looking down into here. This sentence will certainly provide you with a new occasion to apply your great gifts as a psychoanalyst, but still my attitude has not only a psychic but also a very technical background. The more I immerse myself in the measuring problem, the more I see the fine structure constant appearing everywhere, and the impossibility of a rational treatment of the radiation and gravitational problems in the spirit of quantum electrodynamics seems to me more and more evident. As far as the former are concerned, I think that you quite exaggerate the importance of the formal connection between the self-energy paradox and the treatment of the problem of the widths of spectral lines, which finds a natural, although perhaps not elegant, explanation from the simplest premises of atomic mechanics. As far as the gravitational problems are concerned I guess that you have already realized that according to my attitude the answer to the paradoxes treated by Rosenfeld would be a reference to the difference between macroscopic and microscopic possibilities of description, similar to that in electron theory. These very days I have received the proofs from Langevin of the discussion in Brussels, and I am just in the process of finishing my own remarks to be included therein. In a few days I will write to you in more detail about this. Today I only add that now as spring approaches I think as usual of the possibility of a more than welcome little visit from you here, and I should therefore be very happy to hear a few words about your plans for the nearest future.

Yours,
Niels Bohr

P.S. If you have a copy or a proof of the new paper by you and Heisenberg, all of us here would be very interested in seeing it.

BOHR TO PAULI, 15 March 1934
[Typewritten with handwritten postscript]

UNIVERSITETETS INSTITUT
FOR
TEORETISK FYSIK

BLEGDAMSVEJ 15, KØBENHAVN Ø.
DEN 15. Marts 1934.

Kære Pauli,

Dit Brev, som jeg lige har faaet, var i enhver Henseende Anledning til stor Glæde for mig. Først og fremmest sender min Kone og jeg Dig vore hjerteligste Lykønskninger til Dit forestaaende Giftermaal og glæder os meget til snart at skulle have Dig og Din tilkommende Kone paa et lille Besøg hos os. Ugen efter Paaske passer os udmærket, og jeg haaber, at I begge trods Søjlerne vil føle Jer hjemligt paa Carlsberg.

Dernæst var baade Rosenfeld og Klein, der for Øjeblikket er her paa Besøg, saavel som jeg meget interesserede i Dine Bemærkninger og skatter ikke alene Din Munterhed i Udtrykket, men ogsaa Din Saglighed i Indstillingen. Jeg skal naturligvis tage Dine Forslag til Forbedringer af Teksten i mine Bemærkninger under Overvejelse, naar Korrekturen kommer. Bemærkningerne er iøvrigt, som Du ogsaa forstod, kun at betragte som en Skitse til en mere indgaaende engelsk Afhandling over Korrespondenssynspunktets Karakter og Begrænsning, som jeg haaber snart at faa færdig. Jeg sendte dem blot af Sted til Langevin i den foreliggende Form for ikke yderligere at forsinke Trykningen af Solvay Reporten.

Vi var ogsaa alle interesserede i Weisskopfs smukke Arbejde[15] og tror, at der hersker fuld Enighed i Sagen, selv om jeg ikke er klar over Heisenbergs Opfattelse og skulde være glad for at høre Din Mening om hans Bemærkninger i et Brev, som jeg lige har faaet, og hvoraf jeg sender en Kopi.

Trods al Enighed i Sagen vil jeg dog gerne foreslaa en ringe, men maaske ikke uvigtig Ændring i den didaktiske Fremstilling af Situationen. Som jeg antydede i det lille Brev, jeg sendte Dig for nogle Dage siden, har jeg, siden jeg sendte mine Bemærkninger af Sted, tænkt adskilligt over Liniebredde-problemet og mener, at man tydeligere maa betone, hvad man forstaar ved dettes konsekvente korrespondensmæssige Behandling. Jeg ser nemlig ingen dybere Pointe i at gøre en Adskillelse mellem Kvanteelektrodynamikken og en saadan Brug af Feltbegrebet, som det drejer sig om i Weisskopfs Afhandling.

---

[15] V.F. Weisskopf, *Über die Selbstenergie des Elektrons*, Z. Phys. **89** (1934) 27–39; cf. Introduction to Part I, ref. 28.

[474]

Alle disse Metoder har man jo med større eller mindre Ret betegnet som korrespondensmæssige, medens jeg efter den dybere Indsigt, som den senere Udvikling har givet os, gerne vil benytte denne Betegnelse paa principielt mere indskrænket Maade.

Jeg er jo ganske klar over de Farer, som ligger i den mere naive Behandling af Liniebreddeproblemet, som man forsøgte i gamle Dage, hvor man for groft skilte mellem Atomtilstande og Straaling, og jeg forstaar fuldstændigt det Krav til en mere konsekvent Behandling, som gav sig Udtryk i Diracs Teori. Da den sidste Vej, som vi nu ser, imidlertid er ufarbar af principielle Grunde, tvinges vi efter min Mening til at betragte Liniebreddeproblemet som ikke nærmere analyserbart i lignende Forstand som de Interferensfænomener, der optræder ved Elektroner, der kan gaa igennem to Huller i en Skærm. Selvfølgelig er det sidste Problem af en helt anden Natur og den omhandlede Analogi derfor ren dialektisk [didaktisk?], men jeg tror alligevel, at Indstillingen er saavel naturlig som konsekvent.

For en sædvanlig Absorptionslinies Vedkommende, hvor alle Forhold jo ligger klart eksperimentelt belyst, ser jeg saaledes ingen Vanskelighed ved at fuldstændiggøre den simple korrespondensmæssige Behandling af Dispersionsproblemet ved at benytte den klassiske Dispersionsformel med Straalingsdæmpning. Denne Formel danner jo under Hensyntagen til Energibevarelse og Superpositionsprincip et Hele, hvoraf intet Led konsekvent kan udskilles.

For en almindelig Absorptionslinie er vel overhovedet selve Liniebreddens Definition knyttet op til en Forsøgsanordning, der beror paa en Maaling af Dispersionen for streng monokromatisk Belysning. For Spektrallinier, der svarer til Overgange mellem to Tilstande med endelig Levetid, er den klassiske Analogi jo betydelig mere tvivlsom, og navnlig er de eksperimentelle Anordninger, som vilde tillade en entydig Definition af Liniebredden, langt mere kompliceret. Personlig tror jeg imidlertid, at det Resultat, som man efter Din Sprogbrug vel vilde sige, at man har svindlet sig til ved Hjælp af Diracs Teori, er tilstrækkelig simpelt til, at man kan anse det for rigtigt.

Jeg mener, at vi her som ved alle egentlige Korrespondensbetragtninger, indbefattet de Heisenbergske Grundligninger og Diracs Bølgeligning, foruden i Modsigelsesfriheden først og fremmest maa søge Argumentationen i Enkelheden. Man kunde overhovedet være tilbøjelig til som Fysikkens Grundlag at opstille det Postulat, at Naturen er saa lovbestemt, som det overhovedet er muligt at definere paa modsigelsesfri Maade, men jeg skal ikke gaa videre med denne løse Snak, før jeg har Lejlighed til at se dit Minespil paa nærmere Hold.

Det glædede mig ogsaa, at Du forstod Grundstemningen i mine Slutbemærkninger om Energibevarelsen. Jeg er dog siden blevet mere skeptisk med Hensyn til det Haab, som implicit ligger i disse Bemærkninger, nemlig at

Margrethe and Niels Bohr, Carlsberg honorary mansion 1960.

benytte Gravitationsteorien til en Korrespondensudledning af Loven for β-Straaleemissionen. Tanken var den, at en Neutrino, for hvilken man antager en Hvilemasse 0, vel ikke kan være andet end en Gravitationsbølge med passende Kvantisering. Jeg har imidlertid overbevist mig om, at Gravitationskonstanten er alt for lille til at kunne berettige en saadan Opfattelse og derfor fuldt forberedt paa, at vi her virkelig har et nyt Atomartræk for os, der kunde være ensbetydende med Neutrinoens reale Eksistens.

[476]

Jeg vil blot slutte med at sende de hjerteligste Hilsener og Lykønskninger fra os alle og især fra min Kone og mig, der glæder os meget til snart at se Jer heroppe.

Din

Niels Bohr

P.S. Efter at Klein og Rosenfeld har set nærmere på Weisskopfs Afhandling er Spørgsmaalet om en principiel Forskel mellem hans Metoder og Resultat og de tidligere Opfattelser os meget uklart. Maaske kan Du oplyse os derom?

Translation

Copenhagen, March 15, 1934

Dear Pauli,

Your letter which I have just received gave me in every respect occasion for great joy. First and foremost my wife and I send you our most cordial congratulations on your forthcoming marriage and we are very much looking forward to having you and your future wife with us for a little visit here. The week after Easter will suit us splendidly and I hope that you will both feel at home at Carlsberg in spite of the columns.

Furthermore, both Rosenfeld and Klein, who is here on a visit at present, as well as I were very interested in your comments and we appreciate not only the liveliness of your formulation, but also the objectivity of your attitude. When I get the proofs I shall of course consider your suggestions for improvements in the formulation of my remarks. By the way, as you also realized, the remarks are only to be regarded as a sketch for a more thorough treatise in English on the nature and limitations of the correspondence viewpoint, which I hope to finish soon. I only returned them to Langevin in their present form in order not further to delay the printing of the Solvay report.

All of us were also interested in Weisskopf's beautiful paper[15], and I believe that there is complete agreement in this matter, although I am not quite clear as to Heisenberg's view and I should be happy to know your opinion about his remarks in a letter which I just received and of which I send a copy.

In spite of the general agreement in this matter I should still like to suggest a small, but perhaps not insignificant change in the didactic presentation of the

Margrethe and Niels Bohr, Carlsberg honorary mansion 1960.

situation. As I indicated in the short letter that I sent you a couple of days ago, after sending off the remarks, I have given much thought to the problem of line width and I believe that one must stress more clearly what is meant by a consistent treatment of it on the basis of the correspondence principle. Indeed, I do not see any deeper point in making a distinction between quantum electrodynamics and the kind of application of the field concept involved in Weisskopf's paper. All of these methods have with more or less justification been characterized as correspondence methods, while – in accordance with the deeper insight that we have gained through the recent development – I would like to employ this designation in a fundamentally more limited manner.

Of course I am quite aware of the dangers inherent in the more naive treatment of the problem of line width that was attempted in the old days, where one distinguished too crudely between atomic states and radiation; and I perfectly understand the demand for a more consistent treatment as it found expression in Dirac's theory. Since, as we now see, the latter path is barred for fundamental reasons, however, we are in my opinion forced to regard the problem of line width as defying closer analysis in a sense similar to the interference phenomena that occur for electrons passing through two holes in a diaphragm. Of course, the latter problem is of quite a different nature and the analogy mentioned is therefore purely dialectic [didactic?], but I still believe that the attitude is natural as well as consistent.

Thus, as far as a usual absorption line is concerned, where all features are fully accounted for experimentally, I do not see any difficulty for a completion of the simple correspondence treatment of the dispersion problem by using the classical dispersion formula with radiation damping. This formula constitutes, in fact, with due regard to the principles of energy conservation and superposition, a unity from which no part can be separated in a consistent manner.

In the case of an ordinary absorption line I suppose indeed that the very definition of the line width is associated with an experimental arrangement which depends on a measurement of the dispersion of strictly monochromatic light. In the case of spectral lines corresponding to transitions between two states with finite lifetimes, the classical analogy is of course rather more dubious, and in particular those experimental arrangements which would permit an unambiguous definition of the line width, are much more complicated. Personally I believe, however, that the result – which I suppose in your parlance one would say had been obtained through swindle by means of Dirac's theory – is sufficiently simple to be considered correct.

I think that here, as in all other genuine correspondence considerations, including the fundamental Heisenberg equations and Dirac's wave equation, we must base the argumentation on consistency, and above all on simplicity. As far as the basis of physics is concerned, one may even be inclined to advance the postulate that nature is governed by laws exactly to the extent to which it is at all possible to define in a consistent manner. But I shall not continue this idle chatter until I have the opportunity of watching the expression on your face at close range.

I was also happy that you understood the basic mood of my final remarks on energy conservation. Since then I have become more sceptical, however, as regards the hope implicit in these remarks, namely to employ the theory of gravitation for a correspondence derivation of the law of β-ray emission. The idea was that a neutrino, for which one assumes a zero rest mass, could hardly be anything else than a gravitational wave with appropriate quantization. However, I have convinced myself that the gravitational constant is far too small to justify such an interpretation, and I am therefore fully prepared that we are here facing a new atomic feature which might be tantamount to the real existence of the neutrino.

[479]

I shall merely conclude by sending the most cordial greetings and congratulations from all of us and in particular from my wife and me, who are looking very much forward to seeing you here.

Yours,

Niels Bohr

P.S. After Klein and Rosenfeld have looked more closely into Weisskopf's paper, the question of a fundamental difference between his methods and result and the earlier conceptions has become very unclear. Perhaps you could explain this to us.

PAULI TO HEISENBERG, 15 June 1935
[Extract of letter [412] in PWB II]

Zürich, 15. Juni 1935

Lieber Heisenberg!

Über Physik weiß ich nichts Neues. Weisskopf ist am Delbrückschen Problem und die dabei auftretenden Fragen der Subtraktionsphysik werden sehr unschön. – Davon möchte ich aber heute nicht berichten, sondern von zwei *pädagogischen* Fragen, bei denen Du eventuell öffentlich eingreifen könntest.

1. *Einstein* hat sich wieder einmal zur Quantenmechanik öffentlich geäußert und zwar im Heft des Physical Review vom 15. Mai (gemeinsam mit Podolsky und Rosen – keine gute Kompanie übrigens). Bekanntlich ist das jedes Mal eine Katastrophe, wenn es geschieht. "Weil, so schließt er messerscharf – nicht sein kann, was nicht sein darf" (Morgenstern)[16].

Immerhin möchte ich ihm zugestehen, daß ich, wenn mir ein Student in jüngeren Semestern solche Einwände machen würde, diesen für ganz intelligent und hoffnungsvoll halten würde. – Da durch die Publikation eine gewisse Gefahr einer Verwirrung der öffentlichen Meinung – namentlich in Amerika – besteht, so wäre es vielleicht angezeigt, eine Erwiderung darauf ans Physical Review zu schicken, wozu ich *Dir* gerne zureden möchte.

Vielleicht lohnt es sich also doch, wenn ich Papier und Tinte vergeude, um diejenigen durch die Quantenmechanik geforderten Tatbestände zu formulieren, die Einstein besonders geistige Beschwerden machen.

Er hat jetzt so viel verstanden, daß man zwei Größen, die nicht vertauschbaren Operatoren entsprechen, nicht gleichzeitig messen und ihnen nicht

---

[16] C. Morgenstern, *Alle Galgenlieder*, Berlin 1932.

gleichzeitig Zahlwerte zusprechen kann. Aber woran er sich in Verbindung damit stößt, ist die Weise, wie in der Quantenmechanik zwei Systeme zu einem Gesamtsystem zusammengestzt werden. – Man sieht dies am folgenden Beispiel, das er (im wesentlichen) auch heranzieht.

. . .

Überhaupt spukt bei älteren Herren wie *Laue* und *Einstein* die Idee herum, die Quantenmechanik sei zwar *richtig*, aber *unvollständig*. Man könne sie *durch in ihr nicht enthaltene Aussagen ergänzen, ohne die in ihr enthaltenen Aussagen zu ändern.* (Eine Theorie mit einer solchen Eigenschaft nenne ich – im logischen Sinne – *unvollständig.* Beispiel: die kinetische Gastheorie.) Vielleicht könntest Du – bei Gelegenheit der Erwiderung an Einstein – einmal in autoritativer Weise klarstellen, daß eine solche Ergänzung bei der Quantenmechanik nicht möglich ist, ohne ihren Inhalt abzuändern.

. . .

Viele Grüße

Dein W. Pauli

Translation, see p. [251]

PAULI TO BOHR, 18 February 1948
[Typewritten]
See p. [277]

PAULI TO BOHR, 17 August 1948
[Handwritten]
See p. [268]

BOHR TO PAULI , 15 September 1948
[Carbon copy]
See p. [270]

PAULI TO ROSENFELD, 5 July 1949
[Handwritten]

Physikalisches Institut	ZÜRICH 7, 5. Juli, 1949
der Eidg. Technischen Hochschule	Gloriastraße 35
Zürich	

Dear Rosenfeld,

I am writing you today regarding some questions of physics connected with the papers of you and Bohr* on the measurability of fields and charges. (I read once the then existing $(-1)^{st}$ approximation of the second part of your paper. As I do not remember all details of it I shall quote the abstracts which appeared in Pais' book 'Developments in the theory of the electron'[17].) I see, namely, a possibility of a connection of the *properties of the 'vacuum'-charge-fluctuation with the vacuum-polarisation by an external electromagnetic field* with which I was occupied so much during the last months.

According to Schwinger** the vacuum expectation values for products of the current-components on the one hand and its commutator on the other hand can be written [Pauli's formulae are written in units so that $\hbar = c = 1$. The factor e^2 in eq. (I) has been added by me for the convenience of the reader.]

$$\langle \{j_\mu(x), j_\nu(x')\} \rangle_0 = e^2 \cdot \left(2\frac{\partial\Delta}{\partial x_\mu}\frac{\partial\Delta}{\partial x_\nu} - \delta_{\mu\nu}\left[\left(\frac{\partial\Delta}{\partial x_\alpha}\right)^2 + m^2\Delta^2\right] - \right.$$
$$\left. 2\frac{\partial\Delta^{(1)}}{\partial x_\mu}\frac{\partial\Delta^{(1)}}{\partial x_\nu} + \delta_{\mu\nu}\left[\left(\frac{\partial\Delta^{(1)}}{\partial x_\alpha}\right)^2 + m^2\Delta^{(1)2}\right]\right), \quad \text{(I)}$$

"Vacuum"

---

* If you wish to show this letter to Bohr – under the obvious condition that you answer me *before* that – I shall be glad. I don't want to disturb him however in abolishing every Monday everything what was built up the week before of the new Institute's building (Hint for your next article on Bohr; comparison with Penelope).

** See his paper Phys. Rev. **75**, 651, 1949. I use his definitions (in his Appendix) for $\Delta(x), \Delta^{(1)}(x), \overline{\Delta}(x)$ and also $\{A,B\} \equiv AB + BA$, $[A,B] \equiv AB - BA$; $\langle$ is the 'bra', $\rangle$ the 'ket'.

---

[17] The addendum by Wheeler, cf. this volume p. [41].

the argument of the $\Delta, \Delta^{(1)}$-functions is always $x - x'$.

$$\langle [j_\mu(x), j_\nu(x')] \rangle_0 = (-\mathrm{i})4e^2 \cdot \left\{ \frac{\partial \Delta}{\partial x_\mu} \frac{\partial \Delta^{(1)}}{\partial x_\nu} + \frac{\partial \Delta}{\partial x_\nu} \frac{\partial \Delta^{(1)}}{\partial x_\mu} - \right.$$
$$\left. \delta_{\mu\nu} \left(\frac{\partial \Delta}{\partial x_\alpha} \frac{\partial \Delta^{(1)}}{\partial x_\alpha} + m^2 \Delta \Delta^{(1)} \right) \right\} \qquad \text{(II)}$$

The corresponding formulas in quantum electrodynamics (Heaviside units for the field-strengths; you get D, $D^{(1)}$ out of $\Delta, \Delta^{(1)}$ by putting $m = 0$) are

$$\langle \{F_{\mu\rho}(x), F_{\nu\sigma}(x')\} \rangle_0 =$$
$$\left(\frac{\partial^2}{\partial x_\mu \partial x_\sigma} \delta_{\rho\nu} + \frac{\partial^2}{\partial x_\rho \partial x_\nu} \delta_{\mu\sigma} - \frac{\partial^2}{\partial x_\mu \partial x_\nu} \delta_{\rho\sigma} - \frac{\partial^2}{\partial x_\rho \partial x_\sigma} \partial_{\mu\nu} \right) \cdot D^{(1)}(x - x') \qquad \text{(I')}$$

$$[F_{\mu\rho}(x), F_{\nu\sigma}(x')] = (\text{same}) \cdot cD(x - x') \qquad \text{(II')}$$

(The formula (I') is essentially eq. (11) on p. 16, Part I of your paper[18].)
Just as (I') is the field fluctuation and in principle observable, (I) is the charge fluctuation and in principle observable. The latter (I) is identical with Heisenberg's old result after averaging over finite time–space regions and over the thickness b of the boundaries. (NB. That the latter is necessary for (I) but not for (I') is directly caused by the circumstances that in (I) there is a *product* of *two* Δ-type functions, which means higher singularity than in (I').)

Now follow my questions: 1) *I do not agree with your expression 'uncertainty' sub k.)* (See Pais, p. 44) which seems to me *misleading*. The charge fluctuation of Heisenberg is a *real* effect and not an uncertainty and is analogous to the field-fluctuation given by (I'). The 'unavoidable fluctuation' – mentioned sub j. – is no hindrance to consider the result of the measurement of the surface integral – after compensation of the *classical* effects – as the *real* value of the charge in this region, because the result of this measurement is *reproducible*. The situation is completely analogous to the one described in Part I of your paper (see p. 57 above: "Eine nähere Betrachtung zeigt indessen …", [this volume p. [113]].) The result of the measurements characterized by a.) – h.) (Pais, p. 44) has to be considered as the *real charge* value *in spite of the fluctuations*.

---

[18] BR, this volume p. [55]. The formula for the vacuum fluctuations averaged over the space–time domain G.

On the other hand a discussion of the formula (II) and the *uncertainty-relations* claimed by it should not be difficult (particularly for $\mu = \nu = 4$: charge-density in different space–time regions). The discussion of the idealized experimental arrangements will be closely analogous to the one in Part I, §6 [p. [105]]: There will be an incompensable uncertainty due to the parts of the measurements of $\oint E_n \, \mathrm{d}f$ in [the domain] I prop[ortional] to the displacements D^{II} [referring to the test body in the domain II] and vice versa. These will exactly correspond to the right side of (II). I think, that the Part II [the work in progress by Bohr and Rosenfeld] should also contain a consideration of this type, not only a consideration of the charge density-fluctuations.

There is another question which interests me still more. Sub c.) and d.) (Pais, p. 44) you speak of the *pairs produced by an external field*. Could you briefly indicate in your answer *how the charge-distribution of the pairs induced by the external field was computed*. How did you avoid at this place the infinities occurring in the theory of vacuum-polarisation? (*'Self-charge!'* We call this here the 'top nonsense' of the present theory.) I guess, that Bohr had here the right instinct of a physicist, because a "correct" computation will give an infinite result. *Please write this to me as soon as possible.*

I am the more interested in this point as there is a very close connection between the commutator given by the expression (II) above and the polarisation-current induced by an external field. The latter is given by

$$\delta j_\mu(x) = \int K_{\mu\nu}(x - x') A_\nu(x') \, \mathrm{d}^4 x',$$

where $K_{\mu\nu}$ differs (apart from a constant factor) from the commutator $\langle [j_\mu(x), j_\nu(x')] \rangle_0$ only by the famous $\varepsilon(x - x')$, defined as $\varepsilon(t) = \begin{Bmatrix} +1 \text{ for } t > 0 \\ -1 \text{ for } t < 0 \end{Bmatrix}$.

It is the ε which makes the troubles (infinite self-charge, etc.). On the other hand I don't see a possibility to change the right side of (II) – after the usual averaging over space–time regions.

Now I am eagerly looking forward to your answer. (Will you see Bohr in the next time? Is there a chance that Part II will appear?)

We made here great progress regarding particular problems of quantum-electrodynamics, but not with the principle side of the theory. The main obstacle concerning the latter seems to me at present the self-charge.

With many regards
as always
yours
W. Pauli

PAULI TO ROSENFELD, 11 July 1949
[Handwritten]

Physikalisches Institut
der Eidg. Technischen Hochschule
Zürich

ZÜRICH 7, 11 July 1949
Gloriastraße 35

Dear Rosenfeld,

I was glad to obtain your letter of July 7, which clarified a number of points[19]. I am also jubilant, that you [are] just going to Denmark to finish the paper on charge measurements. – The point which has still to be settled is of course an *exact* computation of the charge-distribution of the pairs produced by the field. The expressions of Uehling[20] or of Rose & me[21] should be completely reliable for this purpose.

Your statement "there has never been any consistent theory of positrons and negatons going beyond the trivial first order effects" seems to me too pessimistic. Luttinger and Jost[22] are working out just now the next order corrections (in $e^2/\hbar c$) for the cross section of pair production (with 'renormalisation' à la Schwinger–Tomonaga) and everything is convergent.

What the principles of this new formalism concerns, I become more and more clear that my proposal of 'regularisation' will have only a sense if the masses are *real*. And then it is leading back to the compensation-idea of Pais[23] which is so familiar to Bohr.

In the last days I was thinking again on the self-charge, where certainly no compensation is possible in the first approximation in $e^2/\hbar c$ (from the 'realistic' point of view with respect to the auxiliary masses). But I believe now, that one can seriously try to carry through the other point of view; *the postulate that*

[19] Unfortunately this letter seems to be lost.

[20] E.A. Uehling, *Polarization Effects in the Positron Theory*, Phys. Rev. **48** (1935) 55–63.

[21] W. Pauli and M.E. Rose, *Remarks on the Polarization Effects in the Positron Theory*, Phys. Rev. **49** (1936) 462–465.

[22] R. Jost and J.M. Luttinger, *Vacuum Polarisation und e^4-Ladungsrenormalisation für Elektronen*, Helv. Phys. Acta. **23** (1950) 201–214.

[23] A. Pais, *On the Theory of the Electron and of the Nucleon*, Phys. Rev. **68** (1945) 227–228, and *On the Theory of Elementary Particles*, Kon. Ned. Akad. v. Wet. **19**, no. 1 (1946).

*the self-charge has to be zero could determine the value of $e^2/\hbar c$.**

We are trying now: 1. To check whether the theoretical self-charge is in first approximation really the same for all kind of original charges with the same amount (electrons, Bosons, oil-drops etc). I have some reason to guess that this is actually so.

2. To compute the next approximation (in $e^2/\hbar c$) for the self-charge due to the radiative corrections (one has to retain only terms linear in the *external* field but the electromagnetic radiation field has now to be quantized). One has to look (a) whether in the next approximation the divergence of the self-charge is still logarithmic and (b) whether the sign is opposite to the first approximation.

If the answer to both (a) and (b) is affirmative one could try to determine "in first approximation" the value of $e^2/\hbar c$ by claiming the vanishing of the self-charge. Of course I do not know whether the present theory is good enough to carry through such a programme (even if it is correct in principle), as the result may depend on the mass and spin values of all charged pairs which exist in nature.

But we shall try here something in this direction and I would be glad to hear yours and Bohr's opinion about it.

If the paper is 'finished' please let me know.

Best wishes to you and Bohr and – last not least – to the paper

Yours ever
W. Pauli

BOHR TO PAULI, 15 August 1949
[Carbon copy]

Tisvilde, 15. august [19]49.

Kære Pauli,

Jeg har haft megen dårlig samvittighed over ikke at have skrevet så længe og ikke en gang takket for dine rare og så oplysende breve. Ikke mindst under Rosenfelds besøg her for nogle uger siden har vi begge haft megen anledning til at sende dig mange venlige tanker, idet henvisningen i dine breve

---

* Bohr considered only the fluctuations in this connection in 1946 [perhaps a misprint for 1936?], but I insist on the sharp distinction between fluctuation and self-charge. Heisenberg's fluctuation formula is alright and can not be changed.

til Rosenfeld, til Schwingers arbejde, har været af største betydning for os. Jeg kan desværre ikke berolige dig med at fortælle, at vi har et fuldt færdigt manuskript om måleproblemerne, men et sådant manuskript foreligger nu i meget høj tilnærmelse og vil blive helt afsluttet ved et nyt besøg af Rosenfeld kort efter ferien.

Jeg tror imidlertid, at det vil glæde dig at høre, at vi virkelig har haft stor fremgang med arbejdet. Det har nemlig vist sig, at en omhyggelig gennemgang af målingsmulighederne ved idealiserede instrumenter med alle dertil hørende kompensations- og korrelationsindretninger fører til ubestemthedsrelationer for ladnings- og strømstørrelser i to skarp begrænsede rum–tids-områder, der nøjagtig svarer til konsekvenserne af Schwingers kommutationsrelationer for sådanne størrelser i to rum–tids-punkter. Hele situationen svarer derfor i alle enkeltheder til forholdene ved målingsmulighederne for feltstørrelser.

Ligesom i det gamle arbejde er det også lykkedes os at opklare forskellige tidligere misforståelser vedrørende teoriens konsekvenser, og navnlig har det vist sig, at også ved ladningsfluktuationerne ligger forholdene ganske analogt som ved feltfluktuationerne, idet – i modsætning til hvad Heisenberg antog – man får veldefinerede fluktuationskvadrater selv for skarpt begrænsede rum–tids-områder[24]. Ved mere omhyggeligt at undersøge forholdene ved ladningsmålinger, når man går til grænsen for stadig mindre dimensioner og mindre operationstider for de på randen af sådanne områder anbragte prøvelegemer, viser det sig også at forholdene ikke bliver mere komplicerede, men tværtimod langt simplere at overskue.

Vi har i denne forbindelse også nøje gennemgået Heisenbergs gamle arbejder i det Sachsiske Akademi, hvor han først ved spørgsmålet om energifluktuationerne i et hulrum og senere ved betragtning af ladningsfluktuationerne kommer ind på at betragte uskarpe rumområder[25]. Sådanne betragtninger rummer iøvrigt en tvetydighed, idet det ikke er klart om man tænker på en bestemt forsøgsanordning, der kun kan give en tilnærmet bestemmelse af de pågældende klassisk definerede størrelser, eller man tænker på middelværdier af en række målinger af veldefinerede størrelser. Hvad ladningsfluktuationerne angår, er det derfor lykkeligt at sådanne spørgsmål slet ikke mere kommer i betragtning.

---

[24] This is wrong; there remains always at least a logarithmic divergence, cf. the following letter from Pauli and footnote 4 in N. Bohr and L. Rosenfeld, *Field and Charge Measurements in Quantum Electrodynamics*, Phys. Rev. **78** (1950) 794–798, this volume p. [211].

[25] W. Heisenberg, *Über die mit der Entstehung von Materie aus Strahlung verknüpften Ladungsschwankungen*, Ber. Sächs. Akad. math.–phys. Kl. **86** (1934) 317–322. This volume p. [239].

Hvad energifluktuationerne angår synes jeg iøvrigt, som jeg vist også har sagt til dig for år tilbage, at behandlingen af dette spørgsmål, lige fra Einstein til Heisenberg, rummer misforståelser eller uklarheder. Einsteins oprindelige argumenter for fotonernes lokalisation ved betragtning af energifluktuationerne i en lille del af et større hulrum har jo slet intet klart fysisk grundlag. Hvis det lille område er fuldt adskilt fra det større, kommer vi jo blot tilbage til akkurat de samme slutninger som man lige så godt kunne gøre for det hele område. Det fysiske problem er imidlertid netop følgerne af det lille områdes adskillelse fra det større område og disse vil jo afhænge af, om adskillelsen udføres adiabatisk, hvorved der jo selvfølgelig slet intet kan sluttes om fotonernes oprindelige lokalisation, eller om adskillelsen gøres hurtigt, hvorved energien i det lille område jo kan ændre sig lige så meget som man vil. I Heisenbergs betragtninger er han naturligvis ikke ukendt med denne side af spørgsmålet, men jeg synes at han ganske tilslører problemerne ved at indføre en i rummet uskarp grænse for områderne, idet det jo alene er tiden, der er konjugeret til energien.

Jeg skriver alt dette, fordi Rosenfeld og jeg har haft megen anledning til netop at tænke på de forskellige repræsentationer i den kvantemekaniske formalisme og deres entydige anvendelse for at udtrykke de lovmæssigheder, det drejer sig om ved de forskellige fænomener. I det hele har vi ved det manuskript, som nu er så nær færdigt gjort, den videst mulige anvendelse – på alle punkter – af den fænomenterminologi[26], som har fundet dit venlige bifald.

I denne forbindelse har jeg også fundet fornyet anledning til at fordybe mig i og understrege den principielle dualitet af partikel- og feltbegreberne, som jo bunder i selve begrebernes definition og derfor, trods alle tilkommende komplementaritetsforhold, består lige så vel i kvanteteorien som i den klassiske fysik, således som jeg jo også antydede i min lille slutningstale ved Solvaymødet[27]. Jeg var meget taknemmelig for dine bemærkninger om mit manuskript til denne tale, med hvilke jeg var fuldstændig enig. Jeg har endnu ikke fået nogen korrektur, men har forberedt nogle ændringer i manuskriptet netop for at undgå sådanne misforståelser som du frygtede, og jeg sender indlagt et forbedret manuskript, om hvilket jeg skal være taknemmelig for at høre din mening. Jeg er iøvrigt slet ikke selv tilfreds dermed, idet jeg ikke mindst i tilslutning til de nye målingsundersøgelser er begyndt at fordybe mig langt mere i de principielle betragtninger; men dette er jo fremtidsopgaver og de allerede gamle beretninger

---

[26] The papers by Bohr that are of special relevance in this connection are reproduced in Part II of this volume and are referred to in the Introduction to that part.
[27] Cf. this volume p. [223].

om mødet må jo en gang afsluttes for sig.

I de sidste uger har jeg imidlertid været meget interesseret i en ganske anden sag, nemlig i de meget interessante arbejder over atomkernernes opbygning, der særlig er fremkommet i Physical Review og som jeg har haft megen glæde og nytte af at diskutere med Lindhard, der har været på besøg her ude i Tisvilde[28]. I disse arbejder, særlig af Feenberg[29] og Nordheim[30], synes det at en række fænomener, der hidtil har været ganske uforståelige, simpelt forklares ved antagelsen om, at de enkelte kernepartikler i første tilnærmelse er bundet uafhængigt af hverandre, i lighed med elektronerne i et atom. Det er naturligvis vanskeligt at gennemskue grundlaget for en sådan antagelse, men det forekommer mig at det drejer sig om en ganske ligefrem konsekvens af kvantemekanikken, som man somme tider hidtil har overset som følge af en alt for håndgribelig sammenligning med en klassisk dråbemodel. Jeg er ingenlunde klar over forholdene og finder det navnlig vanskeligt at vurdere hyppigheden af ombytningseffekterne, men jeg har nedskrevet nogle almindelige bemærkninger, som jeg også indlagt sender i håb om en streng kritik. For mig selv er det i det mindste i øjeblikket en befrielse at komme bort fra visse alt for klassiske forestillinger, der har hindret en konsekvent gennemførelse af behandlingen af reaktionsproblemerne, der uden ændringer i de almindelige resultater nu synes at kunne fremstilles langt mere konsekvent.

Jeg skal ikke trætte dig med at fortælle om de mange andre både fysiske og almindelige spørgsmål, som jeg i øjeblikket tumler med, ikke mindst i forbindelse med forberedelsen til en række forelæsninger, som jeg i efteråret skal holde i Edinburgh og som har givet mig megen anledning til at genkalde fælles erindringer fra de gamle dage, hvor vi trods alle menneskelige ufuldkommenheder dog alle somme tider lærte noget. Rent bortset fra, hvad du må mene om min uforbederlige dilettantisme, vil du vist imidlertid forstå, at jeg har haft lidt mere arbejdsro i sommer end det til daglig under bestræbelserne for at genopbygge en ny ramme for instituttet er tilfældet. Jeg trænger imidlertid snart til nogen ferie, som jeg haaber at faa de sidste uger herude i Tisvilde.

Rosenfeld, der efter et brev, som lige er kommet, og hvori han fortæller om gode fremskridt i den matematiske formulering af maaleproblemerne, og som synes at have fundet et dejligt sted i Tyrol for ferien sammen med sin familie,

[28] This part of the letter makes it evident that Bohr in this period was only briefly revisiting the realm of quantum electrodynamics.

[29] Presumably E. Feenberg and K.C. Hammach, *Nuclear Shell Structure*, Phys. Rev. **75** (1949) 1877–1893.

[30] L. Nordheim, *On Spins, Shells and Moments in Nuclei*, Phys. Rev. **75** (1949) 1894–1901.

bad mig, før han rejste, hilse dig saa mange gange og sagde, at han var ked af ikke at kunne træffe dig paa gennemrejsen i Zürich, men at han glæder sig til at være sammen med dig ved kongressen i Como, hvortil jeg desværre ikke kan komme.

Vi sender alle de venligste hilsener til dig og Franca og haaber, at I har fundet en god ferie i sommer.

Som altid

din

[Niels Bohr]

Translation

Tisvilde, August 15, [19]49

Dear Pauli,

I have a very bad conscience about not having written for so long, and not even having thanked you for your good and so informative letters. Not least during Rosenfeld's visit a few weeks ago we both had much cause to send you many kind thoughts since the reference to Schwinger's paper – in your letters to Rosenfeld – has been of the greatest importance for us. Unfortunately I cannot reassure you by reporting that we have a complete manuscript on the measuring problems, but such a manuscript exists now in a very close approximation and will be finished during Rosenfeld's next visit shortly after the vacation.

I think, however, that you will be pleased to hear that we have really made great progress with the paper. It has turned out that a painstaking examination of the possible measurements by means of idealized instruments, with all the required compensation and correlation devices, leads one to uncertainty relations for the charge and current quantities in two sharply limited space–time regions, which correspond exactly to the consequences of Schwinger's commutation relations between such quantities at two space–time points. The whole situation corresponds therefore in all details to the conditions as regards the measuring possibilities for the field quantities.

As in the old paper we have succeeded also in clarifying some earlier misunderstandings concerning the consequences of the theory, and in particular it has turned out that also for charge fluctuations the situation is quite analogous to that for the field fluctuations in that – contrary to Heisenberg's assumption – one gets well-defined mean-square fluctuations even for sharply limited space–time regions[24]. By a more careful investigation of the situation as regards charge measurements in the limit of smaller and smaller dimensions and shorter and shorter durations for the operations involving test bodies placed at the edge of such regions, it also turns out that the situation does not get more complicated, but rather much easier, to grasp.

In this connection we have also checked carefully Heisenberg's old papers in the Saxon Academy, in which he introduces the use of spatial regions with diffuse boundaries, first in the problem of energy fluctuations in a cavity and later in dealing with charge fluctuations[25]. Incidentally, these considerations contain an ambiguity in so far as it is not clear whether one imagines a definite experimental arrangement which can give only an approximate determination of the relevant classically defined quantities, or one considers the average values of a series of measurements of well-defined quantities. As far as the charge fluctuations are concerned it is therefore fortunate that such questions no longer arise.

Moreover, as regards the energy fluctuations, I think, as I have probably told you already years ago, that the treatment of this question, all the way from Einstein to Heisenberg, contains misunderstandings or obscurities. Einstein's original arguments for the localization of photons by considering energy fluctuations in a small part of a larger cavity, have after all no clear physical basis whatever. If the small region is completely separated from the large one, we merely get back to exactly the same conclusions as one could equally well have drawn for the whole region. The physical problem is, however, precisely the consequences of the separation of the small region from the larger one, and these will depend on whether the separation is carried out adiabatically, in which case one evidently cannot conclude anything at all about the original localization of the photons, or whether the separation is carried out rapidly, in which case the energy in the small region can of course change by as much as one likes. In Heisenberg's considerations, this aspect of the question is of course not unknown to him, but I think he obscures the problem completely by introducing a diffuse spatial boundary for the regions, since it is only time that is conjugate to energy.

I write about all this because Rosenfeld and I have had much cause to contemplate precisely the different representations in the quantum-mechanical formalism and their unambiguous application in order to express the regularities involved in the different phenomena. All in all, in our manuscript, which is now so near completion, we have made the widest possible use – on all points – of the phenomenon terminology[26] which has won your kind approval.

In this connection I have also found a fresh opportunity to ponder upon and stress the fundamental duality of the particle and field concepts, which of course arises from the very definition of these concepts and therefore, in spite of all due complementarity relations, exists in quantum theory as well as in classical physics, as I have also mentioned in my little closing talk at the Solvay Conference[27]. I was very grateful for your comments on the manuscript of this talk, with which I agree entirely. I have not yet received any proofs, but I have prepared some changes in the manuscript precisely to avoid such misunderstandings as you were afraid of, and I enclose an improved manuscript, of which I would be grateful to learn your opinion. I am actually not at all satisfied with it, because – not least in connection with the new investigations of measurements – I have begun to get more deeply immersed in fundamental considerations; but this is a task for the future, and the already overdue report of the meeting must after all be finished some time.

In the last few weeks, however, I have become greatly interested in a completely different matter, namely in the very interesting papers on nuclear structure which have appeared, particularly in the Physical Review, and which I discussed with much pleasure and benefit with Lindhard, who was on a visit here in Tisvilde[28]. In these papers, particularly by Feenberg[29] and Nordheim[30], it appears that a number of phenomena, which so far were quite incomprehensible, are simply explained by the assumption that the individual particles in the nucleus in first approximation are bound independently of each other like the electrons in an atom. It is of course difficult to judge the basis for such an assumption, but it appears to me that we are here concerned with a straightforward consequence of quantum mechanics, which has sometimes been overlooked so far because of a too literal comparison with a classical drop model. I am by no means clear about the situation and find it particularly difficult to assess the probability of the exchange effect, but I have written down some general remarks which I also enclose in the hope of severe criticism. For myself, at this moment at least, it has been a relief to get away from certain much too classical ideas which have prevented a consistent realization of the treatment of reaction problems, which can now seemingly be represented far more consistently without any change in the general results.

I am not going to bore you with an account of the many other questions, some concerned with physics and some of a general nature, with which I am now grappling; not least in connection with the preparation of a series of lectures which I am going to give in the autumn in Edinburgh, and which have given me much cause to recall memories we share from the old days, when in spite of all human imperfection we all still learnt something now and again. Quite apart from what you may think of my incorrigible dilettantism, you will nevertheless understand that I have had a little more peace for work this summer than is the case in my day-by-day efforts to build up a new framework for the Institute. However, I will soon need a little holiday, which I hope to get during the final weeks out here in Tisvilde.

Rosenfeld, who, in a letter just arrived, reports good progress with the mathematical formulation of the measuring problems, and who seems to have found a charming place in Tyrol for a holiday with his family, asked me before his departure to send you very many greetings, and said that he was sorry he could not see you when passing through Zurich, but he is looking forward to being together with you at the conference in Como, to which I unfortunately cannot come.

We all send the kindest regards to you and Franca and hope you have had good holidays this summer.

As always,
yours
[Niels Bohr]

PAULI TO BOHR, 21 August 1949
[Handwritten]

Physikalisches Institut
der Eidg. Technischen Hochschule
Zürich

ZÜRICH 7, 21. Aug. 1949
Gloriastraße 35

Dear Bohr,

I was very glad to have your recent letter and was particularly glad to hear that the Part II of the common paper of you and Rosenfeld made such good progress. However, there is one important point regarding the consequences of the quantum-electrodynamical formalism for charge (and current) fluctuations (positron-theory) on which I *cannot* agree with you: According to my opinion Heisenberg was entirely right in his claim of *unsharp* limitations of either the space or the time domain in which the charge (or current) fluctuation is considered – in *contrast* to the case of energy (and momentum) fluctuation of the radiation (where a *sharp* limitation of the space–time region is sufficient to guarantee a *finite* value of the square of the mean value of the field strength over this domain).

First of all I wish to point out, that *Schwinger's formulas* – although more elegant and also more general than the old formulas of Heisenberg – *lead to exactly the same* results for the mean square [of the fluctuation] of the *charge* contained in a finite space–time region as Heisenberg's old formulas. (We have

[493]

checked this here in Zürich some time ago). At least without particular rules to go to the limit of sharp boundaries (rules which are not known to me) the result for such sharp boundaries is *infinite*. The formal reason for it is that mathematical expressions of the type

$$\frac{\partial \Delta^{(1)}}{\partial x_\mu} \frac{\partial \Delta}{\partial x_\nu} \qquad - \text{ or similarly} \qquad \frac{\partial \Delta^{(1)}}{\partial x_\mu} \frac{\partial \Delta^{(1)}}{\partial x_\nu} \text{ or } \frac{\partial \Delta}{\partial x_\mu} \frac{\partial \Delta}{\partial x_\nu}$$

cannot be integrated over a space–time region containing the "light cone" $r^2 - c^2 t^2 = 0$ in a well-defined way because of the singularity of the type $\left(\dfrac{\partial}{\partial x_\mu} \dfrac{1}{r^2 - c^2 t^2}\right)\left(\dfrac{\partial}{\partial x_\nu} \delta(r - ct)\right)$ – or analogous ones for the other cases – at such points*. It is different for the energy fluctuations of radiation where *not the product of two* such functions enters, but only *either* $\dfrac{\partial^2}{\partial x_\mu \partial x_\nu} D^{(1)}$ *or* $\dfrac{\partial^2}{\partial x_\mu \partial x_\nu} D$. It is the product of *two* such functions which frustrates the finiteness of the value of the square of [the fluctuations of] the charge contained in a sharply limited space–time region.

For the evaluation of this quantity after averaging over the boundaries of, let us say, the space of the region considered, it is most convenient to represent Δ and $\Delta^{(1)}$ as integrals over the momentum space and then to perform all integrations over the (x,t)-space, including the one over the position of the boundary, *first* and afterwards the integration over the momentum space. In this way one falls back to Heisenberg's old formulas (as was explicitly verified by us some time ago).

The physical meaning of this procedure is, of course, – as you already explained yourself – that the measured quantity is $\dfrac{1}{V \cdot T} \displaystyle\int_K \mathrm{d}V \int_0^T \mathrm{d}t\, E_r$, where the domain K of the space integration is a spherical shell of thickness b (and of course *not* the average over a series of measurements of well defined quantities).

I write all these critical remarks with the reservation that I did not see the

---

* I use Schwinger's notations for the functions $\Delta^{(1)}(x)$, $\Delta(x)$ which are the invariant solutions of the wave equation $(\Box - m^2)\{\Delta^{(1)} \text{or } \Delta\} = 0$ with the pole and the δ-type singularity for $r^2 = c^2 t^2$ respectively. For $m = 0$ the corresponding functions are denoted by $D^{(1)}$ and D.

details of yours and Rosenfeld's considerations or computations[31] which – according to your letter – were leading to a finite value of this quantity for $b = 0$.

All what I said until now concerns the expectation value of the square of the charge contained in a finite space–time region with sharp boundaries. I think, I can definitely say, that this expectation value must be infinite (unless the boundaries are unsharp) – different from the expectation values of the square of the mean value of the electromagnetic field strength over such a region; moreover that Heisenberg's old formulas are *correct* for this quantity with unsharp boundaries.

Of course, it is an entirely *different* problem to discuss the *indeterminacy relations* for charge and current quantities in these regions. This was not treated by Heisenberg at all, but – as you repeat also in your letter – is contained in formulas derived by Schwinger. I do not know whether in this case the integration over space–time regions with sharp boundaries gives *generally* finite results, because there are still these disagreeable products of two types of singularities, which have to be treated with the greatest care. I shall be glad to discuss the details of this question with Rosenfeld, when I shall meet him in Basel or Como. In any case I have no doubt of the correctness of your statement "that a thorough discussion of the possibilities of measurement for charge and current quantities in finite space–time regions" – both for sharp and for unsharp boundaries – "with all necessary arrangements for compensation and correlation, must exactly correspond to the consequences of Schwinger's commutation-relations for such quantities in two space–time points" and I am very glad that you have reached just this conclusion.

Finally I wish to state expressively my complete agreement with everything what you said in your letter about the adiabatic and the quick separation of the photons contained in a partial volume of a great box.

The improved manuscript of your remarks at the Solvay-congress[32] I read with great pleasure and I agree now completely with everything what you say, which seems to me to give a good review of the present situation.

I also thank you for the other manuscript but at present I do not wish to discuss nuclear constitution at all for reasons of a certain personal spiritual economy. I shall be glad to show this manuscript to all persons who are interested in this moment in these problems.

Hoping that our discrepancy regarding the validity of Heisenberg's old for-

---

[31] From Pauli's letter (29 August 1949), quoted on p. [498], we learn that there was an elementary mistake in Rosenfeld's computations.
[32] Cf. p. [223].

mulas will soon be cleared up I shall come back at a later occasion to the principal questions, with which I am now also busy myself very much. I shall be very much interested to hear your new ideas about it and I hope that my own ideas will get much clearer in the near future*.

Meanwhile our warmest regards to you and your whole family (where and how is Aage?) from both Franca and myself.

As always

Yours,

W. Pauli

BOHR TO PAULI, 29 August 1949
[Typed copy]

UNIVERSITETETS INSTITUT   BLEGDAMSVEJ 15, KØBENHAVN Ø.
FOR   DEN August 29th 1949.
TEORETISK FYSIK

Dear Pauli,

Hearty thanks for your good letter. I was of course deeply interested in what you wrote of the charge and current fluctuations, but Rosenfeld and I have not only explored the matter from the point of view of measurability, but we have also gone through the old calculations of Heisenberg, in which he has omitted the case, where the width of the border is taken to be smaller than h/mc. Actually in connection with one of the earlier frustrated attempts of completing the old paper I had asked Rosenfeld to calculate the fluctuations for the more general case and about a year ago he sent me the enclosed page[33], which I shall be glad to have back again when you have read it. The importance of the point was, however, first appreciated by us, when we went more carefully into the measurement problem in which the fluctuations do not mean a limitation in the measurability of charge quantities, which according to definition depends only on the control of momentum transfer to the test bodies but corresponds

---

* I thought very much on the compensation idea of Pais and you, and the more recent contribution to it of Sakata and his pupils in Japan.

---

[33] A copy, in Bohr's hand, of this page is found in the collection of Bohr's unpublished manuscripts at NBA, in the folder *Field and Charge Measurements*. Microfilmed, Bohr MSS, microfilm no. 19. Reproduced on p. [497].

to the degree to which the non-compensatable effects of pair productions due to the operation of the test bodies can be estimated. I am glad that you will have the opportunity to discuss the mathematical details with Rosenfeld, but so far I believe we are right. The primary endeavour of the investigation is just to clarify the use of space–time concepts within the frame of the present theory, but what is in the background of my mind is of course the ways of further progress. In this respect the duality of the corpuscle and field concepts, which is so emphasized just in the measuring problems might be the essential clue. I have not yet seen the Japanese papers on the compensation problem of which you wrote, but I shall look them up when I return to Copenhagen in a week's time. I need hardly say that I was much relieved, that you are now more satisfied with my concluding comments at the Solvay meeting. The contents of the whole talk is of course very modest, but I am glad, if by your help I have avoided greater misunderstandings of my attitude.

With kindest greetings and best wishes from us all,

Yours ever,

[Niels Bohr]

P.S. As regards your kind question about Aage I can tell, that he is at the moment in Berkeley and seems to enjoy himself and have good progress with his work. In particular he has continued his investigations of the [hyper]fine structure problems together with Weisskopf, but I have not yet heard what results they have arrived to.

Enclosure, BOHR TO PAULI, 29 August 1949
[Handwritten by Bohr]

Copy of Rosenfelds Note from 1948.

The case not discussed by Heisenberg is that in which the breadth of b is smaller than the two other critical quantitites cT and λ_0 (Compton wavelength). In this case we may treat P as small compared with $\sqrt{m^2 + \frac{1}{4}k^2}$. The integration over P then yields

$$\sim \frac{1}{h^2(cT)^2}\sqrt{\lambda_0^2 + \left(\frac{h}{k}\right)^2}.$$

If we further assume $\lambda_0 \ll L$, we see that the integration over k can be carried

[497]

out neglecting λ_0^2 against $\left(\dfrac{h}{k}\right)^2$ under the square root; it then yields $\sim L^2 h^2$ and the final result for the fluctuation $\dfrac{\Delta e^2}{\varepsilon^2}$ is in this case $\dfrac{L^2}{(cT)^2}$.

Summarizing we have:

$$b \quad \text{smallest:} \qquad \frac{L^2}{(cT)^2}$$

$$\lambda_0 \quad \text{smallest:} \qquad \frac{L^2}{(cT)^2} \cdot \frac{\lambda_0}{b}$$

$$cT \quad \text{smallest:} \qquad \frac{L^2}{cTb}.$$

From these formulae we deduce the following rules:

1) If the order of the two *largest* of the three quantities b, cT, λ_0 is reversed, the estimate of the fluctuation does not change.

2) If the order of the two *smallest* quantities is reversed, the fluctuation is multiplied by the ratio of that which was originally larger to that which was originally smaller.

E.g., to pass from case $b < \lambda_0 < cT$ to case $\lambda_0 < b < cT$ multiply the fluctuation for the first case, viz. $\dfrac{L^2}{(cT)^2}$ by $\dfrac{\lambda_0}{b}$.

From the above formulae, it appears that when $b \to 0$, the fluctuations may remain finite, provided the time cT does not tend to zero faster than b. Only if this occurs does the fluctuation become infinite.

Note that if $\lambda_0 \to 0$ i.e. if (e.g.) we go over to the classical theory, the fluctuation vanishes.

PAULI TO BOHR, 29 August 1949
[Handwritten]

Physikalisches Institut
der Eidg. Technischen Hochschule
Zürich

ZÜRICH 7, Aug. 29, 1949
Gloriastraße 35

Dear Bohr,

Many thanks for your last letter and the enclosed page of Rosenfeld, which I am now returning. There is indeed an elementary error on this page for the interesting case $b < cT$ (and $b < L$). In this case there exists always a term $\sim \underline{\underline{\log b}}$

which prevents a finite limit for $b \to 0$. If one assumes $\lambda_0 \ll L$ and $\lambda_0 \ll cT$ (as Rosenfeld also does), one gets for "b smallest"

$$\overline{(\Delta e)^2} \sim \frac{L^2}{(cT)^2} \log \frac{L}{\underline{b}} \qquad \left(\text{and } not \; \left(\frac{L}{cT}\right)^2 \text{ as Rosenfeld states}\right).$$

It is true that Heisenberg did not evaluate his formulas for this case, but Rosenfeld's error is already his first statement that "we may treat P as small compared with $\sqrt{m^2 + \frac{1}{4}k^2}$ " in this case. Actually one has to perform the integration over the whole (infinite) P-space.

The statements of my last letter are therefore correct and it is not possible to restrict oneself to sharply limited space–time regions. – Of course I shall show to Rosenfeld next week all details of our calculations.

On the other hand I agree completely with all your remarks on the *physical* interpretation of the theoretical results, particularly on the distinction between fluctuation phenomena and uncertainty relations. I believe, too, that the duality of the corpuscle and field concept is something very fundamental and may even be "the essential clue". – In this connection I may point to the well-known considerations of Heisenberg in *non-relativistic* wave mechanics, in which he shows the self-consistency of the theoretical results, if the (arbitrarily chosen) situation of the *separation ("Schnitt") between the measured system and the measuring instruments* is shifted from one place to another. Until now a similar consideration was never carried through in the *quantized field theories*. I always felt this "hole" in our theoretical concepts very much. Obviously this problem leads directly to the discussion of the duality of corpuscle and field concepts, as it must be to some degree arbitrary what is considered as measured "field" and what else as measuring "test body" (provided the latter behaves sufficiently classically).

The Japanese papers, I mentioned recently, are partly not yet published (the printing in Japan is lasting more than one year), but were sent to me directly by the authors (Umesawa and others).

I was glad to hear about Aage.

All good wishes to yourself and the whole family (from Franca, too)

Yours ever,

W. Pauli

PAULI TO BOHR, 25 October 1949
[Handwritten]

Physikalisches Institut   ZÜRICH 7, Oct. 25, 1949
der Eidg. Technischen Hochschule   Gloriastraße 35
Zürich

Dear Bohr,

Dr. Jost and I studied yours and Rosenfelds article very carefully and we entirely agree on the principle side of the problem. There is only one minor point which is obviously erroneous, namely the remark on p. 6 "While the commutation rules of the field components are not modified in this approximation ...". Obviously they *are* modified*, as they have to obey now the inhomogeneous Maxwell equation $\dfrac{\partial F_{\mu\nu}}{\partial x^{\nu}} = J_{\mu}$ prop. *e* instead of the vacuum equation $\dfrac{\partial F^{0}_{\mu\nu}}{\partial x^{\nu}} = 0$ (unless a canonical transformation is performed).

Moreover we suspect that for the case of a *sharply* limited space–time region not only the mean square of the charge fluctuation is getting infinite but also the whole distribution law of the fluctuations will become singular**. But we are not sure on this particular point, as, by the way, we find that the case of a *sharply* limited space–time region (with this $\delta(x)$-type singularity of the external field-strengths) is not interesting from a physical point of view. Why you insist so much on this particular limiting case? (I am always glad, if I have one limiting process *spared*).

Jost is writing today (about the same as I do here) to Rosenfeld with a bit more mathematics.

Franca and I intend to go soon to Princeton (to be back in April). We have a boat reservation for Nov. 10 but I am still waiting for the Visum. The State Dept. is quite crazy with physicists now, they should better learn something from England, where everybody can enter without any visum. Some of the younger Swiss physicists which intend to immigrate to the U.S.A. have to wait more than half a year for the necessary papers. If they were chemists or pure mathematicians one would treat them very differently.

I am happy still not to read any newspapers, but *Stern* is here and we often

---

* Of course it is unnecessary to calculate explicitly this modification for the purpose of your paper.
** Maybe Rosenfeld can write something about it to Jost.

laugh together, when he tells me about the nonsense they are writing (particularly in U.S.A.).

I shall say many regards from him to you, he is leaving tomorrow to go back to Berkeley.

Many regards to Born, I wonder how you come along with his 'reciprocity'. All best wishes to yourself.

As ever
Yours
W. Pauli

LÉON ROSENFELD

BOHR TO ROSENFELD, 2 September 1949
[Carbon copy]

[København,] 2. September [19]49

Kære Rosenfeld,

Pauli er sandelig storartet til at svare hurtigt, og jeg fik i gaar et brev fra ham, hvoraf jeg vedlægger en kopi. Som du vil se, opretholder han sin paastand om, at ladningsfluktuationerne er uendelige i skarpt begrænsede omraader. Imidlertid er den logaritmiske formel, han angiver, jo væsentlig forskellig fra, hvad man ville regne med efter de af Heisenberg betragtede tilfælde. Jeg er ikke klar over hans beregning og ved derfor ikke, om der virkelig er en fejl i dine formler. Det er jo ogsaa muligt, at pardannelsen under maalingsoperationen virkelig giver en logaritmisk formel, jeg er imidlertid saa optaget med andre ting, at jeg ikke fuldt kan overse det, men i øjeblikket er du jo saa meget mere inde i den matematiske formulering af saadanne betragtninger.

Med alt forbehold grundet paa manglende kendskab vil jeg imidlertid gerne rejse det spørgsmaal, om det kan dreje sig om et mere dybtliggende punkt. Pauli ved maaske endnu slet ikke, hvor vidt vi er gaaet med subtraktionsnormaliseringerne, uden hvilke jo heller ikke kommutationsrelationerne er fri for logaritmiske uendeligheder, og jeg vil derfor spørge, om det ikke kan være noget lignende, det drejer sig om ved beregningen af ladningsfluktuationerne. Vi maa jo se at finde ud af, hvordan det hele ligger, men det vil jo være skønt, om de definitioner og kompensationer, vi indfører ved maalingerne, skulle danne et fuldstændigt sidestykke til normaliseringsproblemet i formalismen. Jeg

[501]

skal jo være dybt interesseret i at høre, hvad du tænker om det hele, og ikke mindst hvad der kommer ud af diskussionerne med Pauli, og jeg ville være meget taknemmelig, om du fortløbende ville holde mig a jour med sagernes udvikling.

Som du kan forstaa, er min hovedbeskæftigelse i disse dage at tænke nærmere over de filosofiske spørgsmaal, som jeg skal tale om i Edinburgh. Ved at kigge lidt i litteraturen har jeg haft megen anledning til at tænke over forskellige af dine bemærkninger i Tisvilde om de græske filosoffer, og jeg føler mere og mere, at hele den saakaldte filosofiske videnskab fra ældre og nyere tid er en eneste forvirring og staar paa et lignende stade som fysikken før gennembruddet i renæssancen og relativitetsteorien. Jeg skal dog ikke her gaa nærmere ind derpaa, men blot sige, hvor meget jeg glæder mig til at kunne tale nærmere med dig om hele anlægget for Edinburgh-foredragene, naar du kommer til København i oktober.

Vi haaber hele familien fik en rar sommerferie i Tyrol og sender de venligste hilsener fra os alle.

Din,
[Niels Bohr]

Translation

[Copenhagen,] September 2, 1949

Dear Rosenfeld,

Pauli is indeed marvellously quick in answering, and yesterday I got a letter from him of which I enclose a copy. As you will see he maintains his claim that the charge fluctuations are infinite in sharply bounded domains. However, the logarithmic formula he quotes is quite different from what one would expect according to the cases considered by Heisenberg. I am not familiar with his calculations and thus I do not know whether in fact there is an error in your formulas. It is of course also possible that the pair creation during the measuring operation really yields a logarithmic formula, I am, however, so occupied by other matters that I cannot fully see through this question, but at present you are, of course, so much more familiar with the mathematical formulation of such considerations.

However, with all reservations due to my lack of knowledge, I should like to raise the question whether we may here be concerned with a more profound point. Perhaps Pauli does not yet know at all how far we have gone with the subtraction normalizations – without which not even the commutation relations would be free of logarithmic infinities. Therefore I would ask whether it could not be something similar we are confronting in the calculation of the charge fluctuations. Naturally we must find out how all these matters stand, but it would of course be wonderful if the definitions and compensations that we introduce in the measurements should form a complete parallel to the normalization problem in the formalism. I should of course be deeply interested to learn what you think about all this, and not least about the outcome of the discussions with Pauli. Thus I should be most grateful if you would keep me up to date as regards the development of these matters.

As you can understand, my main preoccupation these days is to consider in detail the philosophical questions I am to talk about in Edinburgh. By glancing at the literature I have had much cause to think about some of your remarks in Tisvilde about the Greek philosophers, and I feel more and more clearly that the whole so-called science of philosophy from ancient and modern times is a total confusion and has remained at a position similar to that of physics before the breakthrough in the Renaissance and the theory of relativity. However, I shall not here go into more detail, but only say how much I am looking forward to being able to discuss the whole plan for the Edinburgh lectures with you when you come to Copenhagen in October.

We hope that all the family had a nice summer holiday in Tyrol and we send the most friendly greetings from us all.

Yours,
[Niels Bohr]

ERWIN SCHRÖDINGER

SCHRÖDINGER TO BOHR, 13 October 1935
[Carbon copy]

[Oxford,] 13. Oktober 1935.

Lieber und verehrter Herr Bohr!

Ich habe vor einigen Tagen den Durchschlag Ihres Phys. Rev. Aufsatzes (der Erwiderung auf das Einsteinparadoxon) gelesen. Ich gestehe, dass mich die Lektüre sehr nachdenklich gemacht hat. Aber ich glaube, ich kann mich letzten Endes doch nicht damit abfinden.

Ich will von der Situation beim zweiten Versuch sprechen. Man hatte den Impuls des Zweispaltendiaphragmas sehr genau gemessen, hat dann die zwei Elektronen hindurchtreten lassen, hat dann nochmals den Diaphragmaimpuls sehr genau gemessen. Man hat dann mit *drei* Körpern zu tun. Erstens die zwei Partikel. Drittens das grobe, starre, massige Koordinatensystem (etwa eine eiserne Plattform) von dem aus *alle* Messungen gemacht werden. Ich will dies die Ausgangssituation S_A nennen. Jetzt will ich sprechen von den *möglichen* Situationen, in denen sich *eine* Partikel und ein grober starrer Koordinatenkörper zueinander befinden können, nennen wir diese Situationen S'. In der klassischen Mechanik wird ein S' vollständig beschrieben durch Angabe der Koordinaten und Impulse des Partikels. In der klassischen Mechanik ist es unmöglich, von ein und derselben Ausgangssituation S_A aus zu zwei verschiedenen S' zu gelangen durch Manipulationen, die weder auf die Partikel noch auf den Koordinatenkörper merkliche physische Wirkungen ausüben. In der Quantenmechanik wird ein S' schon durch Angabe der Koordinaten allein oder der Impulse allein vollständig beschrieben (von anderen, komplizierteren Situationen nicht zu reden). Und in der Quantenmechanik *ist* es möglich, von ein und derselben S_A aus zu verschiedenen S' für unseren Koordinatenkörper und, sagen wir, unser Partikel Nr 1 zu gelangen durch Manipulationen, die zwar merklich an Partikel Nr 2 angreifen, am Koordinatenkörper aber nur beliebig geringe Wirkungen ausüben.

Diese Möglichkeit erscheint mir paradox und ich finde, man darf sich damit nicht abfinden. – In der Argumentation könnte Bedenken erregen der letzte Punkt: vielleicht ist es doch nicht eine beliebig geringe Einwirkung auf den Koordinatenkörper, wenn man *von ihm aus* (gleichsam auf ihm stehend) eine sehr genaue Koordinaten- oder Impulsmessung an Partikel Nr 2 vornimmt, wie schwer und massig und solid der Koordinatenkörper auch sein mag. Aber dieser Einwand trifft wohl nicht zu. Denn hätten wir (von der Situation S_A aus) zunächst die Situation S' von Partikel Nr 1 durch *direkte* Messung festgelegt, so würden wir nicht glauben, dass dieses S' etwa abgeändert wird durch weitere Messungen, die wir von derselben eisernen Plattform aus an anderen leichten Partikeln, z.B. auch an Partikel Nr 2 vornehmen. Der physische Einfluss einer solchen Messung auf die Plattform wird also für unbedeutend erachtet. –

Eigentlich wollte ich aber gar nicht von diesem Punkte sprechen, jedenfalls nicht in der Idee, dass Sie mir darauf antworten, sondern von etwas anderem. Sie sprechen immer wieder auf das bestimmteste die Überzeugung aus, dass Messungen durch klassische Begriffe beschrieben werden müssen. Z.B. S. 61 des 1931 bei Springer erschienenen Heftes[34]: "Es liegt im We-

---

[34] N. Bohr, *Atomtheorie und Naturbeschreibung*, Springer, Berlin 1931. In English: *Atomic Theory and the Description of Nature*, Cambridge University Press, Cambridge 1934. The articles quoted are reproduced in Vol. 6, p. [201] and p. [219].

sen einer physikalischen Beobachtung, dass alle Erfahrungen schliesslich mit Hilfe der klassischen Begriffe unter Vernachlässigung des Wirkungsquantums ausgedrückt werden müssen." Und S. 74 ebendort: "die durch das Wesen der Messung geforderte Benutzung klassischer Begriffe". Und auch jetzt sprechen Sie wieder von "the indispensable use of classical concepts in the interpretation of all measurements". Jetzt sagen Sie allerdings bald darauf: "The removal of any incompleteness in the present methods of atomic physics ... might indeed only be effected by a still more radical departure from the method of description of classical physics, involving the consideration of the atomic constitution of all measuring instruments, which it has hitherto been possible to disregard in quantum mechanics."

Daraus könnte man heraushören, dass das, was früher als im Wesen jeder physikalischen Beobachtung liegend, als "indispensable necessity" bezeichnet wurde, doch hinwiederum nur ein bisher glücklicherweise immer noch zulässiges bequemes Auskunftsmittel sei, von welchem man vermutlich einmal abzugehen gezwungen sein wird. Wenn dies Ihre Meinung wäre, würde ich gern zustimmen. Aber der nachfolgende straffe und klare Vergleich mit der Relativitätstheorie lässt mich zweifeln, ob ich Ihre Meinung mit dem, was ich eben gesagt, richtig auffasse. Denn in der Relativitätstheorie, als Gedankengebäude für sich betrachtet, ohne Beziehung zur Quantenmechanik, würde man auf die scharfe Scheidung von Raum und Zeit *bei der Messung* wohl niemals verzichten können. Allerdings wäre es möglich, dass bei der unerlässlichen gegenseitigen Modifikation der beiden Theorien *beide* gezwungen würden, ihre klassischen Eierschalen abzuschütteln – und dass *das* Ihre Meinung ist.

Wie dem auch sei, es müssen ganz bestimmte klare Gründe sein, die Sie zu der wiederholten Erklärung veranlassen, man *müsse* Beobachtungen klassisch interpretieren, das liege durchaus in ihrem Wesen. Sie sagen das immer wenn Sie es sagen, so bestimmt und klar im Indikativ heraus, ohne jedes einschränkende "wohl" oder "dürfte" oder "wir müssen darauf gefasst sein, dass", als ob es das Allerbestimmteste in der Welt wäre. Es muss zu Ihren festesten Überzeugungen gehören – und ich kann nicht verstehen, worauf sie sich gründet.

Es kann doch nicht nur dieses sein (was Sie mir schon 1926 mündlich sehr eindringlich sagten), dass unsere überkommene Sprache und die ererbten Begriffe völlig ungeeignet seien, um die Dinge auszudrücken, denen wir da jetzt gegenüberstehen. Denn das war doch wohl sicher im Laufe der Entwicklung unserer Wissenschaft (und der Mathematik) aus ihren ersten Anfängen bis zu dem Stand am Ende des neunzehnten Jahrhunderts immer wieder und wieder der Fall. Wenn uns der Bruch mit dem Alten jetzt grösser erscheint als je zuvor, so müssen wir doch wohl damit rechnen, dass am Zustandekommen dieses Eindruckes eine gewisse zeitliche Perspektive mitbeteiligt ist, derzufolge uns *die* Entwicklung, die wir selbst mitmachen, als die erheblichere und bedeutungsvollere sich abhebt gegenüber den früheren, die wir nur aus der Geschichte kennen und deren Stadien wir meistens in umgekehrter Reihenfolge kennen lernen. Dabei wird es uns oft schwer uns in die *frühere* Denkweise hineinzuversetzen. Und obwohl die Schwierigkeit, solch einen historischen Schritt in Gedanken wieder *zurück*zumachen, eigentlich am beredtesten dafür spricht, *wie* erheblich er den Pionieren beim ersten Vordringen geschienen haben muss, können wir uns doch manchmal des Gefühls nicht erwehren: "Unglaublich, dass die Leute bis dahin so beschränkt waren!" Darin zeigt sich am besten die zeitlich perspektivische Unterschätzung.

Also ich meine, das Nochnichtangepasstsein unseres Denkens und seiner Ausdrucksformen an die neue Theorie kann doch unmöglich die Überzeugung begründen, dass Experimente immer klassisch, unter Vernachlässigung der wesentlichen Charakterzüge der neuen Theorie beschrieben werden müssen. Es mag ein kindisches Beispiel sein, nur um kurz zu sagen, wie ich es meine: nachdem die elastiche Lichttheorie durch die elektromagnetische verdrängt war, hat man doch auch nicht gemeint, die experimentellen Befunde müssten nach wie vor auf Elastizität und Dichte des Äthers, auf Verschiebungen, Deformationszustände, Geschwindigkeiten und Winkelgeschwindigkeiten der Ätherteilchen bezogen werden.

Verzeihen Sie, dass ich so weitschweifig geworden bin. Meine Idee dabei ist, ob Sie nicht in der ausführlicheren Arbeit, welche Sie in der Phys. Rev.-Note ankündigen, diesen Punkt ganz klarstellen wollten: Warum betone ich immer wieder und wieder, dass es im Wesen der Messung liegt, nur klassisch interpretiert werden zu können? Und vor allem: Ist das ein augenblickliches Sichbescheiden oder können wir irgendwie erkennen, dass wir darüber dauernd nie hinauskommen werden?

Ich möchte Sie so gern wieder einmal sehen und sprechen. Aber die Zeiten sind jetzt so wenig angetan zu Vergnügungsreisen. Bald regt sich in mir der philisterhafte Wunsch, endlich wieder einmal irgendwo dauernd zu sein, d.h. mit erheblicher Wahrscheinlichkeit zu wissen, was man in den nächsten fünf oder zehn Jahren tun wird. Es ist furchtbar philisterhaft und gar nicht in die Zeit passend; vermutlich nur der Reflex der hiesigen Umgebung, die ja noch reichlich ungestört von den Weltläuften abrollt.

Mit wärmsten Grüssen und Wünschen von Haus zu Haus

Ihr ganz ergebenster
[Schrödinger]

[506]

Translation

[Oxford,] October 13, 1935

Dear Mr Bohr,

Some days ago I read the copy of your paper in Phys. Rev. (the reply to the Einstein paradox). I admit that this reading has given me much cause for reflection. But I believe that ultimately I cannot come to terms with it.

I shall consider the case of the second experiment. The momentum of the double-slit diaphragm has been measured very accurately; then the two electrons have been sent through, and again the momentum of the diaphragm has been measured very accurately. We are now dealing with *three* bodies: First, the two particles. Third, the heavy, rigid, massive frame of reference (something like an iron platform) from which *all* measurements are carried out. This initial situation I shall denote by S_A. I shall now consider the [various] *possible* situations in which *one* particle and one heavy, rigid reference body can be found relative to each other; let us denote these situations by S'. In classical mechanics an S' is completely described by the specification of coordinates and momentum of the particle. In classical mechanics it is impossible from one and the same initial situation S_A to arrive at two different S' through manipulations that exert a perceptible physical influence neither on the particle nor on the reference body. In quantum mechanics an S' is exhaustively described solely by the specification of the coordinates, or solely by the specification of the momenta (not to mention other more complicated situations). And in quantum mechanics it *is* possible from one and the same S_A to arrive at different S' for our reference body and, let us say, our particle 1, through manipulations that, while acting perceptibly on particle 2, exert an arbitrarily small influence on the reference body.

To me this possibility appears paradoxical and in my opinion one ought not accept it. – One might feel uneasy as regards the last point of the argumentation: perhaps the influence on the reference body is not really arbitrarily small, when one performs a very accurate position or momentum measurement on particle 2 *from the position of the reference body* (when standing on it, as it were), however heavy, massive and solid this body might be. Still, this objection is hardly tenable. Because if to begin with (starting from the situation S_A) we had ascertained the situation S' of particle 1 through a *direct* measurement, then we could not imagine that this S' could be changed somewhat by further measurements carried out from the same iron platform on other light particles such as particle 2. Thus, the physical influence of such a measurement on the platform is to be considered negligible.

As a matter of fact I did not really want to talk about this point, at least not with the purpose that you should reply, but rather about something else. You have repeatedly expressed your definite conviction that measurements must be described in terms of classical concepts. For example, on p. 61 of the volume published by Springer[34] in 1931: "It lies in the nature of physical observation, that all experience must ultimately be expressed in terms of classical concepts, neglecting the quantum of action." And ibid. p. 74 "the invocation of classical ideas, necessitated by the very nature of measurement". And once again you talk about "the indispensable use of classical concepts in the interpretation of all measurements". True enough, shortly thereafter you say: "The removal of any incompleteness in the present methods of atomic physics ... might indeed only be effected by a still more radical departure from the method of description of classical physics, involving the consideration of the atomic constitution of all measuring instruments, which it has hitherto been possible to disregard in quantum mechanics".

This might sound as if what was earlier characterized as inherent in the very nature of any physical observation as an "indispensable necessity", would on the other hand after all just be a, fortunately still permissible, convenient way of conveying information, a way we presumably sometime will be forced to give up. If this were your opinion, then I would gladly agree. However, the subsequent stringent and clear comparison with the theory of relativity makes me doubt whether, in what I just said, I have understood your views correctly. Because, if we consider the theory of relativity as a conceptual edifice in itself, without any relationship to quantum mechanics, we would presumably never be able to renounce the sharp separation between space and time *in any measurement*. Still, it seems possible that in connection with the unavoidable mutual modification of these two theories, *both* would be forced to shake off their classical eggshells – and that *this* is what you mean.

However that may be, there must be clear and definite reasons which cause you repeatedly to declare that we *must* interpret observations in classical terms, according to their very nature. Whenever you say that, you state it so definitely and clearly, in the indicative, without any reservations like "probably", or "it might be", or "we must be prepared for", as if this were the uttermost certainty in the world. It must be among your firmest convictions – and I cannot understand what it is based upon.

It could not be just this point (about which you talked so insistently to me already in 1926): that our traditional language and inherited concepts were completely unsuited to describe the phenomena with which we are confronted now. Because, in the course of the development of our science (and mathematics), from its earliest beginnings to the situation at the end of the nineteenth century, this was certainly the case over and over again. If the break with the old traditions seems greater now than ever before, then we should take into account that a particular time perspective is responsible for forming the impression that *that* development in which we ourselves take part, stands out as being more important and more essential than earlier ones, which we cite only from history, and whose stages we get to know mostly in reverse order. In fact, it is often difficult for us to imagine *earlier* ways of thinking. And although the difficulty of such a historical step *back* actually speaks most eloquently of *how* significant [the step] must have seemed to the pioneers at their first advances, still now and then we cannot avert the feeling: "Incredible that, up to then, people were so narrow-minded!" Here, the underestimation of the time perspective shows itself most clearly.

Thus I think that the fact that we have not adapted our thinking and our means of expression to the new theory cannot possibly be the reason for the conviction that experiments must always be described in the classical manner, thus neglecting the essential characteristics of the new theory. It may be a childish example [but I use it] only to say briefly what I mean: after the elastic light theory was replaced by the electromagnetic one, one did not say that the experimental findings should be expressed – just as before – in terms of the elasticity and density of the ether, of displacements, states of deformation, velocities and angular velocities of the ether particles.

Forgive my long-windedness. What I mean to say is: whether you couldn't make this point completely clear in the more detailed paper you announce in your Phys. Rev. note: Why do I [Bohr] emphasize again and again that according to the very nature of a measurement, it can only be interpreted classically? And above all: Is this a temporary resignation, or can we somehow recognize that we will never get beyond it?

I should so much like to meet you and talk with you again. But present times are so little suited for pleasure-trips. Suddenly I feel the philistine desire to stay somewhere again for good, i.e. to know with great probability what one will be doing for the next five to ten years. This is terribly philistine and quite out of step with our time, probably just a reflex arising from my present environment which continues truly undisturbed by world events.

With warmest greetings and wishes from home to home,

Yours very sincerely
[Schrödinger]

BOHR TO SCHRÖDINGER, 26 October 1935
[Typewritten]

UNIVERSITETETS INSTITUT
FOR
TEORETISK FYSIK

BLEGDAMSVEJ 15, KØBENHAVN Ø.
DEN 26 Oktober 1935.

Lieber Schrödinger.

Ich danken Ihnen und Ihrer Frau herzlich für die freundlichen Glückwünsche zu meinem Geburtstag, und danke Ihnen selbst für den freundlichen und interessanten Brief, den ich vor einer Woche empfangen habe. Was meine Antwort auf den Einsteinschen Artikel betrifft, glaube ich, dass er schon herausgekommen ist, und wie Sie sehen werden, habe ich an mehreren Punkten versucht die Gedanken etwas klarer darzustellen, und hoffe dass dadurch auch Ihr erster Einwand betreffend die Messanordnung beantwortet wird. Weiter habe ich den Hinweis auf die mögliche Bedeutung der atomistischen Struktur aller Messinstrumente für die Aufklärung der noch ungelösten Schwierigkeiten der Elektronentheorie ausgelassen, weil ich zusammen mit Rosenfeld eben im Begriff bin, eine Arbeit über die Messprobleme der Elektronentheorie abzuschliessen, durch welche diese Frage etwas näher beleuchtet wird[35]. Diese Betrachtungen haben aber keinerlei engere Beziehungen zu den Einsteinschen Paradoxien und zu der Frage der Begrenzung der kausalen Beschreibung der Quantenphänomene. Hier muss ich gestehen, dass ich Ihre Zweifel nicht teilen kann. Meine Betonung der Unvermeidbarkeit der klassischen Beschreibung der Experimente läuft ja letzten Endes auf nicht anderes hinaus als *die scheinbare Selbstverständlichkeit, dass die Beschreibung jeder Messanordnung wesentlich die Anordnung der Apparate im Raume und deren Funktionieren in der Zeit enthalten muss, wenn wir überhaupt etwas über die Phänomene sollen aussagen können.* Das Argument ist ja dabei vor allem, dass die Messinstrumente, wenn sie als solche dienen sollen, nicht in den eigentliche Anwendungsbereich der Quantenmechanik einbezogen werden können. Es hat mich sehr amüsiert, dass Sie bemerkt haben, dass ich, was ich gar nicht selber wüsste, eben an diesem Punkt, und nur an diesem, nicht "dürfte" sage; auch werde ich Ihrem Rat folgen und in der in meinem Artikel angekündeten Arbeit über die Anwendung der Raum–Zeitbegriffe in der

---

[35] Cf. Part I of this volume p. [39].

Atomtheorie möglichst deutlich und ausführlich auf diesen prinzipiellen Punkt eingehen. Sobald diese Arbeit fertig ist, werde ich Ihnen einen Durchschlag zugehen lassen.

Mit den freundlichsten Grüssen und besten Wünschen Ihnen und Ihrer Frau von meiner Frau und

Ihrem
Niels Bohr

Translation

Copenhagen, October 26, 1935

Dear Schrödinger,

My cordial thanks to you and your wife for the kind congratulations on my birthday, and many thanks to yourself for the kind and interesting letter, which I received a week ago. As far as my answer to the Einstein paper is concerned, I believe that it has already appeared, and you will see that on several points I have tried to express the ideas more clearly, and I hope that also your first objection concerning the measuring device has thereby been answered. Furthermore, I have left out the reference to the possible significance of the atomic constitution of all measuring instruments for the solution of the still unexplained difficulties in electron theory. The reason is that together with Rosenfeld I am just about to finish a paper about the measuring problems in electron theory in which this question will be elucidated somewhat more fully[35]. However, these considerations do not have any close connection to the Einstein paradoxes and to the question of limitations in the causal description of quantum phenomena. On this point I must confess that I cannot share your doubts. My emphasis of the point that the classical description of experiments is unavoidable amounts merely to *the seemingly obvious fact that the description of any measuring arrangement must, in an essential manner, involve the arrangement of the instruments in space and their functioning in time, if we shall be able to state anything at all about the phenomena.* The argument here is of course first and foremost that in order to serve as measuring instruments, they cannot be included in the realm of application proper to quantum mechanics. I found it most amusing that you noticed – what I myself had not at all been aware of – that just on this point, and only this one, I do not say, "it might be". I shall also follow your advice in the article on the application of space–time concepts in atomic theory, announced in my paper, to discuss these fundamental points as clearly and in as great a detail as possible. As soon as this paper is finished, I shall send you a copy.

With the kindest regards and best wishes to you and your wife from my wife and

Your
Niels Bohr

VICTOR WEISSKOPF

WEISSKOPF TO ROSENFELD, 2 December 1933
[Handwritten]

[Zürich,] 2. XII [1933]

Lieber Rosenfeld!

Ich trage eben im Seminar hier über die Würmer vor und verstehe eine Stelle nicht. Es handelt sich um die Messung des Gesamtimpulses des Probekörpersystems auf S. 30 der 4. Korrektur.

Es wird da die Methode der Impulsmessung durch den Dopplereffekt beschrieben und dann behauptet, jeder Teilprobekörper bekäme genau die gleiche Verschiebung, deren absolute Grösse aber unbekannt sei. Nun verstehe ich nicht, wieso in der dort beschriebenen Anordnung, tatsächlich jeder Teilkörper dieselbe Verschiebung erleidet. Es ist nämlich durch eine Einfache Modifikation möglich, die dort beschriebene Anordnung dazu zu verwenden, den Impuls jedes einzelne Teil-Probekörpers zu messen: Man muss nämlich nur in den Weg der reflektierten Strahlen ein Spiegelsystem bringen, das das reflektierte Licht jedes einzelnen Probekörpers in eine andere Richtung bringt, sodaß man es *getrennt* Spektral *beliebig genau* untersuchen kann. Natürlich kann die Einführung eines solchen Spiegelsystems, welches nur die reflektierte Strahlung ablenkt, unmöglich die Wirkung des eingestrahlten Lichts auf die Probekörper beeinflußen. Daher müssten auch die *relativen* Verschiebungen der Probekörper unbestimmt sein, wenn man die einzelnen Impulse der Teilkörper messen kann. Messe ich zufällig alle reflektierten Strahlen zusammen, wie es Bohr tut, so ändert das ja nichts an der Verschiebung, sondern stellt, so viel ich sehe, nicht die günstigste Methode zur Messung des Gesamtimpulses dar.

Ich bin felsenfest davon überzeugt, daß Sie und Bohr recht haben, aber ich wäre Ihnen sehr dankbar über nähere Aufklärungen. Wenn Sie es Bohr vorlegen, dann sagen Sie vorher, daß die Frage nicht als Einwand gilt sondern nur paedagogisch gemeint ist!

Mit herzlichen Hilsenern an alle
Københavner und Belgierinnen
aus Ihrer Umgebung!
Herzlichst!

Ihr Weisskopf

Translation, see p. [14].

BOHR TO WEISSKOPF, 5 December 1933
[Carbon copy]

[København,] 5. Dezember [19]33.

Lieber Weisskopf,

Rosenfeld hat mir eben Ihren freundlichen Brief gezeigt, der mich sehr erschrocken hat, nicht weil wir ein schlechtes Gewissen haben, sondern weil ich fühle, dass unser Bestreben, unseren Kampf mit der Sache in Worte zu kleiden, noch weniger Erfolg gehabt hat, als ich fürchtete. Die Ausführung an der betreffenden Stelle kam uns nämlich besonders klar und fast zu umständlich vor.

Die Sache liegt ja so, dass der Verlust der Kenntnis des Orts eines Körpers durch die Bestimmung seines Impulses im allgemeinen gar nicht in der Unverfolgbarkeit des Verlaufs des Stossprozesses liegt, sondern allein in der Unmöglichkeit, diesen Verlauf zugleich relativ zu einem festen raum–zeitlichen Bezugssystem zu fixieren. In unserem Fall kommt letzterer Umstand darin zutage, dass die absolute Zeit der Öffnung der zur Abschneidung des einfallenden Strahlungsbündels benutzten Blende nicht zugleich mit der zwischen dieser Blende und der durchgehenden Strahlung ausgetauschten Energie bekannt sein kann. Dagegen kann unter Annahme einer genügend schweren Blende der Verlauf des Öffnungsprozesses beliebig genau bekannt sein. Die Genauigkeit, die wir bei der Messung des Gesamtimpulses erzielen, ist eben dadurch bedingt, dass wir für diese Messung nur eine einzige Blende zu benutzen brauchen; wäre aber die Kenntnis der Impulse der einzelnen Teilkörper verlangt, so würden ebensoviele unabhängige Blenden wie Teilkörper notwendig sein, und dadurch wurden nicht nur diese Körper verschiedene Verschiebungen erleiden, sondern die Genauigkeit der Bestimmung des Gesamtimpulses wäre um $1/\sqrt{N}$ herabgesetzt, wenn N die Anzahl der Teilkörper ist. Dass man, wie wir gezeigt haben, erreichen kann, dass alle Verschiebungen der Teilkörper gleich sind, und zugleich die Genauigkeit des Gesamtimpulses die maximale, diesen Verschiebungen entsprechende ist, liegt übrigens daran, dass der Gesamtimpuls zur Lage des Schwerpunkts konjugiert, aber mit den relativen Verschiebungen der Teilkörper vertauschbar ist. Diese relativen Verschiebungen kann man ja ausserdem durch ein geeignetes Telegraphverfahren nachher bestimmen und sogar regulieren, ohne den Gesamtimpuls zu ändern. Wie wir in §5 ausdrücklich bemerkt haben, können wir ganz nach Belieben zwischen Spiegelmechanismus und Telegraphenmechanismus wählen.

Ich hoffe, dass diese Antwort Sie befriedigen wird, und wäre dankbar, wenn Sie mir darüber möglichst bald Bescheid sagen wollten. Übrigens brauche ich kaum zu sagen, dass ich für jeden Erschreckung physikalischer oder psychologischer Art nur dankbar bin, und dass Rosenfeld und ich viele solche beim letzten genauen Durchgang der Arbeit erlebt haben, die wir durch eine Anzahl weiterer kleiner Änderungen im letzten Paragraphen dem Leser zu ersparen versucht haben.

Mit vielen freundlichen Grüssen von uns beiden, auch an Pauli, für dessen letzten Brief ich in jeder Beziehung sehr froh war.

Ihr
[Niels Bohr]

Translation, see p. [15]

Niels Bohr, Tisvilde 1961 (Photograph by J. Kalckar).

INVENTORY OF RELEVANT
MANUSCRIPTS
IN THE NIELS BOHR ARCHIVE

INTRODUCTION

The following documents are mainly from the years 1930–1958, which is the period covered by the present volume.

The folders listed below form part of the collection of Bohr Manuscripts in the Niels Bohr Archive. They are microfilmed as part of the AHQP under the designation "Bohr MSS"; the corresponding microfilm number (abbreviated "mf.") is given for each folder.

The titles of the folders have been assigned by the cataloguers, as have all dates in square brackets. Unbracketed dates are taken from the manuscripts.

Numbers in the margin facing an item indicate the pages on which the item is reproduced; they are followed by the letter E if only excerpts are given. Items for which English translations are provided are indicated by the letter T and facsimile by the letter F.

1 *Field Measurements* [1930–1931]

Handwritten [N. Bohr and H.B.G. Casimir], 4 pp., Danish, mf. 12.

Notes and calculations.

2 *Angular Momentum in Radiation Theory* [1932–1939]

Typewritten, carbon copy and handwritten [G. Placzek, L. Rosenfeld and unidentified], 232 pp., Danish, English and German, mf. 13.

Notes, parts of drafts and manuscripts for an unfinished paper on angular momentum in radiation theory by N. Bohr, C. Manneback, G. Placzek and L. Rosenfeld.

3 *Field Measurements* [1932–1933]

Handwritten [N. Bohr], 6 pp., English, mf. 13.

Notes and calculations. Two pages date from Bohr's stay in Pasadena in the spring of 1933.

[183]–[191] E,T 4 *7ième Conseil de Physique Solvay* [22–29 October 1933]

Typewritten and handwritten [L. Rosenfeld and V.F. Weisskopf(?)], 38 pp., Danish, French, English and German, mf. 13.

Manuscript, Danish with French translation, of Bohr's paper in the report of the 7th Solvay Meeting. Notes for the paper.

5 *Numerical Constants of Atomic Theory* [1933]

Typewritten, carbon copy and handwritten [L. Rosenfeld], 8 pp., English, mf. 13.

Notes for unpublished paper with title: "The numerical constants of atomic theory and the limits of the correspondence argument".

6 *Om Positive Elektroner* 27 November 1933

Typewritten and carbon copy, 16 pp., Danish, mf. 13.

Notes for lecture with title: "Om de positive elektroner" (On the positive electrons), delivered before the Danish Physical Society, 27 November 1933.

[338] F 7 *Vorlesungen in Leningrad* 7 May 1934

Carbon copy and drawings [N. Bohr], 34 pp., German, mf. 13.

Shorthand reports of two lectures given in Leningrad on 7 May 1934: (a) "Space and time in atomic theory", at Physico–Technical Institute; (b) "General problems in atomic physics", before the Academy of Sciences. The drawings, which show "Einstein's box" and passage of waves and particles, are dated: Leningrad 7.5.33. At that time Bohr was in the USA, so the date should probably be 7.5.34.

8 *Rum og Tid i Atomfysikken* [1934–1935]
Handwritten with drawings [N. Bohr, O. Klein and L. Rosenfeld], typewritten and carbon copy, 58 pp., Danish and English, mf. 14.

Notes, parts of drafts etc. for unpublished paper on space and time in atomic physics. Four pages dated 5 December 1935, two pages dated 8 December 1935. The drawings show "Einstein's box" and various experimental arrangements, consisting of diaphragms, lenses and photographic plates.

9 *Rum og Tid i Atomfysikken* [21 March 1935]
Carbon copy, 14 pp., Danish, mf. 14.

Title: "Rum og Tid" (Space and Time). Notes for lecture delivered before the Society for the Advancement of Physical Knowledge, Copenhagen, 21 March 1935.

10 *Quantum-Mechanical Description of Physical Reality* 13 July 1935
Carbon copy, 31 pp., English and German, mf. 14.

Manuscript and German translation of the paper in Phys. Rev. **48** (1935) 696–702. Cf. p. [291].

11 *Space and Time, Inst. Poincaré* [18–19 January 1937]
Handwritten [L. Rosenfeld], 6 pp., English, mf. 14.

Notes for lecture on space and time in atomic physics given at l'Institut Poincaré, Paris, 18 or 19 January 1937.

12 *The Causality Problem in Atomic Physics* [April–October 1938]
Typewritten, carbon copy and handwritten [L. Rosenfeld and unidentified], 174 pp., English and Danish, mf. 15.

Notes, drafts and manuscript for the report of the lecture given at the Physical Congress, organized by the International Institute of Intellectual Co-operation in Warsaw 30 May–3 June 1938. The manuscript exists in two copies, one of them containing an additional footnote on p. 1. Part of the material is dated, the dates being in the period 28 April to 17 October 1938. Cf. p. [299].

13   *Analyse og syntese i naturvidenskaben*   3 December 1940
Handwritten [N. Bohr], 2 pp., Danish, mf. 16.

Notes. Outline of lecture with title: "Filosofisk Forening, Lund 3.12.1940. Analyse og syntese i naturvidenskaben". (Philosophical Society, Lund 3.12.1940. Analysis and synthesis in science).

14   *Causality and Complementarity*   [1948]
Typewritten and handwritten [S. Rozental], 17 pp., English, mf. 17.

Parts of drafts of: "On the Notions of Causality and Complementarity", Dialectica **2** (1948) 312–319. Some pages dated between 1 September 1948 and 13 October 1948. Cf. p. [325].

15   *8th Solvay Meeting*   [27 September–2 October 1948]
Carbon copy and handwritten [J. Lindhard], 17 pp., English, mf. 17.

(a) Reports of the discussions following the lectures given by J.R. Oppenheimer, C.F. Powell and R. Serber, and comments on the reports. (b) Draft of Bohr's article in the report of the meeting with the title: "Some General Comments on the Present Situation in Atomic Physics". Cf. p. [223].

16   *Discussion with Einstein*   [1949]
Typewritten, carbon copy and handwritten [Aage Bohr and S. Rozental], 123 pp., English, mf. 18.

Notes, manuscripts and proof for "Discussion with Einstein on Epistemological Problems in Atomic Physics". The two manuscripts and the proof constitute three different versions of the paper, the latter corresponding to the published version. Cf. p. [339].

17   *Relativity Theory and Atomic Physics*   1949
Typewritten, carbon copy and handwritten [J. Lindhard and S. Rozental], 41 pp., English and Danish, mf. 19.

Notes and parts of drafts. Title: "Relativity theory and atomic physics". Unpublished paper intended as contribution to June 1949 edition of Rev. Mod. Phys., dedicated to Einstein on the occasion of his 70th birthday. Dated between 20 April and 15 June 1949.

[197]–[209],
[497] E

18   *Field and Charge Measurements*   [1937, 1947–1950]

Typewritten, carbon copy and handwritten [Aage Bohr, Ernest Bohr, N. Bohr, S. Hellmann, L. Rosenfeld and Mrs. Rosenfeld], 342 pp., English and Danish, mf. 19.

Five folders: (a) "Field and Charge Measurements in Quantum-Theory A. 1937" – notes and an unpublished manuscript; (b) "Field and Charge Measurements in Quantum Electrodynamics – notes from 1937 and 1947–1948. B"; (c) "Notes 1949"; (d) untitled – notes, calculations and galley proofs of "Field and Charge Measurements in Quantum Electrodynamics" (cf. p. [211]), L. Rosenfeld's summary of the paper, published in Physics Today **3**, no. 9 (1950) 35; (e) "Duplicates".

19 *Naturvidenskabens erkendelsesproblem*   19 January 1951
Handwritten [Aage Bohr], 4 pp., Danish, mf. 19.

Outline of lecture on the epistemological problem of natural science delivered before the Royal Danish Academy of Sciences and Letters, 19 January 1951.

20 *Atomerne og den Menneskelige Erkendelse*   14 October 1955
Typewritten, carbon copy and handwritten [Aage Bohr, A. Petersen, L. Rosenfeld], 113 pp., Danish, mf. 21.

Notes etc. for lecture at the meeting of the Royal Danish Academy of Sciences and Letters, 14 October 1955. Part of the material dated between 29 September 1955 and 8 October 1956. Two copies of the draft dated 14 October 1955.

21 *Atoms and Human Knowledge, Nicola Tesla*   [10 July–7 November 1956]
Typewritten, carbon copy and handwritten [A. Petersen], 37 pp., English and Danish, mf. 22.

(a) Tribute at the opening ceremony of the Nicola Tesla Conference, 10 July 1956. (b) Manuscripts. Title: "Atoms and human knowledge". Footnote: "Except for certain points, which have been elaborated, this article presents the content of the author's address delivered without manuscript at the Nicola Tesla Congress, July 1956". Apart from a short introduction referring to the special occasion, the two manuscripts filed under this title are slightly different versions of the manuscript filed under *Atoms and Human Knowledge (I)*. One of the manuscripts is dated 7 November 1956. (c) Notes for a report of the lecture. Dated 27 October 1956.

22 *Atoms and Human Knowledge (I)*   1956
Typewritten, 18 pp., English, mf. 22.

Manuscript of "On Atoms and Human Knowledge", Dædalus **87** (1958) 164–175. Cf. p. [395].

23 *American Academy of Arts and Sciences*   13 November 1957
Typewritten, carbon copy, 2 pp., English, mf. 22.

Manuscript. Title: "Abstract of introductory remarks by N. Bohr at the panel discussion at the meeting of the American Academy of Arts and Sciences, 13 November 1957".

24 *Atomic Physics and Human Knowledge* 1939–1942, 1953, 1957

Typewritten and handwritten [Aage Bohr, Ernest Bohr and A. Petersen], 109 pp., Danish and English, mf. 22.

Various notes related to: "Analysis and Synthesis", "Light and Life", "Position and Terminology in Atomic Physics". Drafts for Introduction and Preface to "Atomic Physics and Human Knowledge" (Wiley 1958), "Postscriptum to Discussion with Einstein" (dated 8 August 1957).

25 *Atoms and Human Knowledge (II)* [1957–1958]

Typewritten, 14 pp., English, mf. 22.

Manuscript. Title: "Atomic Physics and the Problem of Knowledge. Publication in Dædalus, Journal of the American Academy of Arts and Sciences, April 1958". Footnote: "Based upon a Communication presented at the 1401st Stated Meeting of the Academy, November 13, 1957". Does not correspond completely to the cited article in Dædalus.

26 *Quantum Physics and Philosophy* [September 1957–13 June 1958]

Typewritten and printed, 37 pp., English, mf. 23.

Drafts and proofs of "Quantum Physics and Philosophy" in "Philosophy in the mid-Century", 1958 (cf. p. [385]).

27 *Kvantemekanik (efterladte noter)* [22 February 1962–10 May 1962]

Handwritten [E. Rüdinger], 15 pp., Danish and English, mf. 26.

Notes on Stern–Gerlach experiment, spin conservation, Einstein's experiment and quantum-mechanical treatment of molecular rotation. Draft of a letter.

INDEX

Subjects which appear throughout the volume – such as complementarity, measurability/measurement, quantum electrodynamics, quantum mechanics and uncertainty principle – are not listed (but, e.g., charge, classical electrodynamics, energy and momentum are).

When a term on a page is found only in a footnote or in a picture caption, the page number in the index is followed by the letter n or p, respectively.

It is hoped that the cross references will help the reader identify subjects listed under headings which differ from the terms actually used in the text.

Printed and bound by CPI Group (UK) Ltd, Croydon, CR0 4YY

03/10/2024

01040329-0016